Critical Debates in Tourism

ASPECTS OF TOURISM

Series Editors: Chris Cooper, *Oxford Brookes University, UK*, C. Michael Hall, *University of Canterbury, New Zealand*, and Dallen J. Timothy, *Arizona State University, USA*

Aspects of Tourism is an innovative, multifaceted series, which comprises authoritative reference handbooks on global tourism regions, research volumes, texts and monographs. It is designed to provide readers with the latest thinking on tourism worldwide and push back the frontiers of tourism knowledge. The volumes are authoritative, readable and user-friendly, providing accessible sources for further research. Books in the series are commissioned to probe the relationship between tourism and cognate subject areas such as strategy, development, retailing, sport and environmental studies.

Full details of all the books in this series and of all our other publications can be found on http://www.channelviewpublications.com, or by writing to Channel View Publications, St Nicholas House, 31-34 High Street, Bristol BS1 2AW, UK.

Critical Debates in Tourism

Edited by
Tej Vir Singh

CHANNEL VIEW PUBLICATIONS
Bristol • Buffalo • Toronto

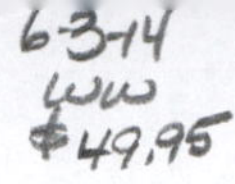

Dedicated To

Dr Naresh Arora
Astrid and Erwin Groetzback

Library of Congress Cataloging in Publication Data
A catalog record for this book is available from the Library of Congress.
Critical Debates in Tourism/Edited by TejVir Singh.
Aspects of Tourism:57
Includes bibliographical references and index.
1. Tourism--Social aspects. 2. Tourism--Environmental aspects. I. Singh, Tejvir, 1930-
G155.A1C743 2012
338.4'791–dc232012021844

British Library Cataloguing in Publication Data
A catalogue entry for this book is available from the British Library.

ISBN-13: 978-1-84541-342-2 (hbk)
ISBN-13: 978-1-84541-341-5 (pbk)

Channel View Publications
UK: St Nicholas House, 31-34 High Street, Bristol BS1 2AW, UK.
USA: UTP, 2250 Military Road, Tonawanda, NY 14150, USA.
Canada: UTP, 5201 Dufferin Street, North York, Ontario M3H 5T8, Canada.

The policy of Multilingual Matters/Channel View Publications is to use papers that are natural, renewable and recyclable products, made from wood grown in sustainable forests. In the manufacturing process of our books, and to further support our policy, preference is given to printers that have FSC and PEFC Chain of Custody certification. The FSC and/or PEFC logos will appear on those books where full certification has been granted to the printer concerned.

Typeset by The Charlesworth Group.
Printed and bound in Great Britain by Short Run Press Ltd.

Contents

Contributors

Editor

Tej Vir Singh is Professor and founding Director of the Centre for Tourism Research & Development (CTRD), Lucknow, India, recognized by the World Travel and Tourism Council (WTTC) for global good practices in travel and tourism human resource development. He is also Founding Editor of the centre's international journal *Tourism Recreation Research,* which won the Golden Page Award in 2003 from Emerald Management Review, UK. A specialist in Himalayan Tourism, Dr Singh has produced about two dozen international books on tourism and many technical papers on tourism development and impacts. He was Senior Fellow Ford Foundation at ICIMOD, Kathmandu (1988). Dr Singh is a fellow of the International Academy for the Study of Tourism (since 1989). He has consulted for the UNEP and also represented UNEP at the ESCAP Conference on Sustainable Tourism Development in LDCs, Pattaya (October 1992). He has chaired several international conferences on tourism and delivered keynotes and valedictories, including the session keynote address on Sustainable Tourism – BOAO Forum for Asia Tourism, Guilin, China (2002). He was awarded, a lifetime honorary professorship of tourism by the Bundelkhand University, Jhansi, India.

Contributors

Gregory Ashworth was educated at the universities of Cambridge (BA), Reading (MPhil) and London (PhD) and has taught at the University of Wales (Cardiff), University of Portsmouth and since 1994 has been Professor of heritage management and urban tourism at the University of Groningen (NL). His research and publishing interests are in the fields of heritage planning and management, heritage tourism and place marketing/ branding mostly in an urban context.

Susanne Becken is Professor of Sustainable Tourism at Griffith University, Australia, and an Adjunct Professor at Lincoln University, New Zealand. Susanne is currently leading a government-funded programme in New Zealand entitled "Preparing the Tourism Sector for Climate Change". Susanne is on the editorial boards of Annals of Tourism Research, the Journal of Sustainable Tourism, Journal of Policy Research in Tourism, Leisure and Events, and the Tourism Review. She acted as a contributing author to the Fourth IPCC Assessment Report. In 2007 and is currently contributing to the Fifth IPCC report. At the UNFCCC conference in Bali, Susanne was an invited member of the United Nations World Tourism Organisation delegation. Susanne has undertaken consultancy work for organisations within New Zealand and internationally, and is working closely with the tourism industry.

Bill Bramwell is Professor of International Tourism Studies in Sheffield Business School at Sheffield Hallam University, UK. He conducts research in the areas of tourism policy and planning, tourism and governance, and tourism and sustainable development. He is Co-Editor of the *Journal of Sustainable Tourism*, Associate Editor of the *Journal of Ecotourism*, and Resource Editor for *Annals of Tourism Research*. He has edited or co-authored books on rural tourism and sustainable development, tourism partnerships, tourism and governance, sustainable tourism in Europe and coastal mass tourism.

Frances Brown is an independent editor, writer and translator. She was Editor of *Tourism Management* for 10 years (1987–1996) and has written and edited a number of books, special issues and articles on tourism, as well as producing material for an NHS Scotland course on tourism and health. She gives occasional lectures in the UK and abroad on publishing one's research and has also published a guide on this topic. Her principal role, however, is as Editor of the journal *Space Policy*.

Ralf Buckley is Director of the International Centre for Ecotourism Research at Griffith University, Australia. He has over 750 publications including 12 books and about 200 refereed journal articles, about half of each in ecotourism. He has worked in more than 40 countries, held nine international professorships and fellowships, and served on eight multilateral and ministerial councils and numerous parliamentary inquiries. He is a longstanding contributor to *Tourism Recreation Research*.

Jim Butcher is a lecturer and writer on contemporary leisure and travel. He has written two influential monographs: *The Moralisation of Tourism* (2003), a hard hitting defence of mass tourism and; *Ecotourism, NGOs and Development* (2007), a critique of the development claims made for ecotourism (both published by Routledge). He has written extensively on tourism and development. Jim has also contributed to debates about tourism in the media. He is a Fellow of the RSA.

Richard Butler has degrees in Geography from Nottingham University and the University of Glasgow and spent 30 years at the University of Western Ontario in Canada, before joining the University of Surrey (1997 to 2005), and then the University of Strathclyde. He has published many journal articles and book chapters, and 16 books on tourism. His principal research interests are the development process of tourist destinations and the impacts of tourism. He is a founding member and former president of the International Academy for the Study of Tourism.

Alison Caffyn is a freelance tourism consultant specializing in tourism development and rural regeneration. She has worked with many communities, local authorities and protected landscapes to develop tourism plans, strategies and partnerships. Alison started her career working in tourism development with tourist boards in two UK National Parks, then spent seven years as a tourism academic at the University of Birmingham. She has also worked at ECOTEC Research and Consulting and as tourism officer for Ludlow and the Shropshire Hills.

Alexandra Coghlan is a Research Fellow at Griffith University's Centre for Tourism, Sports and Services Innovation. She has a BSc (Honours) in Environmental and Marine Biology from the University of St Andrews, Scotland, and a Graduate Diploma in Education as well as a PhD from James Cook University, Australia. She completed her PhD thesis in 2005, studying the volunteer tourism experience. Her primary research interests concern change processes that are facilitated through alternative and nature-based tourism, including transformative travel experiences, environmental education/interpretation, governance issues and tourism partnerships.

Erik Cohen is the George S. Wise Professor of Sociology (emeritus) at the Hebrew University of Jerusalem, where he taught between 1959 and 2000. He has conducted research in Israel, Peru, the Pacific Islands and, since 1977, in Thailand. He is the author of more than 180 publications. His recent books include *Contemporary Tourism: Diversity and Change* (Elsevier, 2004), and *Explorations in Thai Tourism* (Emerald, 2008). He is a founding member of the International Academy for the Study of Tourism. Erik Cohen presently lives and does research in Thailand.

Dennis Conway is Professor emeritus at Indiana University, Bloomington, Indiana. He has co-authored eight books, over 130 articles and chapters on Caribbean urbanization, transnational migration, economic development and alternative tourism models.

Chris Cooper is Pro-Vice Chancellor and Dean of the Business School at Oxford Brookes University, UK. He has more than 25 years experience in the tourism sector and has worked as a researcher and teacher in every region of the world. He held the Chair of the UNWTO's Education Council from 2005–2007 and was awarded the UN Ulysses Medal for contributions to tourism education and policy in 2009. Professor Cooper is currently the Co-Editor of *Current Issues in Tourism*. He has authored a number of leading textbooks in tourism and is the Co-Series Editor of Channel View's influential book series *Aspects of Tourism* and Series Editor of Goodfellows' *Contemporary Tourism Reviews*.

Rachel Dodds is an Associate Professor at the Ted Rogers School of Hospitality and Tourism Management, Ryerson University in Toronto, Canada. She earned her PhD from the University of Surrey, England, in 2005 and has since specialized in sustainable tourism, tourism in islands, tourism policy, corporate social responsibility and climate change issues.

David Fennell is a Professor who teaches and researches mainly in the areas of ecotourism and tourism ethics. He has published widely in these areas, including single authored books on ecotourism programme planning, a general text on ecotourism, tourism ethics and codes of ethics in tourism. Fennell is the founding Editor-in-Chief of the *Journal of Ecotourism,* and is an active member on editorial boards of many academic journals.

Daniel Guttentag is Tourism Business Analyst for Tourism Toronto. He did his Master's in Tourism Policy and Planning, 2010, University of Waterloo, Waterloo, Ontario, Canada. His recently published articles are 'The legal protection of indigenous souvenir products' (*Tourism Recreation Research* 34(1), 2009) and 'The possible negative impacts of volunteer tourism' (*Journal of Tourism Research* 11(6), 2009).

C. Michael Hall is a Professor in the Department of Management, University of Canterbury, New Zealand and the School of Tourism and Hospitality Management, Southern Cross University, Australia; Docent, Department of Geography, University of Oulu, Finland and Visiting Professor, Linneaus University, Kalmar, Sweden. Co-Editor of *Current Issues in Tourism* he has published widely on tourism, regional development, environmental change and gastronomy.

Derek Hall has degrees in geography and social anthropology from the University of London. His PhD was on urban social area analysis (subsequently applied in India). Since the late 1960s he has published in the fields of urban and political geography and tourism development especially in relation to state socialist and post-socialist societies. Previously Head of the Leisure and Tourism Management Department and Professor of Regional Development at the Scottish Agricultural College, he has a number of visiting roles in the UK and Finland.

David Harrison is Professor of Tourism at the University of the South Pacific, Fiji. A sociologist/anthropologist of development, he is especially interested in the economic, environmental, social and cultural impacts of tourism in developing countries, and has carried out research on these in the Eastern Caribbean, South-East Asia, Eastern Europe, Southern Africa and in South Pacific Island countries. He is author of *The Sociology of Modernization and Development* (1988), editor of several books on tourism in developing countries, and has written many journal articles on this topic.

Joan C. Henderson is an Associate Professor at Nanyang Business School in Singapore. Prior to this, she lectured in tourism in the United Kingdom after periods of employment in the public and private tourism sectors. Her PhD thesis was on the subject of social tourism and she also holds an MSc in Tourism and a BSc.Econ (Hons) in Politics and History. Current research interests include crisis management, heritage as a tourist attraction and tourism in the Asia Pacific region.

Carson L. 'Kit' Jenkins is Emeritus Professor of International Tourism in the Business School, University of Strathclyde, Glasgow. He is an economist by background and has taught, researched and consulted in most regions of the world including Europe, Africa, Asia, the Caribbean and Latin America. He has undertaken assignments for most of the international development agencies in over 70 developing countries. As an academic and consultant, Professor Jenkins has specific interests in tourism policy formulation, tourism planning, human resource development, tourism legislation and institutional restructuring.

Dorothea Meyer is Senior Lecturer and director of the Tourism and Poverty Reduction Research Unit at Sheffield Business School, Sheffield Hallam University, UK. Her research interests include tourism as a tool for poverty reduction; the political economy of tourism development; power relations at global-local level; and the role of the tourism private sector in development. Dorothea has published widely and worked over the past decade on approximately 15 research projects related to tourism as a tool for poverty reduction mainly in Sub-Saharan Africa.

Kjell Olsen, Professor Cultural Studies, Finnmark University College, Norway. Olsen is an anthropologist who has worked on topics such as identity and revitalization of indigenous Sámi culture, cultural representations, heritage and authenticity in tourism, local identity, religion and the relationship between indigeneity and the state. Most recent book: *Identities and Ethnicities in a Border Zone* (2010).

Philip L. Pearce has taught in Australian universities for over 30 years. He is the first Professor of Tourism in Australia and is based at James Cook University. He has a First Class Honours degree in Psychology and a Diploma of Education from the University of Adelaide and earned a Doctorate from the University of Oxford studying tourists in Europe. He has held a Fulbright scholarship at Harvard University. In his position at James Cook University he teaches at all levels with a focus on tourist behaviour and experience. In 2008 he won an ALTC award for advancing tourism education and for the supervision of Doctoral level students, having successfully supervised 30 such students. He has 200 publications and eight books on tourism. He was the founding editor of *The Journal of Tourism Studies* (1990–2005) and now reviews and edits manuscripts for other journals on a regular basis. He is a frequent keynote speaker at tourism conferences particularly in Asia. His special interest areas are tourist behaviour, tourism and communities and tourism education and research.

Paul Peeters is Associate Professor at the Centre for Sustainable Tourism and Transport, NHTV University for Applied Science, Breda, The Netherlands. He specializes in the impacts of tourism on the environment in general and climate change in particular. His publications cover a wide range of topics like air transport technology, global and regional tourism and climate scenarios, dynamic system approaches to tourism and tourism transport mode choice and modal shift.

Eliza Raymond has 10 years experience studying, volunteering and working in the field of volunteer tourism. Eliza's Master's Degree focused on good practice in international volunteer programs, and involved researching a variety of different volunteer programs in South America, Africa and Australasia. Following her studies, Eliza ran Global Volunteer

Network's volunteer programs in Peru. She is now based in New Zealand and is the Executive Director of the Global Volunteer Network Foundation.

Lisa Ruhanen is a Lecturer and the Postgraduate Coursework Program Coordinator for the School of Tourism, The University of Queensland. Her research interests include sustainable tourism destination policy and planning, climate change and Indigenous tourism. She has been involved in some 25 international and Australian research and consultancy projects. Dr Ruhanen works closely with the United Nations World Tourism Organization where she is a visiting scholar and consultant.

Regina Scheyvens is Professor of Development Studies at Massey University, New Zealand. Here she combines a passion for teaching about international development with research on tourism and development. In 2002 she published *Tourism for Development: Empowering Communities* and she has written articles on themes such as backpacker tourism, ecotourism, and sustainable tourism. Her most recent book is *Tourism and Poverty* (Routledge, 2011).

Daniel Scott is a Canada Research Chair in Global Change and Tourism at the University of Waterloo. His work focuses on sustainable tourism and the human dimensions of climate change. He has been a contributing author and expert reviewer for the United Nations Intergovernmental Panel on Climate Change Third and Fourth Assessment Reports. He is currently also on the Advisory Committee to the UN coordinated Global Partnership for Sustainable Tourism and the Commission for Climatology of the World Meteorological Organization.

Richard Sharpley is Professor of Tourism and Development at the University of Central Lancashire, Preston, UK. He has previously held positions at a number of other institutions, including the University of Northumbria (Reader in Tourism) and the University of Lincoln, where he was professor of Tourism and Head of Department, Tourism and Recreation Management. His principal research interests are within the fields of tourism and development, island tourism, rural tourism and the sociology of tourism, and his books include *Tourism and Development in the Developing World* (2008), *Tourism, Tourists and Society, 4th Edition* (2008) and *Tourism, Development and Environment: Beyond Sustainability* (2009).

Shalini Singh is Associate Professor in the Department of Recreation and Leisure Studies, Brock University, Ontario (Canada). Singh's research in tourism is largely in the context of developing countries, in general, and India, in particular. Her areas of study include tourism and destination communities, pilgrimages, place and people synergies. Shalini's most recent publication is *Domestic Tourism in Asia – Diversity and Divergence* (Earthscan) and she is presently engaged with a project on the sacred WHS site of Bodh-Gaya. She serves as country representative on the Board of Asia Pacific Tourism Association and is Vice President – publications of the International Sociological Association (RC50–Tourism Group).

Benjamin F. Timms is an Assistant Professor of Geography in the Social Sciences Department at California Polytechnic State University. He studies environmental conservation, land-use and land-cover change, political ecology, and sustainable tourism development in the Caribbean and Central America, with the latter focusing on maximizing economic linkages between tourism and local industries. His publications have appeared in *Tourism Geographies*, *Tourism and Hospitality Research*, *Global Development Studies*, *International Development Planning Review* and *Caribbean Geography*, amongst others.

David Weaver has held previous academic appointments in Australia, Canada and the USA, and is currently Professor of Tourism Research at Griffith University. He has published about 100 refereed journal articles, book chapters and books, and sits on eight editorial boards. He is a Fellow to the International Academy for the Study of Tourism, and is a frequent keynote speaker at international conferences. His specialties include sustainable tourism, ecotourism and destination life cycle dynamics.

Betty Weiler holds a PhD from the University of Victoria (Canada) and is currently Research Professor at Southern Cross University. Her research has centred on the tourist experience, including the role of the tour guide and heritage and nature interpretation. Betty is known for her collaboration with and contribution to the management of protected areas, zoos and heritage attractions. More recently her work has focused on managing visitors and influencing their on-site and post-visit behaviour.

Brian Wheeller is Visiting Professor of Tourism at NHTV, Breda, The Netherlands. He holds degrees in Economics, in Applied Economics, in the Economic Impacts of Tourism, and in American Literature. His doctorate is in Critiquing Eco/Ego/Sustainable Tourism. Brian's current research revolves around the links between travel, tourism and popular culture – in particular literature, art, film and music. He is also interested in humour, the visual and the use of images in tourism and their relevance to tourism education.

Foreword

Critical Debates in Tourism. What a short title for such a 'long' or comprehensive book. While short captions are favored, *menos es más*, few attempt to wordcraft theirs. Movies and songs often use one or two words, so why not articles and books¿ Some may argue that unlike 'artistic' works ours are 'social science' debates, and hence more words are needed to communicate. But Dr Tej Vir Singh's book, with its four-word title, offers more in fewer words – both in label and contents.

The scope of this publication is almost encyclopedic. The perspectives and arguments advanced use what is known to suggest what there is to be known. The book reviews, airs, and debates old and new subjects – tourism planning and development, mass and small-scale options, nature and ecotourism, alternative choices and economic costs, carrying capacity and limits to growth, climate change issues, codes and standards, community-based tourism, empowerment and governance, sustainable development, corporate social responsibility, education and academic responsibility, knowledge management, among others – and brings them into a thought-provoking synergy. Most deliberations are theoretical, some practical, and many both. The volume also echoes seminal questions posed in the 60s and 70s by such scholars as George Young and Louis Turner, updates the debate, reaches out for fresh insights, answers questions, and poses new ones to be addressed in years to come. The discussion solidifies the known retrospectives, for a landmark pathmaking prospective.

Critical Debates in Tourism, in its comprehensive scholarly yet engaging style, makes informed advances. For instance, climate change and tourism (with their structured and structuring impacts on each other) is critically debated with an eye toward the future. This way, the publication connects and adds its voice to new books on this pivotal worldwide issue, allowing one to juxtapose 'carbon footprint' with 'tourism footprint' or 'climate change' with 'culture change'. Even if nature could be revived, culture – when changed – cannot be restored, and if reinstated, it is no longer a living culture but a museumized heritage 'sold by the pound', as Davydd Greenwood put it in the 70s. Again, tourism comes across as a multifarious phenomenon and a complex industry to understand and accommodate.

Other challenging questions of the past, such as 'tourism development for whom¿', have not gone away. The intent of the question was to shift the answer from 'for the tourist' to 'for the host community'. Yes, we continue to conduct research on authenticity and experience, but for whom or from whose point of view¿ Today more is expected of tourism: that it should also contribute to the quality of life of the host community, a proposition which I recently developed into a conference talk, 'A Nice Place to Live is a Nice Place to Visit'.

A good academic book is not an end in itself: it is the beginning to something more, if not to a roadmap. *Critical Debates in Tourism*, with its 14 chapters authored by 35 recognized scholars, is a one-stop reading tome. Dr Tej Vir Singh – Founding Editor of *Tourism Recreation Research* and a Fellow of the International Academy for the Study of Tourism – in this landmark work does not favor add-on or patchwork, but calls for integration. Like any far-reaching book, his leads the reader to many research frontiers. To do this, he sets the stage in the introductory chapter, lets chapter after chapter unfold issues, concepts, and cases, and finally outlines research challenges and opportunities in the concluding chapter. If each scholar follows only one of these inroads, the study of tourism will be going to many new places.

Jafar Jafari
University of Wisconsin-Stout, USA
Jafari@uwstout.edu

Preface

This book was conceived some 15 years ago but it has taken some time to deliver. In fact, initially, it was never the intention to organize this text into book form, for one important reason – that the themes covered were disparate and unconnected. Fortunately, it managed to find a shape. This may sound anecdotal but I would like to share this story with my readers. As an editor of a tourism journal, I received a very interesting and informative paper for my consideration which I forwarded to the referees for assessment; they did not approve it for publication for some valid reasons. For example, 'it [did] not fit into a research format, [lacked] research rigour and [was] generally descriptive and qualitative in tone'. However, the third reviewer found it 'print-worthy and a valuable contribution with high readability'. This set me thinking on the role of an editor. After considering many pros and cons, I created a space for the editor by introducing an exclusive department – *Research Probe*. We believe this new section is an open field where challengers will be asked to state their case clearly and consistently, encouraging discussion and covering some ground usually overlooked by sister publications. I am convinced that the contributions of open discussion to research are far more fruitful than professional sermons which target audiences of believers. True, we should base our scientific proposition on quantitative methods but when adopting an experimental approach for studying humans, we would need qualitative tools.

Thus we encourage debate on perplexing themes such as the relevance of mass tourism; the sudden rise of the new middle class in the developing nations; the impact of mass culture on travel consumption, as well as other issues which demand deeper theoretical probing. As these probes into research appeared in the journal, more and more scholars with special expertise entered into the fray and the interest shown by readers soared. Many of my friends suggested that the contributions to *Research Probe* should be compiled in book form for the benefit of research scholars and students. I am glad that Channel View Publications realized the value of this work and readily agreed to publish it. And, here is the book: *Critical Debates in Tourism.*

Much credit should be given to the contributors and lead authors who strongly supported this proposal, giving top priority to the challenge – especially as some of the probes were anything up to 10 years old. In many cases they had to re-do the work within a given time frame. I am particularly grateful to lead authors who gave 'Context' and included 'Concluding Remarks' in their chapters. I remember that some of them were anxious after being struck by natural calamities. Erik Cohen was overtaken by sea-waters and yet revised his chapter from a hotel in Bangkok; C. Michael Hall was fighting with the fury of an earthquake and living the life of a semi-nomad, yet he was the first to submit his chapters; so was Susanne Becken who triumphed over catastrophe after catastrophe but never missed her commitments. The book has contributions from about 34 authors from different parts of the world and has a variety of themes. Tourism is personified and makers of tourism are made responsible for the good and bad of tourism. The number of areas covered is vast, from fast, mainstream tourism to slower types of tourism.

I owe a word of thanks to my colleagues at the Centre for Tourism Research and Development who assiduously accomplished this task on time. Special thanks also go to Masood A. Naqvi who continued to work on the book, even when his son was seriously ill. I appreciate Parun Nanwani's commitment in finishing her assignments, on time. Thanks are also due to Sagar Singh for reading the book to help get it ready for print. My heart goes to Dr Naresh Arora, Dr Samir Singh, Dr Salil Singh and Dr R.K. Singla who looked after my health during the preparation of this book. And how can I forget to thank Ralf Buckley who suggested a suitable title for this work? Sincere thanks to Jagdish Kaur, my wife, who took care of me when I worked too hard on the book. My special thanks to Elinor Robertson and Sarah Williams of Channel View Publications who facilitated publication of this book. Lastly, I am grateful to Professor Jafar Jafari for writing a foreword to this book.

Lucknow, India
February 28, 2012

Tejvir Singh

Introduction

Tej Vir Singh

When ye entered; ye defiled my land, and made mine heritage an abomination.
King James Bible (Jeremiah 2:7)

Tourism is a many splendoured phenomenon. If I were to name, in one word, what the best and worst thing in the world is, my unequivocal answer would be – tourism. Made up of strange paradoxes, it offers experiences that are magnificent, spectacular, languorous, horrific, good, bad and ugly – it is an experience industry. Given such superlatives, contradictory, often damning, opinions exist about tourism. Poets, politicians, prophets, scholars and even the man in the street have something to say about the tourism industry. A few examples may be interesting to cite here. Ulysses, in Lord Tennyson's famous verse, after the Trojan War, says: 'I cannot rest from travel: I will drink life to the lees. . .'.

Lord Macaulay's (1913) dislike for tourism is well-known. When he witnessed the tourism movement increasingly growing in England, he characteristically remarked, 'this nation of shopkeepers is turning into a nation of innkeepers'. Tourism scholars have made serious efforts to study the diverse impacts of tourism but most of them have found that tourism has more of a negative than a positive influence. A few of them have pronounced it to be a 'new form of colonialism' (Turner & Ash, 1975), so much so that Boorstin (1964) called it 'Vulgar'; others thought it to be an environment-defiler – using the environment to the extent of using it up. Many an island has been transformed from gregarious green to a cluster of sky-rise hotels. A few also believed that tourism development promoted anomie and culturelessness in the destination society.

Not many favoured the idea of tourism development in the way Archer *et al.* (2004) did: 'tourism seems to be more effective than other industries in generating employment and income particularly in peripheral regions where tourism can make the most'. For the Third World, it is considered an appropriate tool of development as it promises to create employment opportunities because it is labour-intensive; it also earns foreign currency, direly needed for infrastructure development, and stops depopulation in the mountain regions, and the like.

An obvious question that arises – is tourism really so bad? Or is it more sinned against than sinning? And if it really sins why, after all, do nations, multi-nationals, corporations and other stakeholders invest money in it? Today, practically all nations are involved in the tourism business in one form or another. Some countries are entirely dependent on tourism, whereas the world's top 20 most tourism-dependent destinations are all islands. While it is true that less-developed peripheral regions are more dependent, it is also true that many islands with a significant tourism sector, particularly those in the Caribbean and Mediterranean, enjoy a higher than average income and an advanced level of socio-economic development (Sharpley, 2012). One simple answer to this question is that governments and civil societies are duty bound to arrange necessary facilities for the welfare of their people and visitors, such as the construction of roads, provision of drinking water, eateries and inns as well as some medical facilities on travelling routes. This was done in classical times, with vestiges of these social amenities existing even today. In the East *sarai* (travellers' rest houses) and *khanqah* (hospices) feature in travellers' tales. Roman and Grecian folklores bear witness to this fact. That it promotes socio-economic development, in its wake, is also the main purpose of tourism development, and this is discussed elsewhere.

Tourism is the environment – economic, physical, social, psychological, cultural and political. It affects and is in turn affected by the environment. Any study of tourism must examine this complex relationship in order to find out what the impact is. To understand other vital influences on tourism it is advisable to give a brief overview of the history of tourism.

The Tourism Story

We all know that travel and tourism is part of man's nature. To wander and seek out new places is as old as human civilization itself. In the past, man must have designed strategies so that he could move faster to reach his destination. But at that early stage of history there were only a few people who travelled – those who had the means to be able to as well as those who possessed a brave heart, for the journey was not only tedious but also filled with danger.

For the moment we must leave our heroic traveller trotting along on horseback though and jump to the post-war era – a time when there was better awareness of place and speedier means of transportation and communication. The tourism story begins now.

Today's tourism is essentially a by-product of Western culture, propelled by forces of globalization and capitalism. Thomas Cook, the architect of modern tourism, ushered in the railway era from 1840 onwards and connected up the urban industrial heartland of the UK. He arranged visits to

The Great Exhibition, in 1851 in London, in which about 6 million visitors participated (Page, 2007). Beside his other remarkable feats, he connected the coastal resorts that had emerged in the late 18th century. This virtually led to the birth of mass tourism, a topic which we shall discuss at length in this book.

Tourism got a boost when industrial workers began to receive financial assistance and paid holidays under the *Front populaire* programme in France (Ghimire, 1997). Because of industrialization, people had enough disposable income and flexi-time to begin to travel. This sizeable middle class of the West has been the biggest consumer of tourism. Interestingly, similar trends can be seen in the Third World, where the emergence of new middle class tourists forms the biggest tourism market. For the lower classes, social tourism – tourism for all – worked well, though it is more feasible in the case of domestic tourism. Other factors that encouraged the movement of tourists, besides bank holidays, were institutions such as the European Grand Tour – a rite of passage which encouraged the rich, aristocratic and privileged classes to follow prescribed travelling routes for the purpose of developing their personality through learning, education, culture and philosophy by enjoying all that nature has to offer. Much of this tradition can be seen to have been revived in the practice of Volunteer Tourism, where young people from the West are carefully selected by sponsoring organizations to experience life in Third World nations.

A considerable growth in land-based transport, particularly the growth in popularity of cars, has added to the volume of tourists. While cars offered flexibility and ease of access, they nonetheless created pressure on the roads, and traffic-jams and snarl-ups are now common sights in big cities and resorts. Buying a car may not be difficult but constructing new roads is a challenging task. This is one of the unwelcome effects of mass tourism.

Drifting Towards the East

Around the 1960s the Hippie phenomenon happened, particularly in the developing world. This flow of drifters from the West is metaphorically phrased 'nomads from the affluence' by Cohen (1973). Said to be satiated by material wealth, they desired to taste life in this part of the planet. More and more visitors flowed into the Third World so that they could enjoy the offerings of nature and culture, people and places. Since these drifters were from outside the established tourist circuit, they followed the way of life of the local people and dressed like them, ate their food and shared their accommodation. This cultural camaraderie was appreciated by the locals, but it added little to the local economy. They were the complete opposite of mass tourists (Cohen, 1972).

Boom and Bust

The drifters gradually gave way to the leisured classes, businessmen and pleasure seekers, who arrived en masse from different climes and countries, with their diverse cultures and different attitudes. Over time their numbers multiplied and created a boom and bust scenario that has continued to give a bad name to tourism. Host societies, mainly from developing countries, welcomed them with euphoria, as they subscribed to the common notion that tourism is a painless type of therapy which can cure all economic and social ills. They believed that touristic development was non-invasive, innocuous and an industry that would not cause any harm. It would not only give them bread, but butter also. The governments in these developing nations took it as a blessing on their economy as they considered it to be an invisible export industry, giving them foreign currency for the development of basic infrastructure which improves their quality of life.

Given such a dream who would not be tempted to realize it, and who would not rush to experience the exotica of the Third and Fourth Worlds, with all their offers of sun, sand, sex and sea, beside the mystique and charms of orientalism? Streams of visitors from the North left their homes in search of this unusual 'pleasure-dome' to be discovered in the South – another world, so unlike and so different, in its cultural and bio-physical manifestations, than theirs. Besides other important growth factors (improved transport technology, package holidays, the internet, etc.) globalism was the most powerful factor in the rapid growth of the tourism industry. Resilient to negative forces such as economic recession, natural hazards and terrorism, it progressed exponentially, with developed nations such as France, Spain, Italy and the USA in the lead. Recently China has emerged as the biggest challenger (see Table 1).

In the 1970s, tourism burgeoned all over the globe in one form or another. Some hot-spots, such as attractive beaches in the Mediterranean, Alpine Resorts, Himalayan pilgrimage-centres and unique historic ruins, were like honey-pots to visitors. A kind of lemming syndrome was witnessed, with people moving towards the East in search of novelty, genuine authenticity or serendipity. In 1950, international tourist arrivals reached 25 million, peaking with 528 million in 1995; in 2011 this figure jumped to 980 million (UNWTO). Nations are on the move, breaking their geographical boundaries. Tourism growth trends go on increasing unabated, abridging the questionable gap of tourist arrivals between the developed and developing nations. Surprisingly, some developing countries have leap-frogged. Table 2 shows that international arrivals in Asia and the Pacific have more than doubled between 1995 and 2010. This is the same with countries in the Middle East and Africa. There has been an appreciable increase in tourism revenues, estimated to have reached US$ 919 billion in 2010 as compared with US$ 851 billion in 2009 (see Figure 1). This phenomenon of over-crowding was

Table 1 International tourist arrivals – Top 10 countries

Rank	*Figures in Millions*		*Change (per cent)*
	2009	2010	10/09
1. France	76.8	76.8	0.0
2. United States	55.0	59.7	8.7
3. China	50.9	55.7	9.4
4. Spain	52.2	52.7	1.0
5. Italy	43.2	43.6	0.9
6. United Kingdom	28.2	28.1	-0.2
7. Turkey	25.5	27.0	5.9
8. Germany	24.2	26.9	10.9
9. Malaysia	23.6	24.6	3.9
10. Mexico	21.5	22.4	4.4

Source: UNWTO (2011)

Table 2 International tourist arrivals by region (in millions)

Region	*1995*	*2000*	*2005*	*2010*	*Change 1995–2010*
Europe	304.1	385.6	439.4	476.6	57%
Asia/Pacific	82.0	110.1	153.6	203.8	149%
Americas	109.0	128.2	133.3	149.8	37%
Africa	18.9	26.1	35.4	49.4	161%
Middle East	13.7	24.1	36.3	60.3	340%

Source: UNWTO (2011)

given the name of 'mass tourism' – not a pleasant word in tourist vocabulary where tourism is made to sin for man's avarice and greed for wealth; where more is less.

Culture of Affluence Versus Culture of Poverty

By the 1980s, host communities in the Third World became disillusioned with the negative consequences of the tourism industry. Prior to this the World Bank and UNESCO had grown anxious to know whether the adverse effects of tourism were real or apocryphal: in order to uncover the reality of

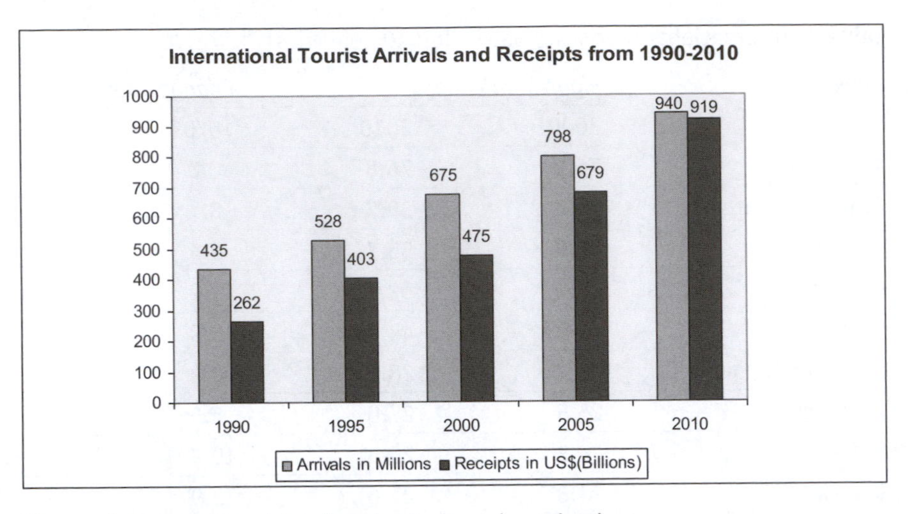

Figure 1 International tourism (arrivals and receipts)
Source: UNWTO (2011)

the situation, the World Bank organized an international seminar in 1976. The report, which had the title: '*Tourism: A Passport to Development*', edited by de Kadt (1979), found that, rather than economic consequences, the non-economic consequences of tourism were serious and had imperceptibly gone un-noticed – particularly social and cultural impacts. Because most of the visitors from the West belonged to the wealthier classes, while those who were visited were poor, there was a conflict between the culture of affluence and the culture of poverty. Quite a voluminous amount of literature has appeared on mainstream mass tourism. Sociologists have accused tourism of many ills, including acculturation, xenophobia, homogenization, modernization, transformation of traditional art forms, inauthenticity and modification of indigenous architecture, and so on. As transport and communication technology improved, the volume of tourists increased tremendously and mainstream mass tourism was found guilty of having many adverse impacts – which tourism was only partly responsible for (Wall & Mathieson, 2006:168).

Quest for New Horizons

Tourism became a major industry in the 19th and 20th centuries, with the first passenger jet service launching in 1958 and bringing about the start of modern mass tourism (Eadington & Smith, 1992). Tourism providers, not content with what had been discovered in nature and culture so far, never stopped seeking new horizons for tourism – the search for new ways to experience pleasure leading them on. They established undersea lodges, built hotels at the tops of high mountains, reached Antarctica, conducted land tours and cruises in the Arctic, and so on. In Antarctica about 22,000 visitors enjoyed the beauty of nature in 2007 (Splettstoessor *et al.*, 2004). This figure soared to 33,824 in the 2010/11 season (MercoPress, 2011). And, with the sky no longer the limit, the search for new skies has continued; space tourism has already come into existence with Dennis Tito's visit to the International Space Station in 2001. Despite the prohibitive costs of such a visit though, more and more tourists have already booked their own trips into space. Such 'space' vacations shouldn't sound so surprising, with the moon or Mars probably becoming the wonder-destinations of tomorrow (Singh, 2004).

Harnessing Tourism

With all its complexities and inherent contradictions, tourism throws down the gauntlet to resource managers to help shape it in such a way as to have only minimum negative impacts. It is often difficult to winnow out the gold from the sand. It is, therefore, a challenge to isolate the principal cause of change or impacts; 'whether changes or impacts are directly attributable to tourist development or whether [the] tourist is only one among a number of agents of change' (Wall & Mathieson, 2006). It is not uncommon; instead it's rather fashionable to vilify tourism for faults that are often indeterminable. There is nothing wrong with tourism when it depends so much upon the dexterity of tourism makers. For example, tourism injects new money into an economy; whether this money stays there or drains out in leakages should be the responsibility of planners and developers. Similarly, sound tourism development strategy is capable of protecting, conserving and even preserving the environment for the enjoyment of residents and the visiting community and, for future generations. But it is also true that due to poor planning or the effects of mass tourism, it can degrade and defile that same environment – tourism destroys tourism. An astute planner should know that a kind of symbiosis exists between tourism and conservation, and he should work with this in mind. Best practices in tourism exemplify that tourism's innocence does not diminish with sustainable growth management approaches. Thus, we can ensure the positive or negative effects of tourism through sound policy goals, proper planning and appropriate development strategies.

Given the fast pace of life, the hustle and bustle, the din and clamour of the modern age, tourism has to stay with us because of its positive effects and the quality of life it promises. This beautiful planet still retains some splendid environments – extraordinary natural features, beautiful odours, the spirit of solitude – things that man has not been able to spoil. Nature has bequeathed her bounties to humankind. Man is privileged to enjoy these delights and tourism can make that possible. Tourism should be treated carefully and cautiously so that our future generations may be proud of this legacy. If mass tourism is part of the problem it should also be part of the solution. Let tourism and conservation be partners in organizing 'soft tourism' and not be rivals. Mankind needs tourism for its well-being, welfare, amity, peace and brotherhood, as well as for socio-economic reasons; it is inevitable.

A Threatened Future

With the world population exploding, the new middle class rising, leisure time increasing, transport and communication getting faster and cheaper, more and more people will want access to tourist destinations. This would threaten managerial acumen, and pose a crisis worst compounded. All those who are working *for* tourism must be on their guard to meet this challenge, either to change current mass tourism practices or invent a new and more appropriate form of tourism that is in harmony with its environment (Poon, 1993: 3). With the democratization of tourism, it does not seem possible that a reduction in the volume of tourist traffic can be managed, nor is it possible to invent any new form of tourism that is manageably small in size, follows principles of equality, respects the environment, subdues consumerism (responsible consumption), emphasizes positive human values, cares for local communities, lives better with less and uses renewable resources. It would not be out of context to quote M.K. Gandhi (1948) in reference to the world's finite resources which must be consumed parsimoniously: 'the world has enough for every one's need but not for every body's greed'. This truth has to be brought home and will be a real challenge for the coming generation.

The Mantra of Sustainable Development

The Brundtland report, *Our Common Future* (WCED, 1987), introduced a magical phrase, 'sustainable development', to tourism vocabulary that brightened the development concept. Having contested definition, it advocates the wise use and conservation of resources in order to maintain their long-term viability (Eber, 1992: 1). In essence, sustainable tourism involves the minimization of negative impacts and the maximization of positive impacts, emphasizing what is now pronounced the 'triple bottom line' – caring for the integrity of the physical ecosystem, economic viability and social responsibility. Sustainable development in tourism takes into account the present generation's needs and cares for the needs of the future generation. The concept is essentially intergenerational but is difficult to put into practice. In other words, sustainability means the consumption of tourism should not exceed the ability of the host destination to provide for future needs. This ideology has inspired many worldwide tourism organizations to develop guidelines, regulations for ethical consumption, sustainable auditing, UNWTO's global code of ethics for tourism, the tour operator's initiative for sustainable tourism development and the like. Much has been written on the concept of sustainability that confuses its real import, nonetheless its impact has been far-reaching and it has become a buzzword in development parlance. Planners, public and private, became conscious of the importance of long-term planning for the well-being of resident communities. Surprisingly, mass tourism has expressed an interest in the local environment and tourists seem willing to practise prescribed codes of conduct during their travels. Mak and Moncur (1995) report that tourists were seen paying fees for the maintenance of a destination's environment and readily accepted limits on the number of visitors in order to avoid overcrowding.

It is encouraging to note that green consciousness has been further promoted by sustainability. With the mellowing of this ideology one can believe that one day much derided mass tourism may shake hands with sustainability. Mowforth and Munt (1998: 102) sceptically observe that 'mass-packaged tours may be just as sustainable as some of the new forms of tourism'. Cohen (1989), in his thought-provoking essay 'Alternative tourism – A critique', observes 'even mass tourists are becoming more circumspect and reflective and hence more careful not to hurt local feelings'. He advises reformation of the more unsavoury aspects of contemporary tourism and to make sincere efforts for a more 'appropriate' mass tourism.

Sustainable Mass Tourism

Scholars have diverse opinions about the validity of the sustainability concept; a few consider it illusive, ambiguous and hard to put into practice. Others are so frustrated that they want to abandon its practice. I sought opinion from the scholars involved in sustainability research. David Weaver (1.1) argued that the mass tourism industry is moving towards sustainability, but more as a 'paradigm nudge' than a 'paradigm shift'. He admitted that academics should be more serious in their action-research and move beyond 'superficial environmentalism'. Ralf Buckley (1.2) questioned that the mass tourism industry should be more involved in translating the concept into result-oriented practices through codes, guidelines, certification and awards schemes. Brian Wheeller (1.3) summarily rejected the concept itself for the inherently unsustainable nature of transportation. Bill Bramwell (1.4) accepted the ambiguity aspect of the concept but would not advise the abandoning of it, halfway. It needs critical evaluation for any positive change or modification. It requires 'knowledge, thoughts and understanding' to enhance its practical use. Bill suggested some ideas that can help enhance its practicability. Such as, 'political, economy approach', 'ecological modernization' based on 'responsible capitalism'. Brundtland (1987) in her report *Our Common Future* observes that sustainable development is the need to integrate economics and ecological considerations in decision-making. She emphasizes that economical and ecological concerns are not necessarily in opposition.

No fruitful consensus could be arrived at except that the concept of sustainable tourism is still in its incipient stage and needs more research. In the meantime efforts should be directed to make 'big' better and beautiful by the joint efforts of the industry and academia.

Tourism and Consumerism

One of the most undesirable facts that comes with mass tourism is that it is closely linked to rampant consumerism, which often makes tourism appear to be sinful, bad and even abominable. The worst side of mass-consumerism is that it perpetuates consumer culture. Richard Sharpley (2.1) examines this aspect of mass tourism – the massive and often irreversible consumption of tourism resources. He states that consumerism promotes materialism in a bad sense – too much love for wealth, insatiable thirst for material goods and services, excessive self love – 'a shallow cultural value that impoverishes the human spirit and fuels narcissistic self-absorption' (Pearce *et al.*, 2011: 107).

This materialistic consumption of tourism can be a destructive force both for tourists and for the destinations. He asserts that to achieve sustainability, tourism may be consumed as a means of experiencing the beautiful sights and sounds that Earth has to offer, liberally. He prescribes 'ethical consumption' and responsible holidays for 'good' tourism whereas over-consumption of vulnerable and non-renewable resources is 'bad'. This is not to deny the value of consumption as it is a driver of economic growth.

Michael Hall (2.2), citing from UNEP (2001), suggested four strategic elements for efficient consumption: first, *dematerialization* (green growth), second, *different consumption patterns*, third, *appropriate consumption*; fourth, *conscious consumption*. He calls this definition the 'slow approach'. For a better tomorrow, he approves the concept of Voluntary Simplicity (choice out of free will) where a tourist limits expenditure on consumer goods out of his or her own free will. He argued that the happiness index should not be equated with GDP. Sustainable consumption is concerned as much with quality of experience as it is with quantity. Admittedly, wealth and riches do not buy happiness. It would be interesting to quote from the *Voice in the Wilderness:*

Money will buy a bed but not sleep;
Books but not brains;
Food but not appetite;
Finery but not beauty;
A house but, not a home;
Medicine but not health;
Luxuries but not culture;
Amusement but not happiness
Religion but not salvation;
A passport to every where, but heaven.

Aristotle is more axiomatic in his statement, when he said 'virtue lies between excess and want'. Wants have no limits, and it is the same with consumption. Hall questions, 'do we really have to travel so much to be happy' and leave behind our abominable ecological foot-prints to further pollute our good Mother Earth? If tourists practise austerity and avoid conspicuous consumption, they would certainly contribute to the sustainability of resources that are scarce. Thus tourists have a definite role in changing consumption behaviour. Hall cites anti-consumption groups that are working to reduce ecological footprints. Finally, he sums up that tourism is not value-neutral. There is no such thing as 'good tourism' or 'bad tourism' – much depends upon who benefits and who does not. It matters how one manages and runs the show. This is a people's industry – 'of the people, by the people and for the people'. It should therefore be people-centric in its policy structure, character, spirit, growth and developmental processes.

Joan Henderson (2.3) believes that not all consumerism has a shady side. Opinions about its 'goodness' and 'badness' are determined by personal, economic, political, socio-cultural, geographical and historical contexts. Good tourism, she elaborates, is generally linked to sustainability, whereas sustainability itself is marked with ambiguities – that is the rub. Mass consumption in fragile ecologies can harm the ecosystems. Hence goodness eludes the grasp of the industry. Defenders of consumerism argue that it propels economies, creates jobs, pushes growth and enhances the standard of living, provided it is practised with a sense of ethics and responsibility.

When tourism was heavily criticized by critics, it re-invented itself and emerged in different forms – generally under the umbrella term of 'Alternative Tourism'. This mainly happened in the 1970s, in the developing countries of the South. Such projects were small-scale and low-key in nature and involved the participation of the local population (Eadington & Smith, 1992). The 'Tourism in Discovery Project' (Bilsen, 1987) firmly established that tourism can be benign if it is locally-owned, locally-managed, locally-catered for, using locally-designed architecture and traditional technology and if it protects community values and assets. The success of this form of tourism has given birth to a wide network of support groups in both developing and developed countries. Known to the world under different names (ecotourism, rural tourism, farm tourism, craft tourism, green tourism, volunteer tourism, etc.) it is considered sustainable, respects the 'triple bottom line' and offers authentic experiences. Hence, small is beautiful.

Small is Beautiful

Is it really beautiful? David Harrison (3.1) has several reasons not to wholly agree with Schumacher's ideology in relation to tourism. First, small or mass are different faces of tourism and form part of the overall tourism system of demand. However, UNESCAP considers small-scale business operations more appropriate for tourism enterprises. UNWTO found that 80% of small and medium enterprises practise sustainability using the motto to 'conserve and not consume resources'. Yet, it would not be fair to state that one is bad and another is good, when they are often interdependent. It sounds an overstatement to equate smallness with sustainability; neither do all tourists prefer small scale accommodations. David Weaver (3.2) concurs with Harrison that well-managed large-scale tourism can also be beautiful; to him one that has fewer negative impacts on the environment and fewer leakages from the economy is beautiful. Richard Butler (3.3) retorts 'many people seem to enjoy being a mass-tourist'. Elsewhere, Weaver (2001) chimes with Butler that small-scale alternative tourism may actually be inappropriate and unsustainable under certain circumstances while large-scale or mass tourism may be sustainable under other circumstances.

To Butler size matters and he subscribes to the concept of carrying capacity. An overcrowded beach would look ugly whereas an empty one may be beautiful. He attempts to unfold the phrase 'small is beautiful', and admits that beauty is hard to explain; the nearest words that come to mind are 'benign, appropriate, responsible and sympathetic' – words which are seldom used for large-scale tourism.

The three authors agreed that in demand and supply of tourism 'one size does not fit all'. Big or small, sustainability is the soul of things that makes them beautiful. Harrison mused about whether beauty itself is sustainable!

Tourism: A Community Industry

Tourism is a community industry where local people matter – their good-will, active participation and willing cooperation – because they are part of the product (Murphy, 1985). It is the local community which bears the cost of providing the tourist infrastructure and hence they should be real beneficiaries, which in many cases doesn't happen. Murphy further states that, the 'tourism industry uses the community as a resource, to sell it as a product and in [the] process affects the lives of everyone' (1985: 165).

'Tourism has become too important to be left to tourist experts' (Seekings, 1982) leaving behind local residents as they are generally considered novice, or not knowledgeable enough to be consulted.

Things are changing now and community participation has become popular in many areas of society, but the progress is no better than of tokenism or at best a part of democratic agenda. Jim Butcher (4.1) argues that community participation is accepted uncritically and therefore its efficacy is poor. Phrases such as 'empowerment' and 'control' are invoked often in inverse proportion to any real power or prospect of meaningful development. In particular 'power' is always (including in Pretty's well known typology of empowerment) situated *within* the community rather than between nations or classes. He argues that the radical view of participation being part of the construction of an 'alternative' new world order is at best wishful thinking.

David Weaver (4.2) argues for a pragmatic rationale for community participation. Community-based tourism (CBT) has had a measure of success but time-limited assistance makes it dependent on funding agencies and hence the concept of 'empowerment' becomes farcical. He suggests a more practical partnership-based tourism model which would be more effective and likely to promote realistic empowerment.

Shalini Singh (4.3) argues that community participation is vital to the life of a community. It must be fostered for the progress and development of the destination community, despite its many inherent shortcomings, and for the well-being of the society. She presents two case studies to illustrate the strength and cohesiveness of the community by active participation in development projects. She argues that local communities should not remain at the mercy of the global tourism industry.

The first two authors are critical of community participation and are disappointed about the realities on the ground. According to them the concept leaves much to be desired and they ask for a fresh look into the problem with pragmatism as a guiding principle – 'put the last first' says everything.

It is commonplace to articulate the shady side of tourism (bad and ugly); but, except for its economic benefits, only a few have bothered to discover its sunny-side. Of late, appreciable literature on these subjects has appeared that washes the sinning hands of tourism – this is probably as a result of the loud voices of environmentalists, NGOs, the UNWTO, the green revolution and other pressure groups. More positive angles have focused on how tourism has cared for visitors' health, wellness, well-being and welfare. Tourism is making efforts towards the reduction of poverty (Scheyvens, 2011). There are stimulating papers in tourism research journals and even special issues are devoted to these critical themes. The UNWTO has set up a separate research cell, called ST-EP (Sustainable Tourism-Eliminating Poverty), although some scholars prefer the term Anti-Poverty Tourism

(APT). Popularly pronounced as PPT, Pro-poor Tourism is now in the spotlight, and scholars are enthusiastic about finding ways and means for a tangible solution through tourism – one of the world's largest industries, alongside widespread support from big development agencies. Not much has been achieved so far. In fact the problem is too huge to be solved in a short time – under-development, widespread inequalities and poverty are rampant worldwide, and continue unabated.

Tourism and Poverty

Does tourism reduce poverty? Four scholars, Regina Scheyvens, Dorothea Meyer, David Harrison and Paul Peeters, were consulted for their opinion as to whether PPT projects and schemes have borne some substantial results or if it is just 'window dressing'. Scheyvens admits that tourism as a sector fits nicely with pro-poor growth because it is a labour intensive and informal sector, based on the natural and cultural assets of the poor. Unfortunately, the concept has got into the hands of big development agencies that have a vested interest. She fears the concept may become a fad like other trends before. At best, it is a programme of minor reforms, where tourism is largely a passive beneficiary of the development. Considering the overall situation, PPT presents an example of 'tokenism [rather] than [being] transformational'. It is an industry that makes efforts to satisfy market interest and not the poor. Ashley and Haysom (2006) commented that tourism enterprises have to move beyond philanthropy and focus their policy on equity rather than growth.

Meyer concurs with Scheyvens that it would be unwise to think that stakeholders would place the PPT agenda as a top priority, setting aside their profit motives – nor would it be included in mainstream tourism. In fact, it needs the cooperative efforts of the public and private bodies along with local partners. The fact of the matter is that major players are still concerned with profit maximization. Nevertheless, recently, private tourism stakeholders have displayed greater social responsibilities and efforts to reduce poverty (Scheyvens, 2011: 143). There is a need for a well-considered shift in approach in order to achieve positive results in poverty-alleviation.

David Harrison (5.3) raises some significant questions such as, 'Are small benefits sufficient and sustainable?' PPT does not take into account the environmental, social or cultural impacts of growth-oriented strategies of tourist development; the poor might only be passive participants; business is there to make a profit, not to service the poor. Harrison concludes that any kind of tourism can benefit the poor, provided governments are actively committed to the welfare of citizens. Elsewhere, Scheyvens (2003) asks a probing question about who should manage tourism in the host community. Paul Peeters (5.4) observes that current practices of running tourism are incompatible with sustainable development because of the effects from growing greenhouse gas (GHG) emissions. Hence, present PPT will not help in reducing poverty. He suggested a refocusing on domestic tourism in order to re-distribute wealth in poorer countries.

The concept of PPT is relatively 'untried and untested' without any road-map or best practice (Roe *et al.*, 2004). Some lessons, drawn from the experiences of scholars, may be summed up briefly. PPT needs diversity of action from micro-to macro-level. PPT should be incorporated into the tourism development strategies of government or business. PPT works best where the wider destination is developing well; poverty impacts work better in more remote areas; PPT strategies often involve the development of a new product which should be integrated with the mainstream.

More humane forms of tourism work best for the well-being of destination communities, such as volunteer tourism (where guests care for the hosts), accessible tourism (for the disabled), tourism for the elderly (senior tourism), medical tourism, ethical tourism, health and wellness tourism, spiritual tourism, education tourism and the like.

Volunteer Tourism

Volunteer tourism (VT) has emerged as one of the most benign forms of sustainable tourism. It benefits the volunteers, the host community and the local environment. It is ideally small in size ('small is beautiful'). VT comes closer to the Boorstinian traveller and is the antithesis of mass tourism.

Volunteer tourists immerse themselves in the host culture: they dress like their hosts, enjoy their indigenous food, try to learn their language, teach English to the host children, construct their houses and repair their paths and do as many things as they can for the good of the community. McIntosh and Zahra (2008: 179) state that with VT, 'a new narrative between the host and guest is created. This is mutually beneficial'. Volunteer tourists donate their time, energy and money to improving the quality of the host community. This presents an enviable example of a cross-cultural relationship. They are often metaphorically called ambassadors of goodwill. They experience enhanced awareness, increased confidence, self-contentment and reject materialism – it is a discovery of the self (Wearing, 2001). With so much righteousness and altruism attached to voluntourism, scholars have found fault in the activity while Jim Butcher (6.2) suspects, 'it may not be as good as it seems' because it encourages a dependency model for VT projects, which are short-lived. Considering its limitations, Butcher doubts its promised benefits. He cryptically observes that VT today is a personal and lifestyle strategy rather than making any difference to the world. Daniel Guttentag (6.1) argues that the concept needs more analysis and field experience to pass any judgement – good or bad. Better results can be achieved if VT-sponsoring organizations and the opinion-leaders of the host community sit together to discuss issues and arrive at grounded solutions.

It is eminently desirable that VT organizations should take the responsibility of guiding their volunteers before sending them to host countries (Guttentag, 2009).

Eliza Raymond (6.3) suggests a 'bottom up' approach. This means that VT organizations should find out the needs of the host community and then match these with the right kinds of volunteers instead of getting a volunteer 'dumped on the project'.

The three scholars agreed that the VT phenomenon should be researched from the perspective of the host community and VT-organizers. Not much is known about the impacts of VT on the host community – how do they react to the presence of foreign visitors because most VT destinations are located in remote and far flung areas, marked by insularities, isolation and cultural sensitivities?

Tourism and Welfare

The tourism industry put on a personable countenance when it embraced welfare, ethics and responsibility. It made it easier to achieve sustainability. This chapter is contributed by Derek Hall and Frances Brown, Jim Butcher and David Fennell. Hall and Brown (7.1) focus on the tourism industry's response to welfare and ethical issues and examine the nature of responsibility. They suggest a welfare-centred approach, which is applicable to both mass tourism and alternative tourism, and which places tourism within its wider social approach. They elaborate on the concept of Corporate Social Responsibility (CSR), and how it forges a stronger connection between business and society and improves the business environment (Hall & Brown, 2006). This will also benefit tourism industry stakeholders as well as the local community besides improving the quality of life of the workforce.

Jim Butcher's main point is that the discussion of welfare has been cut adrift from, and is often viewed as antithetical to, economic growth. The questioning of mass tourism as lacking in ethical credentials, and the talking up of small scale niches, is indicative of this. He argues that a particular ethical standard (typically characterised by being small scale, local and organic) is too often put forward as a universal ethical imperative when it comes to what is good for communities. Therefore attempts to benchmark must always presuppose a prior view of 'good' development. Butcher wants to problematize this ethical climate, rather than benchmark it.

David Fennell (7.3) laments that the ethical philosophy has been delayed in reaching the tourism discipline, while the business motto should be 'We choose to be ethical if ethics pays'. He admits the concept of ethics is hard to define, even harder to practice but is essential to the industry. He writes that the philosophy needs more serious deliberation than a 'surface approach'. In fact, a 'critical understanding of "deep ethics"' is needed.

Tourism Education

Education and training enhance the quality of human resources. This chapter (Chapter 8) addresses the contents of these mutually exclusive terms. Chris Cooper (8.1) wants to know the state-of-the-art of the contemporary tourism industry. He is anxious to know whether or not the current educational system is alert to the demands of the 21st century, particularly the equipment of the 'knowledge economy'. He emphasizes a perfect relationship between tourism phenomenon, tourism knowledge and tourism curriculum. Cooper laments that, the educational system of tourism is still living in the past, with a questionable gap between educators and the tourism sector, between educators and researchers. This hiatus hinders the much needed transfer of knowledge. Lisa Ruhanen (8.3) concurs with Cooper that tourism education has marginally changed during the last 40 years and that this gap should be urgently filled in order to prepare a skilled and knowledge-based workforce for the tourism industry. She urges there should be innovative and meaningful change for this fiercely fast moving and progressive industry as it faces new challenges (e.g. climate change, terrorism). Brian Wheeller (8.2) takes the debate on a different slant. He argues with Cooper that the two – education and training – are not the same, as 'education has a far broader remit and deeper responsibility than mere vocational training'. He comments that, higher education is to 'think' rather than to 'do'. Brian is against the idea of universities preparing students for the private sector, and for menial jobs. All three authors focus on the right curriculum that responds to the demands of the industry.

Post Colonialism

Derek Hall and Frances Brown (9.1) discuss the significance of educational morality based on ethical responsibility and its need to come into the mainstream of tourism to gain better competitive advantage. Diverting on Hendry's concept, of bi-moral society, they question, 'Who is tourism and training ultimately for?' Quoting Fennell, they suggest, that an 'understanding of self-interest' is a 'first order need' for our consideration of ethics in the practice of tourism. They illustrate the phenomenon of 'self-interest' by comparing the roles of 'Northern/Western' consultants, working in less-developed countries and developed societies. They are concerned that the tourism academy is divided between the 'intellectual pursuit' of conceptual thinking and critical analysis and 'acquisitionism'. They assert, referring to 'aid-delivery', 'too much is often consumed by consultants and not enough is delivered to the needed assistance in teaching and infrastructure' – a 'boomerang aid'. They narrate a sordid practice of recruiting high fee-paying students and exploiting them – a quasi colonial process. Universities have shifted from more altruistic goals to a policy of self-interest – Finland and Sweden are singular exceptions though, who, for a long time, have imparted free education for all at the point of delivery. Hall and Brown wish that the tourism academy could return to a more traditional, ethically-based education morality!

Michael Hall (9.2) takes a broader view, based on the globalization of higher education that has spread new-liberal ideas with respect to governance and education policy. He discusses the concepts of 'academic capitalism', 'the entrepreneurial university' and 'new managerialism'. Academic capitalism briefly means an 'encroachment of the profit motives into the academy'. Contemporary tourism faces the challenge of changing the structure and institutional arrangement. This is a huge task.

Erik Cohen (9.3) rejoins that it is impractical to suggest that by introducing 'ethics' to the curriculum, the problem will be solved. He believes that ethical conduct can be learned from the personal conduct of educators and from practical training. In the absence of an accepted professional code of ethics only 'responsible restraint of self-interest' would help long-term sustainability of the tourism industry.

Authenticity–Inauthenticity

This chapter looks at the contemporary debate on the much discussed subject of authenticity – a key driver for most tourist experiences. A simple word, authenticity refers to the 'genuine, unadulterated real-thing', but it becomes complex when seen from a tourist perspective. It is often problematic to identify what is authentic and what is not. Globalization, societal changes and technological progress, modernity and standardization of tourism products have severely affected their authenticity. Boorstin (1964) argued that most tourists thrive on pseudo-events. Later MacCannell (1999) asserted that inauthenticity still feeds on tourists' curiosity; they survive on contrived attractions. It is a polysemic concept with many connotations. For example Brown's (1996: 39) tourist quest for the authenticity of the Other and a 'quest for self' sits alongside Wang's (1999) 'constructive authenticity' as 'verisimilitude'. It all began with MacCannell's stage-authenticity which Kjell Olsen (10.2) elaborates critically. Erik Cohen (10.1) goes into detail on subjective and objective authenticity. Philip Pearce (10.3) points to 'mundane authenticity'. Cohen observes that 'contemporary tourism appears to be moving into a "post authentic age"'. Olsen argues that authenticity could be seen as a 'Western grand idea' but that this is not applicable to non-Western culture, where it does not hold good. In a post-tourist era, the concept of authenticity is succumbing to the law of change – 'nothing is authentic' (Robinson & Wiltshier, 2011: 56). Cohen (1988), in contrast, suggests 'emergent authenticity' (where inauthenticity can become authentic over time). To Pearce 'authenticity still matters'. Cohen, Olsen and Pearce continue this debate in this chapter.

Heritage Tourism

So often voices are heard that say tourists destroy the heritage they have come to experience. Such a resonance generally comes from the managers of heritage resources and sites, sometimes by politicians and the locals themselves; 'localites' are distressed by the presence of mass tourists, originating from diverse cultures and with diverse behaviours. Tourists generally are not iconoclasts but, on the contrary, add to the financial gains of the local economy and generate some employment. Gregory Ashworth (11.1), in a dramatic vein, argues both *for* and *against* five serious accusations levelled at tourists and the tourist industry (e.g. heritage is trivial, inappropriate motives/ behaviours, damage to heritage, more worthy uses of heritage are being crowded out, etc.). The World Monument Fund identifies three major threats to heritage sites: political conflicts, climate change and tourism. To most of the charges, Ashworth's answer is inadequate planning and poor management and still poorer surveillance. Ashworth in his verdict expects the answer to two major questions: 'the distinctiveness of the tourists' and the 'distinctiveness of the heritage'.

Brian Wheeller (11.2) argues that more than tourists, 'the forces of commodification often distort heritage to match tourists' demand'. It is difficult to say who is responsible for the distortion of heritage – tourists or locals? If tourists are responsible for the destruction of heritage, Wheeller quips, are they responsible tourists? And responsible to whom?

Michael Hall (11.3) holds that the linkage between heritage and identity is crucial to understanding the importance of heritage – their preservation essential because 'they are part of our identity'. Heritage is a resource – and a resource can be moulded into anything that is useful or beautiful; hence Hall favours the 'creative destruction' of heritage.

Nature-Based Tourism

Tourism put its best face on with ecotourism around the 1980s. It promised conservation and preservation of the environment, as well as the possibility of soft and benign recreation/tourism education and enhancing the well-being of the local community. The idea was delightful but could not be contained into a well-defined concept. Since it happened in a natural setting, a variety of nature-based tourism activities sprang up, such as wildlife tourism, adventure tourism, fishing, bird-watching, cycling, and so on. Often nature-based activities and ecotourism are viewed synonymously, which ecotourism researchers do not accept. A team of three specialists in the field were asked for a discourse on the problem.

David Fennell (12.1) argued for a clearer distinction between ecotourism and other forms of 'natural resource-based tourism'. The statement had the support of ecotourism thinkers. Seen historically, 'nature tourism' was a label given to what we call ecotourism today and that included many outdoor recreation-based experiences. Over time, because of a lack of any precise definition for ecotourism, the term was misused by commercial operators leading to 'green washing'. Buckley, Coghlan and Weiler believe that ecotourism is distinguishable more by its ideology orientation than by its attributes. Fennell asserts that a 'good definition leads to good policies and good practices'. He laments that some of the most quoted definitions of ecotourism are narrow and passive and it is regrettable that such definitions have been the basis of policy formulation by government and the industry that made ecotourism a misnomer. Ecotourism, as an acme of tourism-forms, should be strengthened in theory and practice so that human beings may continue to enjoy the work of both nature and man. Whatever the name is, ecotourism or nature-based tourism, the naturalness of the area where it occurs must remain green and attractive, and its high aesthetic amenity values maintained (Hall & Boyd, 2005).

Tourism and Climate Change

Nature is not only benign and beautiful but it can be disastrous and dangerous too. We have taken too much from the Earth and our survival is now at stake. Our nemesis appears as different types of natural calamity (earthquakes, floods, global warming, tsunamis and the recession of glaciers). Most of this is attributed to climate change, mainly caused by the accumulation of greenhouse gases (GHG) in the atmosphere. CO_2 breeds extreme weather-events and catastrophes, though opinions vary on the origins of climate change and its impacts. It is a huge global concern. Scientists, researchers, academics, climatologists and the media are all involved in establishing the precise reasons for this climatic havoc. The United Nations has set up an international panel on climate change in 1988 to report on this threat to our climate, besides other important non-governmental organizations. Four specialists in this field of research were consulted to discuss the issue.

Michael Hall (13.1), the lead author, provides a thematic snapshot of this emergent phenomenon by focusing on the relationship between tourism and climate change research.

He elaborates on key issues such as a brief on policy context for tourism and 'climate change', tourism and IPCC assessment and Stern's significant review, major current research themes, an account of research on regional basis and lastly, the status of current knowledge on tourism and climate change. Hall goes in detail to identify knowledge gaps, especially at a regional level, that provide a shortfall when formulating effective policy. His data reveal that with lower carbon footprints, tourism has been a modest sinner.

Susanne Becken (13.2) regrets that tourism is often under-reported in some important documents on climate change, whereas tourism research has made appreciable contributions in reducing the impacts. She affirms that there are areas where tourism has the potential to show leadership. High quality research and sound policies will increase the 'Voice of Tourism'.

Ralf Buckley (13.3) provides a framework to identify research on: artificial climates, climatic variability and climatic means. Because of the knowledge still required, Buckley believes that the coastal resorts industry has responded to climate change more slowly than tropical coral reefs. 'The top-most single priority research would be the social and environmental consequences of extreme weather-events at destinations, and the flow-on effects for tourism'.

Daniel Scott (13.4) briefly sums up climate change research and events with his comments on their strengths and weaknesses. He is desirous to know how far these discourses and research findings have found their way into the research community. He discovered that research into 'adaptation' has been unsatisfactory while tourism operators seem to be optimistic; but it shouldn't be taken as simple or easy to come by – mainstreaming climate change in decision-making is not only important but unavoidable.

All in all, the experts reached consensus that high quality, interdisciplinary field research in cooperation and partnership with other sister organizations should bear some fruitful results in helping to fix policy goals and provide sound planning processes and strategies for establishing climatic equilibrium

Slow Tourism

Modern man is obsessed with living the fast life; speed characterizes his routine. He is tense and over-scheduled and constantly running a race against time. 'He has no time to stand and stare'. Tourism (mass tourism) is often accused of this by many scholars. A new form of tourism – 'slow tourism' – has therefore been created which is an antidote to the fast life. Chapter 14 is devoted to this interesting theme.

Slow tourism was born in a peace-loving city, Bra, in south west Italy. The Bra community loved the slow pace of life, organic living and so started a 'slow food' movement (1989) to celebrate this. They protested against the McDonaldization that came in their way and so brought the concept of 'slow' into the heart of their lives. 'Slow food' was the starting point for other concepts such as slow tourism, slow movement, slow cities and slow travel (Heitmann *et al.*, 2011). Slow tourism mocks the disharmony and disquietude of modern, hurried and worried life. Dennis Conway and Benjamin Timms (14.1) attempt to distinguish between slow travel and slow tourism, though they admit that the two terms are used synonymously.

Alison Caffyn (14.2) advocates the practice of slow tourism and its implementation. She cites an example of Cittaslow, the first slow city in Ludlow (UK). There are about 154 slow cities in Europe. Slow food festivals attract thousands of gastro-tourists. Caffyn believes that while slow tourism is key, transport (low carbon) and distance are crucial to the proper understanding of the phenomenon. A ride on a horse-driven cart is preferable to a motorized vehicle. Rachel Dodds (14.3) focuses on aspects of sustainability. The slow tourism concept is closer to carbon friendly travel – the staycation (also homecation) was highlighted. Given the positive qualities of slow tourism, all travellers cannot be slow tourists and mass tourism with all its frailties can't be done away with. Most important is to make tourism sustainable – big or small.

Tourism: A Patriotic Obligation

This was an exceptionally long journey beginning with the illustrious Thomas Cook (1840) and ending with the daring of Dennis Tito's fast flight on Soyuz (2001). I am fatigued, I must 'slow down' – relax or staycate for another trip. I had glimpses of this beautiful planet through a prism-glass of tourism; 'much have I seen and known; cities of men and manners, climate, councils', yet my heart is hungry – I am a part of all that I have met, yet all experiences are an arch through which 'gleams that untraveled world' – a world so large and time so short. Tourism makes it possible through its many faces to witness the glory and the gloom. That is why I said *tourism is the best*. One can see through the chapters of this book, how tourism has been shaped to match the needs of tourists – if you love crowds, become a mass tourist; if you are adventurous, become a mountain climber; if you love nature, become a green or ecotourist; if you love solitude, go to a wilderness somewhere; if you want to see some tribal life, go to a deep forest, and so on. Tourism offers a varied menu – you can make your own choice about what to try. If tourism tastes unsavoury or bitter, improve the recipe. Humans are the makers of tourism – good or bad. It is everybody's concern – the architect, planners, developers, policy-makers, residents, visitors – all who want to see tourism – it is our responsibility to make it something to relish and enjoy.

Our tourism resource managers ought to be anxious to know the subtle nuances of making tourism 'the best' it can be. Brumeister pinpoints a few 'hows': how to create possibilities for solitude without creating loneliness; how to create an architecture that is part of the landscape; how to avoid the noise that goes with any concentration of people, when more concentration is unavoidable; how to create possibilities for relaxation without excluding stimuli that are desirable; how to avoid a decline in beauty when, in reality, for most people, an empty beach is more beautiful than a full one, a primitive winding street with trees more beautiful than a highway with a gas-station, and; how to avoid the pollution that the sector of industry catering to tourism causes directly or indirectly – for tourists (cited in Singh *et al.*, 1982). These are disturbing 'hows' but a good tourism provider has to answer them and put them into practice. This book touches upon a few relevant questions and seeks to find reasonably practical answers from the experts.

The Why of the Book

Readers might also like to ask, 'what is new about this book?' There is a plethora of literature, touching on these intriguing problems, which publishing houses the world over are ceaselessly churning out. Much research work is being done in university tourism departments, institutes and research centres and these occasionally appear in print. Despite this mass of research, we are failing to answer the questions that this book has posed. There may be many reasons for this but to my mind, we have scarcely trained our students to think against the grain and not become lost in the labyrinth of definitions. For example tourism very organically evolved in the form of 'ecotourism' some 20 years ago, and there are a few examples of best practice models to follow. Scholars are still struggling to refine its definition though and are asking questions about whether it is a product or principle, rhetoric or reality?

Ever since Hector Ceballos-Lascurain defined the term ecotourism in 1987, definitional debates and discourses have never stopped. We now have a plethora of them – so many that we're almost at the point of confusion. A universally accepted definition is needed to deliver quality-ecotourism otherwise the term loses its meaning (Donohoe & Needham, 2006). Similarly, the concepts of sustainable development and carrying capacity have been rendered almost meaningless because of over-definition. These substance-loaded phrases have become buzzwords, dogmas and stereotypes in research usage, having lost much of their pith and marrow. We have considerable good literature on the negative consequences of mass tourism but we have a few empirical studies on the positive influence of mainstream tourism. I found it interesting to read David Weaver's (2001) essay 'Ecotourism as mass tourism: Contradiction or reality?' in *The Cornell Hotel and Restaurant Administration Quarterly* where he explores the possibilities for making mass tourism sustainable. He states: 'The conventional tourism industry is learning from its past mistakes and realizes, like TUI, that its future, profitability and viability depend on adopting an attitude of enlightened self interest with respect to environmental and socio-cultural sustainability.'

He argues that there is considerable evidence that the mainstream industry made significant progress towards sustainability during the 1990s. Such penetrative probes are needed to understand this highly dynamic industry. There are problems that demand innovative approaches, sophisticated tools and intensive field work. Climate change and tourism is a hot research topic for serious scholars to work on and it is awaiting the findings of comprehensive research carried out by scholars.

This book has a large canvas, covering the problem areas in tourism and the subtle research-nuances in diverse environments. Methodologically, at first, problems of critical significance were identified and then a scholar, having significant expertise in that subject, was invited to discuss the problem and all its pros and cons. This core piece was further probed, laid bare by three or four invited critics; thus building a comprehensive critique of the theme. The lead author of the probe would finally give concluding remarks or elaborate whether or not the main argument raised by them was upheld, had failed or needed a new direction and approach for further research.

Thus the chapters in the book collectively present a platform for minds to meet and unravel research-conservatism and stereotypes in tourism. It will encourage scholars to think laterally and probe the consistency of critical notions and research trends whose heuristic value is often taken for granted.

Content-wise the book is multidisciplinary, embracing disciplines like geography, anthropology, sociology, economics, education, biology and humanities. The book may not provide comprehensive answers to some chronic problems, which need innovative approaches, but I am sure it will provoke the reader to think about possible solutions. With this book, planners, policy-makers and resource managers of tourism will be equipped with the knowledge to make more realistic projects.

References

Archer, B., Cooper, C., and Ruhanen, L. (2004) The positive and negative impact of tourism. In W.F. Theobald (ed.) *Global Tourism* (3rd edn). Amsterdam: Elsevier.

Ashley, C. and Haysom, G. (2006) From philanthropy to a different way of doing business: Strategies and challenges in integrating pro-poor approaches into tourism business. *Development Southern Africa* Vol. 23(2): 265–280.

Bilsen, F. (1987) Integrated tourism in Senegal: An alternative. *Tourism Recreation Research* 13(1): 19–23.

Boorstin, D. (1964) *The Image: A Guide to Pseudo Events in America*. New York: Harper and Row.

Brown, D. (1996) Genuine fakes. In T. Selwyn (ed.) *The Tourist Image: Myth and Myth Making in Tourism* (pp. 33–47). Chichester: Wiley.

Brundtland, G.H. (1987) *Our Common Future*. Oxford: Oxford University Press.

Cohen, E. (1972) Towards a sociology of international tourism. *Social Research* 39(1), 64–82.

Cohen, E. (1973) Nomads from affluence: Notes on the phenomenon of drifters-tourism. *International Journal Comparative Sociology* 14(1&2): 89–103.

Cohen, E. (1988) Authenticity and commoditization in tourism. *Annals of Tourism Research* 15: 371–386.

Cohen, E. (1989) Alternative tourism – A critique. In T.V Singh, H.L. Theuns and F.M. Go (eds) *Towards Appropriate Tourism: The Case of Developing Countries* (pp. 127–142). Frankfurt: Peterlang.

de Kadt, E. (1979) *Passport to Development*. New York: Oxford University Press.

Donohoe, H.M. and Needham, R.D. (2006) Ecotourism: The evolving contemporary definitions. *Journal of Ecotourism* 5(3): 192–210.

Eadington, W. and Smith, V. (1992) Introduction: The emergence of alternative forms of Tourism. In V. Smith and W. Eadington (eds) *Tourism Alternatives: Potentials and Problems in the Development of Tourism* (pp. 1–12). Philadelphia: University of Pennsylvania Press.

Eber, S. (ed) (1992) *Beyond the Green Horizon: A Discussion Paper on Principle of Sustainable Tourism*. London: Tourism Concern and WWF.

Gandhi, M.K. (1948) *Gandhi's Autobiography: The Story of My Experiments with Truth*. Washington, DC: Public Affairs Press.

Ghimire, K. (1997) *Emerging Mass Tourism in the South*. Geneva: United Nations Research Institute for social development.

Guttentag, D.A. (2009) The possible negative impacts of volunteer tourism. *International Journal of Tourism Research* 11(6): 537–551.

Hall, C.M. and Boyd, S. (2005) *Nature-Based Tourism in Peripheral Areas: Development or Disaster?* Clevedon: Channel View Publications.

Hall, D. and Brown, F. (2006) *Tourism and Welfare: Ethics, Responsibility and Sustained Wellbeing*. Walling Ford: CABI Publishing.

Heitmann, S., Robinson, P. and Povey, G. (2011) Slow food, slow cities and slow tourism. In P. Robinson, S. Heitmann and P.U.C. Dieke (eds) *Research Themes for Tourism* (pp. 114–27). Wallingford: CABI Publishing.

Macaulay, T.B. (1913) *Warren Hastings*. London: David Salmon.

MacCannell, D. (1999) *The Tourist: A New Theory of the Leisure Class*. New York: Shochen Books.

Mak, J. and Moncur, J.E.T. (1995) Sustainability and tourism development: Managing Hawaii's unique tourist resource – Hanauma Bay. *Journal of Travel Research* 33(4): 51–56.

Mowforth, M. and Munt, I. (1998) *Tourism and Sustainability: New Tourism in the Third World*. London: Taylor and Francis.

McIntosh, A.J. and Zahra, A. (2008) Journey for experience: The experience of volunteer tourists in an indigenous community in developed nations – A case study of New Zealand. In K. Lyon and S. Wearing (eds) *Journey of Discovery in Volunteer Tourism* (pp. 166–181). Walling Ford: CABI.

MercoPress (2011) Press release from South Atlantic News Agency about Antarctica Visitors. June 14, 2011.

Murphy, P.E. (1985) *Tourism: A Community Approach*. New York: Methuen.

Page, S.J. (2007) *Tourism Management: Managing for Change* (2nd edn). Oxford: Butterworth-Heinemann.

Poon, A. (1993) *Tourism, Technology and Competitive Strategies*. Walling Ford: CABI.

Pearce, P.L., Filep, S. and Ross, G. (2011) *Tourists, Tourism and the Good Life*. London: Routledge.

Robinson, P. and Wiltshier, P. (2011) Community tourism. In P. Robinson, S. Heitmann and P.U.C. Dieke (eds) *Research Themes for Tourism*. Wallingford: CABI Publishing.

Roe, D., Goodwin, H. and Ashley, C. (2004) Pro-poor tourism: Benefiting the poor. In T.V. Singh (ed) *New Horizons in Tourism: Strange Experiences and Stranger Practices* (pp. 147–161). Wallingford: CABI Publishing.

Scheyvens, R. (2003) Local involvement in managing tourism. In S. Singh, D.J. Timothy and R. Dowling (eds) *Tourism in Destination Community* (pp. 229–252). Wallingford: CABI.

Scheyvens, R. (2011) *Tourism and Poverty*. New York: Routledge.

Seekings, J. (1982) Tourism for the public good. In T.V. Singh (ed) *Studies in Tourism Wildlife Parks Conservation* (pp. 284–90). Delhi: Metropolitan Book Co.

Sharpley, R. (2012) Island tourism or tourism on island. *Tourism Recreation Research* 37(2) (forthcoming).

Singh, T.V. (2004) Tourism searching for new horizons: An overview. In T.V. Singh (ed) *New Horizons in Tourism: Strange Experiences and Stranger Practices* (pp. 1–10). Wallingford: CABI.

Singh, T.V., Kaur, J. and Singh, D.P. (eds) (1982) *Studies in Tourism Wildlife Parks Conservation*. Delhi: Metropolitan.

Splettstoessor, J., Landau, D. and Headland, R.K. (2004) Tourism in the forbidden lands: The Antarctica experience. In T.V. Singh (ed.) *New Horizons in Tourism: Strange Experiences and Stranger Practices* (pp. 27–36). Wallingford: CABI Publishing.

Turner, L. and Ash, J. (1975) *The Golden Hordes – International Tourism and Pleasure Periphery*. London: Constable.

UNEP (2001) *Consumption Opportunities: Strategies for Change*. Geneva: UNEP.

UNWTO (2011) Tourism highlights. Madrid: UNWTO: 4.

Wall, G. and Mathieson, A. (2006) *Tourism: Change, Impacts and Opportunities*. London: Pearson.

Wang, N. (1999) Rethinking authenticity in tourism experience. *Annals of Tourism Research* 26(2): 349–370.

WCED (1987) *Tokyo Declaration, 27 February (Our Common Future)*. Oxford: Oxford University Press.

Wearing, S. (2001) *Volunteer Tourism: Seeking Experiences that Make Difference*. Wallingford: CABI.

Weaver, D. (2001) Ecotourism as mass tourism: Contradiction or reality? *The Cornell Hotel and Restaurant Administration Quarterly* 42(2), 104–112.

Chapter 1
Mass Tourism and Sustainability: Can the Two Meet?

David Weaver, Ralf Buckley, Brian Wheeller and Bill Bramwell

Context

Weaver initiates this probe by arguing that the mass tourism industry is moving toward sustainability, but more as 'paradigm nudge' (i.e. the opportunistic adjustment of capitalism) than 'paradigm shift' (i.e. a fundamental change of worldview). Largely, this occurs as a response to normative consumerism, which at best displays superficial environmentalism that claims concern about the environment but an unwillingness to adopt inconvenient behaviours. Finally, he argues that academics are not actively leading any effort to move beyond superficial environmentalism. Buckley responds by corroborating and elaborating the paradigm nudge argument through his overview of codes, guidelines, certification schemes, awards and other practices. He adds that there are innovative businesses that should be emulated, and enabling environmental legislative structures that should be implemented. Wheeller responds by rejecting the idea of sustainable mass tourism, due to the inherently unsustainable nature of transportation. He laments the continuation of some scholars to give credibility to this 'charade'. Bramwell concludes by reminding us of the inherent complexity of 'sustainable tourism' as both a theoretical and practical construct, which engenders both confusion and conflict. Critical assessments and normative responses are both warranted to effect positive change.

1.1

Towards Sustainable Mass Tourism: Paradigm Shift or Paradigm Nudge?

David Weaver

From The Margins to the Mainstream

There is perhaps no conceptual theme so dominant in the contemporary tourism literature as 'sustainability'. Yet, almost none of this literature existed prior to 1990, as a quick search of the search engine *leisuretourism.com* will attest. Equally remarkable is the extent to which the vocabulary of sustainability has been formally adopted by conventional tourism-related corporations and organizations over the same time period. High profile examples illustrative of this institutionalization include the establishment of the Sustainable Development of Tourism Department at the United Nations World Tourism Organization, the *Blueprint for New Tourism* manifesto of the World Travel and Tourism Council, and the explicit sustainability focus of the UNEP Tourism Programme. Formal sustainability entities have also been established within major corporations such as Marriott, British Airways and TUI, while initiatives such as the International Hotels Environment Initiative (IHEI), the Ocean Conservation and Tourism Alliance (involving the ICCL), and the Tour Operators Initiative for Sustainable Tourism Development all indicate official sector-specific engagements with the concept since the early 1990s.

But what substance underlies this rhetorical flourish? To the degree that this issue has been empirically investigated, it appears that the tourism sector's engagement with sustainability is neither broad nor deep. A lack of breadth is apparent in the prominence of just a few corporations such as those mentioned above, and by the fact that segments of the tourism industry such as the travel agency sector still have almost no involvement (Weaver, 2006). The absence of depth is evident in the dominance of selective practices such as recycling, energy use reduction and some types of community outreach that expend few resources but often yield substantial cost savings and/or positive publicity (Enz & Siguaw, 2003). The increasingly ubiquitous hotel bathroom signs (made of course from recycled plastic or paper) that exhort guests to re-use bath towels and face cloths in order to 'Save Mother Earth' are frequently the extent of a company's efforts in this regard. While it can be argued that such practices, unless they support corporate green-washing activity, are not without some ecological and social benefit, they are still essentially cosmetic and reflect no fundamental change in the underlying assumptions that inform the actions of the typical tourism corporation. Thus, one often hears of a corporation's commitment to 'smart *growth*' or 'sustainable *development*', but almost never of any decision to actually *curtail* growth or *cancel* a development in favour of ecological or social considerations.

This pessimistic assessment is reinforced by the current state of play with regard to the primary mechanisms that demonstrate the degree to which particular companies or products adhere to the criteria of sustainability. Codes of conduct are abundant, but these are regarded by some as inadequate by themselves due to their vague, voluntary and self-regulated character (Mason, 2007). Awards address some of these problems by relying on third party adjudication, but are still hindered by the sometimes questionable protocols through which certain products or businesses are nominated, assessed and rewarded (Weaver, 2006). Their emphasis on high profile annual awarding ceremonies and the common practice of including the sponsor name in the title (e.g. the *British Airways Tourism for Tomorrow Awards*), moreover, imply a central publicity motive that affiliates awards with superficial practices such as recycling and energy reduction (Font & Tribe, 2001). Finally, certification-based ecolabels in principle represent the gold standard of quality assurance, based as they are on the formal recognition of products that adhere to rigorous standards of practice as assessed by qualified third-party verifiers (Black & Crabtree, 2007). The limited duration of such recognition (usually one year) better ensures that product quality is maintained so that certification status can be renewed. According to Buckley (2002), effective ecolabels require sufficient 'guts', or substantive criteria, and 'teeth' – that is, effective enforcement to ensure the exclusion of non-qualifying products.

If the widespread adoption of effective ecolabels is the litmus test of sustainable tourism, then contemporary tourism does not receive a passing grade. Aside from relatively effective but regionally- and sector-specific ecolabels such as Australia's EcoCertification Programme (focused on ecotourism and nature-based tourism) (Thwaites, 2007) and the Blue Flag Programme (focused on European beaches) (Gössling, 2006), there are no ecolabels that have made a substantial penetration into the global tourism industry as a whole. Green Globe, which is the best known aspirant to such a universal ecolabel (Parsons & Grant, 2007), reported on its website as of mid-February 2007 just 88 Certified Members (the highest level of certification offered by the ecolabel) and 108 that were Benchmarked (an intermediate status promoted as a preparatory stage towards Certified status). The distribution of this miniscule membership, moreover, is geographically skewed, with North America accounting for just one member at either of these two levels and all of Europe just seven (three Certified, four Benchmarked). Other problems associated with Green Globe have been discussed by Honey and Rome (2001).

Mirroring the Mainstream

Kuhn's (1970) idea of paradigm shift offers a useful framework for contextualizing the above synopsis of the tourism status quo with regard to sustainability. In this perspective, the contradictions and anomalies inherent in laissez-faire conventional mass tourism, encapsulated in the self-defeating dynamics of the destination life-cycle model (Butler, 1980), led to recognition in the early 1990s of the need for, and possibility of, sustainability across the entire spectrum of tourism activities, and not just within the parameters of the small-scale 'alternative tourism' that was initially proffered as the best solution to resolving those contradictions. Rather than evidence of tourism's failure to engage sustainability in a meaningful way, the status quo accordingly can be seen as evidence that tourism is in the early stages of a slow but inevitable transformation indicative of a paradigm shift. The level of current involvement, though shallow and modest in an absolute sense, would have been unimaginable just two decades ago and represents tangible progress. The genius of sustainable tourism as a vehicle for paradigm shift is its adaptability, embodied in Hunter's (1997) suggestion of 'strong' and 'weak' paths of sustainable development, the former being appropriate for ecologically and/or socio-culturally sensitive destinations and the latter appropriate for destinations already heavily urbanized or otherwise modified.

However, there is another, much less deterministic, way of interpreting this clash of paradigms, based on the observation that confrontation between competing models is usually characterized by mutual appropriation; that is, supporters of each model consciously or unconsciously incorporate those complementary elements of the competing paradigm that make their own paradigm stronger and, therefore, more competitive; whichever prevails is a synthesis of the two combatants. Thus, advocates of the green paradigm embrace scientific method and are not completely hostile to free markets, thereby gaining support from those who might otherwise be hostile to its basic assumptions about the relationships between humans and the natural environment. Similarly, advocates of the existing paradigm qualify support for free markets, growth and development with green adjectives such as 'smart' and 'sustainable' to win support from moderate environmentalists and support strategies of 'strategic self-regulation' that serve to pre-empt further government intervention (Maxwell *et al.*, 2000). In this scenario, an actual shift from one paradigm to another is not inevitable and it is conceivable that the old paradigm will prevail – synthesised into a softer and greener adaptive entity, but basically the same old paradigm nonetheless. This is the scenario of 'paradigm nudge' described in the title of this paper.

Whether the current confrontation is ultimately assessed as part of a paradigm shift or a paradigm nudge will, in this commentator's opinion, depend on the actions of the general public. In very broad terms, consumer surveys reveal that societies in the more developed countries resemble a bell curve wherein about one-quarter of adults (these ratios vary from country to country) are 'non-environmentalists', including some who are hostile towards environmentalism but most for whom the environment does not register as a serious personal or social concern (PollingReport.com, 2007). In contrast, approximately one-fourth of society at the opposite pole has 'environmentalist' tendencies in the sense that they are concerned about the environment and are more or less willing to modify their lifestyle and sacrifice accordingly. 'Activists' generally account for just a small minority even within this group and are the leading edge in inducing increased sales of products such as organic food and fuel-efficient vehicles. In the centre, about one-half of society is a mainstream consisting of 'veneer environmentalists' who express concern about the environment but are unwilling to make substantive personal concessions for the sake of the latter. It can be argued that the veneer sustainability of the corporate world – and government – is simply a reflection of and response to the veneer environmentalism of society more generally (and not just pervasive greenwashing, as some aver). The lack of corporate interest in expensive certification-based ecolabels, by this logic, is because these quality assurance mechanisms currently have almost no resonance among tourists, and thus provide certified products with no significant competitive advantage over their non-certified competitors. Change will, therefore, have to be driven by consumers rather than the industry.

For contemporary mass tourism, the situation may even be worse than in other sectors in terms of actual consumer agitation for 'green' corporate behaviour and products, even though concerns about the environment and preferences for environment-friendly tourism experiences are consistently elicited from consumer surveys (Chafe, 2007). In part, this is because contemporary alternative tourism activities such as 'voluntourism', farm vacations and homestays already siphon away many of the relatively hard-core environmentalist travellers who would otherwise serve as vehicles of this agitation. In addition, it is very difficult to obtain credible information on the web and through other media about sustainable mass tourism products, which are essentially invisible to the consumer. The relative infrequency of leisure travel also impedes the establishment of environmental purchasing habits (Hjalager, 1999), while the vast majority of potential mass tourists have no awareness at all of the underlying issues and theories associated with unsustainable tourism, and have been exposed to no tourism-specific *cause célèbre* equivalent to climate change that would stimulate consumers to significantly modify their travel behaviour. Finally, tourism-dedicated organizations such as the UK-based Tourism Concern and the Ecumenical Coalition on Third World Tourism lack critical mass to function as effective lobbying and publicity entities, and are largely unknown among consumers (Turner *et al.*, 2001).

Beyond the Status Quo

How, then, can the consumer mainstream be induced to move beyond veneer environmentalism with regard to its mass tourism expectations and behaviour? Greater public awareness and activism is required, and in this regard conventional mass tourism needs its Rachel Carson (1962) – a charismatic (and media-savvy) representative who can attract and sustain the public's attention, raise awareness and induce widespread change in expectations and behaviour. To ignite the latent environmentalism of veneer environmentalists, this outreach should coalesce around an issue such as climate change (probably including natural disasters such as Hurricane Katrina) that is already capturing the public imagination and implicates tourism-related activity such as aircraft greenhouse gas emissions. Residents of major tourist regions and resorts might constitute a particularly receptive target audience because of their vested interest in the well-being of the tourism industry. A more radical idea that could complement the public relations effort and represent a tangible outcome of the latter is the introduction of a certification protocol for *tourists* to work in tandem with those being introduced to tourism products. While such an idea would require a great deal of thought as to criteria, implementation and expectations, the basic idea involves a reciprocal arrangement based on the principle of enlightened self-interest whereby certified tourists agree to patronize qualifying certified products and behave in certain ways in exchange for discounts and other privileges.

As stated earlier, the environmentalism bell curve in society is a phenomenon primarily of the more developed countries, and this is notable to the extent that residents of those regions continue to account for the large majority of international and domestic tourists (UNWTO, 2006b). Attention must be paid, however, to the growing proportion of tourist arrivals originating in less developed countries in which environmental awareness and activism – much less recognition of the negative impacts of tourism – is more incipient. China is illustrative, with over 1.2 billion domestic tourist trips reported for 2005, a 100 million increase over the previous year (China National Tourist Office, 2006). In addition, it is anticipated that China will produce 100 million outbound tourists by 2020, compared with 31 million in 2005 (including visits to Hong Kong and Macau) (UNWTO, 2006a). Chan (2001) reports that strong environmentalist sentiments solicited from residents of Beijing and Guangzhou – that is, those in the urban vanguard of modernization – did not correspond with a pattern of strong green agitation or purchasing. Prospects for attaining sustainable mass tourism in China and other rapidly evolving less developed countries, therefore, appear grim in the foreseeable future, although exposure to industry best practice in ADS (Approved Destination Status) countries such as Australia may stimulate interest domestically. Government in China is also better positioned to force changes in industry, although the priority currently is more on sustained economic growth than a clean environment. Ironically, some of the most tangible evidence of substantive if inadvertent change in the more developed countries is industry-led, manifested in the decision by many major property insurers in the USA to cancel policies and/or raise premiums in hurricane-prone regions along the Atlantic and Gulf coasts (Insurance Information Institute, 2007). However, the same free market forces that are inducing such self-correction could just as easily induce accelerated development in other areas or in the same areas under different circumstances (i.e. more rigorous building standards), and thus appear to be an unstable mechanism for sustained attention to sustainable mass tourism in either more or less developed regions.

Conclusion: The Role (or Non-Role?) of Academics

By way of recapitulation, the basic contention of this probe is that the adoption of practices affiliated with the ubiquitous rhetoric of sustainable tourism is neither broad nor deep within the conventional mass tourism industry, and that, at least in the more developed countries, this veneer sustainability on the supply side mirrors the pervasive veneer environmentalism that characterizes the demand side of tourism. As such, it indicates a 'paradigm nudge' that will only progress into a 'paradigm shift' if the consumer mainstream is sufficiently motivated to demand and patronize only genuinely sustainable products and destinations within conventional tourism. The emergence of a charismatic spokesperson and the introduction of a demand-side certification programme are suggested as two ways to induce this transformation, though it is recognized that the effort is only partial if it does not somehow take into account the burgeoning tourist markets of the less developed world.

It is fitting to conclude this probe by considering the role that could be played by the academic community. Formidable intellectual capital, opportunities to influence future managers and other sector leaders through teaching and advisement, and a core occupational mandate to conduct research and pursue useful knowledge are all assets that confer this community with enormous potential to lead such change. The reality, however, is that academics within the field of tourism studies are engaged in sustainable tourism more on the theoretical and descriptive than empirical level. Rigorously conducted research suffers from paucity in many areas of core relevance, including the assessment of the credentials of certified tourism products, indicator identification and their appropriate thresholds, indicator monitoring, measuring consumer attitudes specifically about tourism and weighing these against actual tourism behaviour, and effectiveness and impact of industry codes of conduct and awards, to name a few. Equally unfortunate is the failure of that literature which does exist to inform or have meaningful influence over the actions of the tourism industry, although this problem applies well beyond sustainable tourism and says much about the deeper long-standing dysfunctional relationship between academics and practitioners within tourism. These are indicative of a deeper malaise that may not be overcome soon enough, and in the meanwhile one hopes, at least, that the community of tourism academics can produce its Rachel Carson.

1.2

Is Mass Tourism Serious About Sustainability?

Ralf Buckley

Is mass tourism making a serious paradigm shift towards sustainability or is it merely a nudge? This issue may be examined from a range of perspectives (Weaver, 2007), including: treatments in the academic literature; environmental management performance in the industry itself; tools such as ecolabels and awards which are intended to improve sustainability; the relative roles of consumers and corporations; the different perspectives of tourists from different nations; and the interplay between academic analysis and industry practice. These are considered in turn below.

Sustainability in the Tourism Research Literature

Sustainability is a major theme in the academic tourism literature, but these analyses are largely conceptual rather than practical (Saarinen, 2006). Remarkably little has been published about the actual practices of commercial tourism corporations and operators in reducing their environmental impacts; and even less about the ecological significance of any such reductions, the reasons why they have been undertaken, the benefits gained by the companies concerned, the financial scale of the measures compared to company turnover, and similar practicalities. Academic tourism journals, and their referees, seem to look down on such submissions as too descriptive. This, however, applies not only to sustainability and environmental management, but arguably also to academic analyses of the tourism industry more generally. This is unfortunate, because it also means that commercial operators pay little attention to the academic literature. To a certain degree this may reflect differences in approach between the social sciences and the natural sciences, though not entirely. The natural sciences demand real-world observations as well as theory, but so do some of the social sciences.

Environmental Management Practices in the Industry

Environmental management practices which have actually been adopted within the mainstream tourism industry are generally only those which save money, e.g. by reducing energy or water consumption; or which provide opportunities for public relations exercises; or both. There are a small number of companies which have indeed made significant investments in conservation and environmental management (e.g. Buckley, 2003, 2006). In general, however, and especially in the mainstream tourism accommodation, transport and large-activity sectors, there is a great deal of publicity about rather small measures. Some companies boast about environmental measures which in fact merely represent compliance with applicable environmental law, or with conditions of development consent or operating permits. Some adopt (and publicize) rather small-scale energy or water conservation, or materials recycling programmes, as a way to cut costs without reducing guest satisfaction. Some claim green kudos from programmes which simply encourage guests to donate to environmental causes, without any financial contribution from the company itself. Some invest a miniscule proportion of annual revenue, as low as hundredths of a percent, in heavily publicized measures which they use to apply for environmental awards, which then provide political capital when the company wants to negotiate permits for much larger-scale operations or expansions which do have significant environmental impacts.

Environmental Codes and Guidelines

There are numerous manuals, guidelines and codes of conduct for different components of the tourism industry and different types of tourism activity, ranging from manuals produced by the International Hotels Environment Initiative over a decade ago to the *Green Guide* series for outdoor tour operators (International Centre for Ecotourism Research, 2007). Some protected area management agencies also provide their own codes of conduct. The Great Barrier Reef Marine Park Authority (2007), for example, has quite a detailed and comprehensive set. Some of these codes are indeed very vague or basic. Some, however, contain quite detailed and comprehensive sets of operational instructions (Garrod & Fennell, 2004). Even for these, however, there do not seem to be any rigorous tests to determine whether tour operators actually do change their practices in response to such codes. Where current compliance with existing codes has been tested, it has generally proved to be quite poor, even where the codes are enshrined in legally enforceable regulations and even when operators know they are under observation (Scarpaci & Dayanthi, 2003; Waayers *et al.*, 2006).

Environmental Award Schemes

Over recent decades a number of environmental award schemes have been established in the tourism sector (Font & Buckley, 2001), and some of these are still operational. The better-known awards are highly coveted within the tourism industry, and are featured heavily in advertising materials. Presumably this indicates that these awards, or more probably their parent organizations, are recognized by tourists themselves and are, hence, valuable in marketing. The three best known are probably the World Legacy Award from National Geographic, the Ecotraveler Award from Condé Nast and the Tourism for Tomorrow Awards, established by British Airways but now run by the World Travel and Tourism Council. All of these depend very heavily on the goodwill and voluntary services of relevant experts as judges and assessors. Some include site assessments for shortlisted applicants, others do not.

Tourism Ecocertification Programmes

Ecocertification schemes, in contrast, seem to be largely ignored by individual tourists, and most such schemes have extremely low penetration of their potential target markets. As noted by Weaver (2007), the much-hyped and supposedly comprehensive Green Globe scheme has in fact been adopted only by an infinitesimally small proportion of its target market, despite very considerable investment for over a decade. The World Hotel-link scheme, a recent competitor to Green Globe which originated from a World Bank project, has achieved very much greater membership in a short period of time using a franchise-style design. It uses a customer feedback system to provide social and environmental information, which may not be rigorous but is at least transparent. The Green Globe system will supposedly certify any operator which is marginally (5%) above mean environmental performance for the industry sub-sector concerned on the basis of energy, water and waste criteria – but since these data have not been published, Green Globe certification means rather little from the perspective of an environmentally-concerned consumer.

There are very few large-scale tourism ecocertification programmes that have achieved reasonably high penetration of their target markets. The Blue Flag programme for water quality at European bathing beaches is a government-funded destination-quality certification scheme which is of immediate significance for the health and enjoyment of every individual beachgoer, so it is not surprising that it is widely adopted and supported. The others certify environmental performance of tour operators, which appears to be of much less concern to most tourists, but which is of considerable interest to the management agencies of protected areas where tour companies may want to operate. Ecotourism Australia, in particular, states openly that its aim is to gain preferential access to protected areas for its ecocertified commercial members, and it is this which has made this scheme popular amongst Australian tour operators. The certification criteria for both Blue Flag and the Ecotourism Australia programme are publicly available.

Consumers and Corporations

The degree to which consumer consciousness is driven by corporate marketing or corporate marketing reflects consumer concerns, is a complex issue which goes well beyond environmental management in the tourism industry. Certainly, however, evidence to date suggests firstly that environmental management measures by most tourism operations, and the labels which certify them, are rather small-scale and shallow; and secondly, that most tourists pay rather little attention to these ecocertification labels. Whether the ecolabels are weak because they know tourists don't care, or whether the tourists don't care because they know the ecolabels are weak, however, is difficult to deduce. There are indeed some commercially viable and quite large-scale tourism operations which have excellent environmental performance and also make significant contributions to conservation (Buckley, 2003, 2006); but since they also provide a very high quality product and service, it is not clear whether their clientele is attracted by environmental concerns. The most likely scenario seems to be that commercial tour operators use ecocertification programmes merely as a means to negotiate preferential access to protected areas; that retail tourists generally expect a basic level of good environmental management from all tourism operations, certified or not; and that only the top-ranked environmental awards, not certification programmes, are recognized by retail consumers and have any significance in marketing.

Newly Industrialized Nations

The suggestions above are derived principally from the observable behaviour of international tourists from the developed Western nations. International tourists from developed Eastern nations such as Japan and Korea do not necessarily behave in the same way as those from European and Anglophone nations. In addition, the numbers of both domestic and international tourists from the heavily populated and newly-industrialized nations such as China, India and Brazil is increasing very rapidly in response to changing economic and social conditions in those countries. And their attitude towards environmental management tourism remains unknown. Weaver (2007) suggests that possibly, exposure to ecocertified tourism products in countries such as Australia may change the attitude of tourists from countries such as China. The Chinese government, however, is not relying on such an indirect effect. Instead, it is establishing its own national ecotourism standards, and has sent Chinese ecotourism experts to Australia to glean relevant information (Zhong *et al.*, 2007).

Information Flow Between Industry and Academia

Finally, Weaver (2007) questions whether or not academic debate on any of these topics has any real influence on the tourism industry itself. A good question indeed, and one which is itself worthy of research attention. Equally significant, perhaps, is whether academic tourism researchers pay any attention to what is happening in the industry. Recent publications listing the most-cited tourism journals and the academics who have published in those journals most often reinforce the impression of a small clique of mutually cross-citing authors with a strong focus on social sciences, particularly on human motivations, perceptions and emotions. Articles which analyze the structure of tourism products, the geographic locations where they operate, the ways in which they dispose of human waste, what they say in their marketing materials, or how much money they make, are few and far between. It seems that the editors and referees of most academic tourism journals consider such topics too descriptive. It is hence not surprising that engineers, economists and geographers and natural scientists working on tourism prefer to publish in their own disciplinary journals (Tribe, 2006); nor that even the highest-ranked tourism journals barely register in the overall scale of academic publishing (Moodie, 2007).

Conclusions

Overall, therefore, Weaver (2007) is surely correct in the claim that the mainstream mass tourism industry has done very little to improve its sustainability, except for fuel-saving and other cost-cutting measures with incidental environmental benefits; and that so-called voluntary measures to improve environmental management, such as ecocertification, are no more effective in the tourism industry than any other (*cf.* Gunningham & Grabosky, 1998). I would add two further conclusions. Firstly, despite this overall trend there are indeed a small number of commercial tourism operations with excellent environmental performance, which do indeed deserve credit. And secondly, the most effective way to improve environmental performance in tourism, as in any other industry sector, is to improve environmental legislation which governs it.

1.3

Sustainable Mass Tourism: More Smudge than Nudge – The Canard Continues

Brian Wheeller

Arguing that there is any case to be made, practically, for sustainable mass tourism is, at least to my way of looking at things, missing the point. Totally. As, of course, is similar discussion on 'conventional' sustainable tourism. All tourism involves travel: all travel involves transport: no form of transport is sustainable: so how on earth can we have sustainable tourism? Whether it be the (mythical) 'conventional' variety as we now know (or rather, don't know) it, or (Weaver's) latest fictitious pipedream (namely, the nebulous sustainable mass tourism), the notion of effective sustainable tourism must surely, almost by definition, be unsustainable.

But it is not just the insurmountable transport problem that mitigates, or rather negates, any possibility of effective implementation of this theoretical white elephant. Whatever its compass, for sustainable tourism (of any ilk) to have real practical credibility 'we must exit in fantasy-land and contextualize the sustainability debate within the wider arena of power, economics, greed, racism and hypocrisy – a "whole-istic" not "hole-istic" approach. That is, one where all the relevant issues are tackled head on, rather than on a partial, selective, cherry-picking basis in which the difficult/impossible issues are conveniently dispatched into a black hole and quietly forgotten' (Wheeller, 2005a).

Instead, what have we here? With Weaver's (2007) article, yet again, we are once more skirting around the fundamental (and I argue intractable) questions, distracted and diverted by what in effect are irrelevances. Whether the 'smoke-sgreen' is deliberate, a complete oversight or just naivety is open to conjecture, but continued discussion at a theoretical level is little short of further obfuscation. And yet that is precisely what we are presented with here. Ostensibly it is about implementation but, in fact, once again it is little more than rhetoric without substance: superficially very appealing, but totally impractical. Yet another theatrical theoretical masquerade. In just the same way that sustainable tourism did (and unfortunately still does), the concept of sustainable mass tourism sounds wonderful, further massages our 'do-good' sensibilities and in so doing easily seduces willing 'believers'. Seemingly less exclusive and more democratic in nature, in actual effect sustainable mass tourism is just as spurious and devoid of delivery as the 'parent' from which it has mutated. In the late nineties I voiced my concern as to how ecotourism 'has metamorphosed into the equally deceptive oxymoron of mass ecotourism' (Wheeller, 1997: 48) – a mere marketing ploy by industry, with little/no concomitant environmental benefits. I have similar, I suspect well-founded, fears for the current incarnation under discussion. No doubt, if the ideals of sustainable mass tourism were something that could reach fruition, then it would, in all probability, be marvellous. As would, of course, the realization of conventional sustainable tourism. But the chasm here between theory and practice cannot, I contend, be bridged (see Wheeller, 2005a).

Sustainability somehow assumes an innate goodness in mankind, an optimism that I do not share. The news each evening must surely be enough to at least question the validity of such a positive (naïve) view of humanity. Despite it being one of the root problems, sustainability (at whatever level) fails to embrace the reality of pervading corruption of humanity – the evil in us all – be it individual, societal and/or institutional. Weaver's (2007) probe is another example of this burgeoning canon that chooses to ignore the blindingly obvious. And the canard continues.

Look around. We live in a world of avarice and greed: of short-term perspectives, of a 'what's in it for me' mentality, vested interest and conflict. Yet the prerequisites of (tourism) sustainability apparently are those of care and compassion, harmony, concern for others, a selflessness of spirit and a long-term, considerate perspective. Clearly, the complete antithesis of what, unfortunately, we have in reality. I suspect it is going to take more than another Rachel Carson to rectify this glaring, but oft ignored, anomaly. To me, there is a need for a fundamental shift in (Western) culture: from one of 'taking' to one of 'giving'. And this, at least for as far as I can see, is never going to materialize. Everything in our current popular culture points diametrically in the opposite direction. Sustainability is overwhelmed and swamped by the hegemony of the 'Me' generation.

Weaver's argument does not seem to be based on the premise that sustainable tourism has been so successful in its limited target group that it should be extended to embrace everybody. Rather, it is being suggested that the reason why sustainable tourism hasn't been successful is that it is too restrictive and that if it was more egalitarian. . .albeit hardly a word the mass tourist might use. . .then it would, somehow (miraculously¿) become more effective¿ So it is not working now but, if we convert more tourists and more of the tourist sector, it suddenly will¿ Admirable intentions, misplaced logic.

Admittedly though, at a superficial level, there is some sense in this. For years I have argued that, *along with all its other deficiencies*, sustainable tourism's ineffectiveness is continually exposed by its restriction to such a minute segment of tourism. Given this, broadening its base would appear reasonable. But my further caveat has always been *even if it had been successful* within this limited arena its impact would be severely restricted in the global tourism arena. And, as Weaver indicates, evidence of real success even at this 'elite' level is hard to find. The notion is futile: but not (only) because it is based on exclusivity but simply because it is based on a completely false premise – sustainability itself. To somehow advocate widespread adoption of the phenomenon, while perhaps laudable and certainly seductive at the abstract level, is surely doomed at the practical. As far as effective solutions to tourism impacts are concerned, we are, once again, on the road to nowhere.

Similar erroneous reasoning is later adopted with regard to Weaver's suggestion of somehow introducing a 'certification protocol for tourists', a move which he regards as 'radical'. Forlorn, would be putting it kindly: absurd, far nearer the mark. As Weaver outlines in the probe, certification has not really been effective at the industry supply level. On the basis of this failure and (I would say absolute) ineffectiveness, the argument once again suggests that it may well be successful if applied, presumably on a mass scale, to a different target group – this time, tourists.

But who are we to pontificate when tourism education has in turn metamorphosed into the tourism education business with the inevitable, concomitant shifts in priorities (Wheeller, 2005b). It could be argued that we in academia are moving further away from tourists in the real world... our study field more, not less, isolated as our publication driven research interests take us, onwards and upwards, to new frontiers ...witness the titles of some recent articles in our journals, or convoluted conference topics. Despite protestations to the contrary, the gap is getting ever wider and deeper, exemplified by recent calls for the more widespread use of the term 'mobility' rather than tourism. At least in this respect it is readily acknowledged here that Weaver's attempts to embrace the 'mass' is an admirable (if, unfortunately, ultimately futile) effort.

As Humpty Dumpty, presumably before his fall from grace, scornfully, but in the context of 'all things to all men' sustainability, so presciently put it: 'When I use a word it means just what I choose it to mean – nothing more nor less' (Carroll, 1872[1991]: 135). But in this arena, slippery shenanigans and spin are not, unfortunately, restricted to interpretation of definition: they underpin 'logic' as well. A classic example here is the reading of recent (questionable) surveys purporting that Western travellers are cutting back on their flights in the interest of the environment. Even if this were true (which I doubt) how can this, and the 'so-called' green taxes and carbon-footprint offsetting, be used as evidence of success in changing our attitude to the environment when cut-price airlines generate/satisfy latent demand; airport expansion is ubiquitous; new routes are de-regulated and opened up, and the overall number taking flights escalates at unprecedented rates¿ The current Boarding Pass for the low-cost carrier Cebu Pacific reads 'It's time everyone flies. Cebu Pacific is committed to giving you quality service at the lowest airfare possible and now, with our brand new Airbus fleet, we believe it's time every Juan flies'. Surely this puts into context the absurdity of the concluding session of a recent international tourism conference which drew solace from the 'fact' that a few academics were (apparently) giving up international travel. As if Durable tourism, yes: sustainable tourism, no. Never, as the saying goes, in a hundred, never in a thousand, never in a million years.

Predictions as to the reticence of tourists from emerging donor countries to adopt a sustainable ethos were (confidently) made more than 15 years ago: 'It seems inconceivable that tourists from countries new to the international tourist scene will behave sensitively or sympathetically. It is naïve and unrealistic to expect otherwise. For them it is a new experience and in the circumstances they will want as much from it as possible at as little cost (to themselves) as possible.... This may seem cynical but, unfortunately, there is a close correlation between cynicism and reality' (Wheeller, 1992: 104).

Subsequent behaviour by tourists fanning out from former Eastern Bloc countries would surely go some way to vindicating this view. And new donor countries continue to emerge. Turner and Ash's (1975) pleasure periphery knows no bounds. Off the beaten track, by degrees, inevitably becomes off the beaten trek – as the combination of Conspicuous Consumption and Successive Class Intrusion take their spatial toll. Western concepts of carrying capacity (shouldn't that be caring capacity?) might well be severely tested – and found wanting – by a gross underestimate of the huge numbers of outbound Chinese tourists. Although there are uncertainties growth, on a truly massive scale, is looming.

Where Weaver is on more solid ground is his assessment of tourism academics' role in industry. Do we hold any sway at all? I am unconvinced as to the extent, if any, that 'liberal' academia has in influencing the realms of either tourism policy or vital aspects of the tourist industry behaviour and mores. Vocational training yes, affecting corporate strategy for the greater good – well, no. And similarly, at the level of resource management, I'm skeptical of the contribution of tourism academics. Despite some concerted efforts, I would suggest there is a void between academia and tourism industry practice/tourism policy. This unhealthy situation is further compounded by the insular remoteness of (tourism) academia from the general public and, *de facto*, tourists. If, as tourism academics, we are ever to leave our ivory towers, it is these 'distances', rather than our more conventional spatial usage of the term, that surely should concern us far more than is the case at present.

My cynical views of eco/ego/sustainable tourism have not mellowed over the years. On the contrary, they have hardened as, alarmed, I have become increasingly pessimistic in the (dismal) light of the burgeoning optimism of others as to the potential, always potential, of sustainable tourism. Weaver's paper is another perfect example of this charade, avoiding the real issues yet still having faith in a concept which clearly isn't working. Even though he expresses doubts as to its effectiveness, my interpretation of his message is that we should continue to try to implement (the theoretical concept of) sustainable tourism while, disturbingly, once more continuing to ignore its inherent impracticality. In so doing, my interpretation is that his argument continues to fudge the real issues: hence, more smudge than nudge. The counter plea here then, in ever more stringent tones, is the necessity to contextualize eco/ego/sustainable tourism within reality: I repeat, to exit fantasy-land. Some things are worth saying twice.

1.4

Critical and Normative Responses to Sustainable Tourism

Bill Bramwell

The concept of sustainable tourism is often criticized as being difficult to pin down. This criticism is based on the view that with sustainable tourism it is hard to know exactly what one should work towards and also to ascertain whether or not progress is being made. The discussion here considers reasons why the idea of sustainable tourism is thought to be illusive, and it is argued that – despite its difficulties – the concept should be central to our approaches to tourism.

One reason for the difficulty with sustainable tourism is that opinions about it vary from person to person. People's opinions vary according to their interests, accumulated experiences and their beliefs about the world, with those opinions being constructed through social and political processes that are embedded within social relations. As a result, there tends to be multiple and contested conceptualizations and discourses within sustainable tourism. The complexity of the concept also arises because it must be applied across geographical or spatial scales and over periods of time. Thus, policies directed at a global scale may have unintended local consequences, and actions applied at a local scale can result in unforeseen global outcomes. Difficulties may also occur over periods of time, because policies applied at one moment in time potentially can have delayed effects that are not planned, or they can lose their effectiveness due to the subsequent changes in related circumstances. Sustainable tourism's complexity also stems from the varied contexts found in different places. Thus, the approach to this concept adopted in a national park environment may need to be rather different to that taken in a long-established mass tourism resort.

The complexities of sustainable tourism mean it is understandable that some people are frustrated because they do not know how best to work towards it. In addition, it is a common complaint that in practice any progress towards implementing sustainable tourism has been, at best, very limited. Indeed, many would argue that the industry continues to expand with insufficient controls, limited self-regulation and few changes in people's consumption habits. Gössling and Hall (2006: 305), for example, contend that 'for all the writing about sustainable tourism since the mid-1980s, the global evidence clearly suggests that in terms of damage to the environment things have got worse, not better'. For many people the lack of substantial progress in applying this idea is a major challenge. Yet the contention here is that these difficulties should not lead us to abandon the concept. Rather, it is argued that it has great importance as a basis for critical evaluations of tourism development, as a framework for moral and normative judgments and as a focus for mobilization.

Critical Responses

It should come as little surprise that sustainable tourism is complex, that people hold differing views about it and that it has become the subject of intense debate. This is in a sense its strength; it is not a fixed or inflexible concept and it does require knowledge, thought and understanding. After all, this is also the case for other really important organizing concepts in society, including central ideas such as 'democracy' and 'social justice'. We should not abandon these notions just because they are difficult to define or strongly contested. Instead, we should focus attention on them because they are so important, and this should involve critically evaluating what they entail.

A first contention, therefore, is that the complexity and importance of the idea of sustainable tourism means we ought to subject it to detailed critical assessments. Any such critical evaluation, however, will be influenced by the theoretical perspectives that inform our analysis. A political economy approach, for example, would emphasize how sustainable tourism is shaped by specific economic and social contexts – notably by the material influences on relations between actors – and by the actors' interactions with nature and cultural heritage. This context of social relations and of human–environment interactions means that sustainable tourism will always be complex, that it will vary between places and that it will change over time. Because of this complexity and fluidity, sustainable tourism needs to be conceptualized as an on-going transition rather than as an end-state that can be achieved in specific locations at particular points in time. Miller and Twining-Ward (2005: 18) note in relation to sustainable tourism that: 'actions to create a simultaneous sustainability are likely always to fall short, [as] understandings change and co-evolve with the place in which actions are taking place'.

From a political economy perspective, many widely-held conceptualizations of sustainable tourism are considered to be affected by the character of advanced capitalism, including the pressures for capital accumulation and the encouragement to consumption habits that promote increased production. This means that often there is an emphasis on short-term over longer-term priorities, and on economic over social and environmental considerations. A political economy approach encourages us to examine the economic interests, patterns of social relations, and beliefs and opinions of different social groups as they relate to sustainable tourism. It also encourages us to consider how these social groups seek to encourage others to support their opinions or to ensure that their views prevail. This includes examining instances where more influential groups adopt the discourses of sustainable tourism in order to give their actions a perhaps unwarranted aura of environmental or social sensitivity (Hannigan, 2006: 55). Indeed, Escobar (1996) argues that capitalist development today is routinely sheathed in such seemingly beneficial discourses as 'sustainable development' so as to obscure the exploitation of people and environmental resources.

A political economy perspective could facilitate critical assessments of the values that underpin different approaches to sustainability. One approach that has gained in importance in some policy and industry circles is called 'ecological modernization'. This approach assumes that business can become more competitive through producing high-value, high-quality 'environmentally friendly' products and through environmental improvements based on technological innovation. In tourism, this approach might include the development of such products as agri-tourism, eco-lodge accommodation and energy-efficient hotels. The logic behind ecological modernization is that further economic growth can be reconciled with environmental protection because that growth is based on 'greener' business practices and products and services. This is considered to be realistic because it is thought that business can profit by protecting the environment, such as by reducing energy costs and attracting 'caring' customers (Bramwell, 2005; Hajer, 1995). But critics who adopt a political economy view might assert that, while this approach concedes that environmental problems are a structural outcome of capitalist society, it has a potentially misleading political message that capitalism can be made much more 'environmentally friendly' through modest reform rather than through more radical change. They might argue that ecological modernization is based on an overly optimistic faith in 'responsible capitalism' as well as in technology to provide environmental 'solutions'. And many might argue that generally it offers a weak version of sustainability.

Normative Responses and Mobilization

A second contention is that sustainable tourism should form an important basis for our moral or ethical stances and for our normative views about how tourism should develop. But in our so-called 'postmodern' times there is often doubt, suspicion and even downright hostility around any appeal for a relatively permanent or stable set of values, such as around an idea like sustainable tourism. Nowadays, the belief that universal truths are both discoverable and applicable is often held to be an error of the 'Enlightenment' and of the totalizing and homogenizing tendencies of the 'modernism' that it supposedly generated. There is an obvious sense in which this questioning certainly is both proper and imperative. It is vital, for example, to remember that at times in the past the strong advocacy of fixed views has crumbled into tyranny or dissolved into violence, corruption and injustice (Harvey, 1996: 342). However, it is argued here that, while we should recognize the impossibility of knowing the exact parameters of sustainable tourism, this does not take away our need to make some judgements about this concept. There is no option but to examine the concept critically, to articulate our own values about it and then to stick to those values for at least a time. There is a need for a moral or ethical stand, not least to invoke the precautionary principle (Fennell & Ebert, 2004).

A third contention is that there should also be a commitment to secure change and to encourage the mobilization of others who will work towards that change (Bramwell & Lane, 2006; Smith, 2000: 2). That mobilization can involve theoretical academic research, as there can be diffuse and useful connections between that research and more practical tourism-related activities. It can also involve activities such as advocacy through the public media, and research and support work to assist interest groups and social movements in society. Some may see the latter activities as political in character, but that is true of many aspects of our work.

If we do not articulate our values through normative stances and through mobilization, then we are unable to strive for the changes that we believe are necessary for a better society and for an improved environment (Harvey, 1996: 12). If there is no search for common discourses, even though these are transitory, then arguably it is difficult to challenge the overall qualities of a social system and its environmental problems. As Macbeth (2005: 980–1) argues in relation to sustainable tourism: 'without a clearly articulated and utilized ethical stance, it will simply serve the interest of short-term development and profit takers'. An 'anything goes' stance tends to result in a focus solely on economic growth rather than on a wider set of environmental, cultural and socio-economic goals; it also often assists the already dominant groups in society, and not the poor or politically weak.

Our judgements about sustainable tourism should be based on experience, critical evaluation and reflection, with these providing knowledge and understanding. The judgements need to be based on our general understanding and values, and also on the consideration of specific circumstances and the views of others, including the marginalized as well as the powerful, in each case. And our appreciation of the existence of multiple and disputed truths, of the shifts in opinions over time, and of the fallibility of judgements, should teach us that we need to reflect critically on our value positions. While holding opinions, we should also be tentative and receptive to alternative perspectives because it is important to acknowledge that one day, maybe very soon, we may change our views.

This discussion has contended that sustainable tourism's importance means that it should be subject to critical assessments, that we ought to adopt normative stances towards it, and that we need to strive for associated changes in tourism activities. But it is vital to do all these things in ways that are fully self-critical, reflexive and dialectical in nature.

Concluding Remarks

The contributors to this debate exhibit attitudes toward the prospects of sustainable mass tourism that range from cautiously hopeful to the relentlessly cynical. These diverse attitudes may reflect differing opinions about the basic nature of human-kind, but also the continuing ambiguity of the concept of 'sustainability'. How do we know when a particular tourism product or activity is 'sustainable'? One can argue that we will never possess sufficient knowledge to assess the unintended consequences through space and time of activities that we might well reasonably deem at present to be sustainable. This uncertainty is compounded in mass tourism, where industrial scales and modes of activity seem more likely to militate against sustainable outcomes. As to what has actually been achieved, all the contributors concede a limited engagement based on the opportunistic adoption of practices such as recycling and linen re-use signs that incur little corporate risk in terms of attendant costs and benefits. This pragmatic paradigm 'nudge', which leaves little room for the higher-risk adoption of certification protocols with real 'guts' and 'teeth' (Buckley, 2002) certainly does invite accusations of the sustainability debate as a 'fantasyland' and other expressions of cynicism. Even to the more hopeful, one cannot help but be dismayed at the continuing chasm that separates academics from practitioners, and the explosive growth of tourism in non-traditional markets such as China and India where concerns about sustainability are likely to rank for the time being far below motivations such as status enhancement and curiosity.

Yet, for the Don Quixotes among us, there is still hope. Look closely enough and you will find examples of innovative sustainability practice in all areas and scales of the tourism industry, stimulated by various external factors. One of these, expressed in the debate over 'peak oil', is the rapid convergence of conventional and unconventional energy prices, prompting a near revolution in the adoption of solar, wind and other alternative energy supplementation by hotels and other tourism facilities. Secondly, and notwithstanding the emergence of powerful reactionary voices, the climate change issue has thrust the environment into the limelight of public opinion like no other previous issue, and at very least has prompted some governments to pursue policies of mitigation and adaptation. Further evidence of tangible biophysical change, such as rising sea levels, will accelerate such efforts and perhaps move more individuals from the 'superficial' to the 'genuine' environmentalist cohort. Third, the ongoing global financial crisis continues to spawn a public emoscape of anger and fear over laissez-faire capitalism that at the time of writing was playing out in the spreading Occupy Wall Street movement, which may demand more than 'nudge'. Finally, the movement of the internet into a position of social centrality greatly facilitates the dissemination of good (and, alas, bad) practices so that their diffusion through the sector is accelerated. Equally important is the facilitation of online communities focused on environmental or social issues of one sort or another, some of which are sub-communities spawned by the more popular travel sites. These are all promising developments, and ones that sustain hope in the prospects for sustainable mass tourism.

Discussion Questions

(1) Is 'paradigm nudge' a pragmatic and realistic way of eventually achieving sustainable tourism outcomes, or a form of 'greenwashing' that prevents such outcomes from being achieved?
(2) Is the idea of 'sustainability' fundamentally compatible with the continued growth of the tourism sector?
(3) (a) Why do so few tourism-related businesses participate in environmental and social certification schemes? (b) Should they be forced to do so?
(4) (a) Can small-scale 'alternative tourism' function independently from mass tourism, or is it necessarily an appendage of the latter? (b) How does this relationship positively or negatively influence the sustainability of alternative tourism?
(5) (a) Is meaningful movement toward tourism sustainability ultimately dependent on pressure and advocacy from the general public? (b) If so, how can this involvement be effected?
(6) What more should academics be doing to effectively advance the cause of sustainable tourism?

References

Black, R. and Crabtree, A. (2007). Achieving quality in ecotourism: Tools in the tool box. In R. Black and A. Crabtree (eds) *Quality Assurance and Certification in Ecotourism* (pp. 16–22). Wallingford. CABI.

Bramwell, B. (2005). Mass tourism, diversification and sustainability in southern Europe's coastal regions. In B. Bramwell (ed) *Coastal Mass Tourism: Diversification and Sustainable Development in Southern Europe* (pp. 1–31). Clevedon: Channel View Publications.

Bramwell, B. and Lane, B. (2006). Policy relevance and sustainable tourism research: Liberal, radical and post-structural perspectives. *Journal of Sustainable Tourism* 14(1): 1–5.
Buckley, R. (2002). Tourism ecolabels. *Annals of Tourism Research* 29: 183–208.
Buckley, R.C. (2003). *Case Studies in Ecotourism*. Oxford: CABI.
Buckley, R.C. (2006). *Adventure Tourism*. Oxford: CABI.
Butler, R. (1980). The concept of a tourist area cycle of evolution: Implications for management of resources. *Canadian Geographer* 24: 5–12.
Carroll, L. (ed) (1872 [1991]). *Through the Looking Glass*. Manchester: World International Publishing.
Carson, R. (1962). *Silent Spring*. New York: Houghton Mifflin.
Chafe, Z. (2007). Consumer demand for quality in ecotourism. In R. Black and A. Crabtree (eds) *Quality Assurance and Certification in Ecotourism* (pp. 164–195). Wallingford: CABI.
Chan, R. (2001). Determinants of Chinese consumers' green purchase behavior. *Psychology and Marketing* 19: 389–413.
China National Tourist Office (2006). China Tourism Statistics. Online at: http://www.cnto.org/chinastats.asp. Accessed 19 February 2007.
Enz, C. and Siguaw, J. (2003). Revisiting the best of the best: Innovations in hotel practice. *Cornell Hotel and Restaurant Administration Quarterly* 44(5/6): 115–123.
Escobar, A. (1996). Constructing nature: Elements for a post-structural political ecology. In R. Peet and M. Watts (eds) *Liberation Ecology: Environment, Development, Social Movements* (pp. 46–68). London: Routledge.
Fennell, D.A. and Ebert, K. (2004). Tourism and the precautionary principle. *Journal of Sustainable Tourism* 12(6): 461–479.
Font, X. and Buckley, R.C. (2001). *Tourism Ecolabelling*. Oxford: CABI.
Font, X. and Tribe, J. (2001). Promoting green tourism: The future of environmental awards. *International Journal of Tourism Research* 3: 9–21.
Garrod, B. and Fennell, D.A. (2004). An analysis of whalewatching codes of conduct. *Annals of Tourism Research* 31: 334–352.
Gössling, S. (2006). Tourism certification in Scandinavia. In S. Gössling and J. Hultman (eds) *Ecotourism in Scandinavia: Lessons in Theory and Practice* (pp. 63–75). Wallingford: CABI.
Gössling, S. and Hall, C.M. (2006). Conclusion. Wake up...This is serious. In S. Gössling and C.M. Hall (eds) *Tourism and Global Environmental Change: Ecological, Social, Economic and Political Interrelationships* (pp. 305–320). London: Routledge.
Great Barrier Reef Marine Park Authority (2007). *Onboard*. Online at: www.gbrmpa.gov.au/Onboard. Accessed 30 July 2007.
Gunningham, N. and Grabosky, P. (1998). *Smart Regulation*. Oxford: Clarendon.
Hajer, M. (1995). *The Politics of Environmental Discourse: Ecological Modernisation and the Policy Process*. Oxford: Oxford University Press.
Hannigan, J. (2006). *Environmental Sociology*. London: Routledge.
Harvey, D. (1996). *Justice, Nature and the Geography of Difference*. Oxford: Blackwell.
Hjalager, A. (1999). Consumerism and sustainable tourism. *Journal of Travel and Tourism Marketing* 8(3): 1–20.
Honey, M. and Rome, A. (2001). *Protecting Paradise: Certification Programs for Sustainable Tourism and Ecotourism*. Washington DC. The Institute for Policy Studies.
Hunter, C. (1997). Sustainable tourism as an adaptive paradigm. *Annals of Tourism Research* 24: 850–867.
Insurance Information Institute (2007). Catastrophes: Insurance Issues. Online at: http://www.iii.org/media/hottopics/insurance/xxx/. Accessed 19 February 2007.

International Centre for Ecotourism Research (2007). *Green Guides*. Online at: www.griffith.edu.au/centre/icer. Accessed 30 July 2007.

Kuhn, T. (1970). *The Structure of Scientific Revolutions* (Second edition). Chicago: University of Chicago Press.

Macbeth, J. (2005). Towards an ethics platform for tourism. *Annals of Tourism Research* 32(4): 962–984.

Mason, P. (2007). 'No better than a band-aid for a bullet wound!' The effectiveness of tourism codes of conduct. In R. Black and A. Crabtree (eds) *Quality Assurance and Certification in Ecotourism* (pp. 46–64). Wallingford: CABI.

Maxwell, J., Lyon, T. and Hackett, S. (2000). Self-regulation and social welfare: The political economy of corporate environmentalism. *Journal of Law and Economics* 43: 583–617.

Miller, G. and Twining-Ward, L. (2005). *Monitoring for a Sustainable Tourism Transition: The Challenge of Developing and Using Indicators.* Wallingford: CABI.

Moodie, G. (2007). The publishing environment. Sydney. Faulkner Workshop, 10 February.

Parsons, C. and Grant, J. (2007). Green Globe 21: A global environmental certification program for travel and tourism. In R. Black and A. Crabtree (eds) *Quality Assurance and Certification in Ecotourism* (pp. 81–100). Wallingford: CABI.

Pollingreport.com (2007). Environment. Online at: http://www.pollingreport.com/enviro.htm. Accessed 19 February 2007.

Saarinen, J. (2006). Traditions of sustainability in tourism studies. *Annals of Tourism Research* 33: 1121–1140.

Scarpaci, C. and Dayanthi, N. (2003). Compliance with regulations by swim-with-dolphins operations in Port Phillip Bay. Victoria, Australia. *Environmental Management* 31: 342–347.

Smith, D.M. (2000). Moral progress in human geography: Transcending the place of good fortune. *Progress in Human Geography* 24(1): 1–18.

Thwaites, R. (2007). The Australia EcoCertification Program (NEAP): Blazing a trail for ecotourism certification, but keeping on track? In R. Black and A. Crabtree (eds) *Quality Assurance and Certification in Ecotourism* (pp. 435–463). Wallingford: CABI.

Tribe, J. (2006). The truth about tourism. *Annals of Tourism Research* 33: 360–381.

Turner, L. and Ash, J. (1975). *The Golden Hordes*. London: Constable.

Turner, R., Miller, G. and Gilbert, D. (2001). The role of UK charities and the tourism industry. *Tourism Management* 22: 463–472.

UNWTO (2006a). The Chinese outbound tourism market. Online at: http://www.world-tourism.org/newsroom/Releases/2006/november/chineseoutbound.htm. Accessed 19 February 2007.

UNWTO (2006b). *Yearbook of Tourism Statistics: Data 2000 – 2004*. Madrid.

Waayers, D., Newsome, D. and Lee, D. (2006). Observations of non-compliance behaviour by tourists to a voluntary code of conduct: A pilot study of turtle tourism in the Exmouth region. Western Australia. *Journal of Ecotourism* 5: 211–220.

Weaver, D. (2006). *Sustainable Tourism: Theory and Practice*. London: Butterworth-Heinemann.

Weaver, D. (2007). Towards sustainable mass tourism: Paradigm shift or paradigm nudge? *Tourism Recreation Research* 32(3): 65–69.

Wheeller, B. (1992). Is progressive tourism appropriate? *Tourism Management* 13(1): 104–105.

Wheeller, B. (1997). Here we go, here we go, here we go eco. In M.J. Stabler (ed) *Tourism and Sustainability: From Principles to Practice* (pp. 39–49). Oxon: CABI.

Wheeller, B. (2005a). Ecotourism/egotourism and development. In C.M. Hall and S. Boyd (eds) *Nature-Based Tourism in Peripheral Areas* (pp. 309–318). Oxford: Elsevier Science.
Wheeller, B. (2005b). Issues in teaching and learning. In D. Airey and J. Tribe (eds) *International Handbook of Tourism Education* (pp. 309–318). Oxford: Elsevier Science.
Zhong, L.S., Buckley, R. and Xie, T. (2007). A Chinese perspective on tourism ecocertification. *Annals of Tourism Research* 34(3): 808–811.

Further Reading

Bramwell, B. (ed) (2004). *Coastal Mass Tourism: Diversification and Sustainable Development in Southern Europe*. Clevedon: Channel View Publications.
Buckley, R.C. (2009). *Ecotourism Principles and Practices*. Oxford: CABI.
Buckley, R.C. (2010). *Conservation Tourism*. Oxford: CABI.
Buckley, R.C. (2011). Tourism & environment. *Annual Review of Environment & Resources* 36: 397–416.
Lansing, P. and De Vries, P. (2007). Sustainable tourism: Ethical alternative or marketing ploy? *Journal of Business Ethics* 72: 77–85.
Liu, Z. (2003). Sustainable tourism development: A critique. *Journal of Sustainable Tourism* 11: 459–475.
Sheldon, P., Knox, J. and Lowry, K. (2005). Sustainability in a mature mass tourism destination: The case of Hawaii. *Tourism Review International* 9: 47–59.
Weaver, D. (2012). Organic, incremental and induced paths to sustainable mass tourism convergence. *Tourism Management* 33(5). (forthcoming).

Chapter 2
Consumerism and Tourism: Are They Cousins?

Richard Sharpley, C. Michael Hall and Joan C. Henderson

Context

Tourism and consumerism are intimately linked. Contemporary society has come to be defined by consumerism inasmuch as consumption has become a key process in identity creation, whilst tourism is a pervasive form of such consumption. Moreover, on a mass scale, both consumerism in general, considered by many to be the pursuit of happiness through materialism, and tourism in particular are widely criticised; mindless consumption is seen to be manifested in 'bad' tourism. This research probe and its responses explore the phenomena of, and the relationship between, consumerism and tourism. It is argued that both are subjected to value judgements and may be variously defined and, as a consequence, generalizations should be avoided; not all consumerism is 'bad', nor is all mass tourism (or tourists). Moreover, consumerism, both generally and in the tourism context, is essential to contemporary economies. Therefore, the question should, perhaps, be how to encourage more ethical consumerism. Or, as one response to the probe suggests, it is not a question of 'good' and 'bad' tourism, but why tourism? That is, it should be asked why the consumption of tourism (and, implicitly, consumerism more generally) is seen as a path to greater happiness.

2.1

Does Consumerism Necessarily Promote Bad Tourism?

Richard Sharpley

Since the advent of rail travel in the mid-19th century first heralded the opportunity for large numbers of people to travel cheaply and in relative comfort and security, criticism has been directed at those who participate in what has come to be referred to, usually disparagingly, as mass tourism. John Ruskin, for example, lamenting the development of the railways, wrote that the train 'transmutes a man from a traveller into a living parcel' (cited in Buzard, 1993: 33) whilst, towards the end of the 1800s, the novelist Henry James famously described the then new breed of tourist as 'vulgar, vulgar, vulgar'.

Over half a century later, as mass travel began to evolve on an international basis, the debate was subsequently re-ignited by Daniel Boorstin (1964). In his seminal essay on the 'lost art of travel', he suggested that modern tourism reflects the ephemeral, pseudo nature of contemporary (American) society, that in comparison to the traditional traveller actively seeking novelty, knowledge and adventure, the modern tourist is the passive, satisfied recipient of contrived, meaningless experiences. In so doing, Boorstin not only maintained the elitism of many 19th century commentators with respect to the modern tourist; he also established a position that has come to underpin numerous themes and debates within the contemporary study of tourism, from conceptualizations of authenticity of tourist experiences to issues related to sustainability, responsibility and ethics. In particular, however, he fuelled the ongoing criticism levelled against mass tourism, criticism that, by the mid-1990s, was to reach almost apocalyptic proportions. For example, Poon (1993: 3) argued that 'the tourism industry is in crisis. . .a crisis of mass tourism; for it is mass tourism that has brought social, cultural, economic and environmental havoc in its wake', whilst similarly, Croall (1995: 1) wrote of mass tourism as a 'spectre haunting the planet'. In short, mass tourism came to be seen as 'bad' tourism.

Recently, a more measured approach has been in evidence, exemplified perhaps by the World Tourism Organization's assertion that sustainable tourism is not an alternative, niche form of tourism development, but an approach applicable to all forms of tourism, including mass tourism (UNEP/WTO, 2005). Nevertheless, explicit within the long-held and continuing criticism of mass tourism is the belief or perception that the 'problem' lies with tourists themselves. Whilst 19th century commentators were overtly critical of the new 'tourists' (as opposed to 'travellers'), attention has focused more recently on the solution to the problem. Poon (1993), for example, prophesied the emergence of the 'new' tourist – though there is still limited evidence that this prophecy has been fulfilled – whereas others have extolled the virtues of being a 'good' (Wood & House, 1991) or 'responsible' tourist (Goodwin & Francis, 2003), although these concepts have themselves long been criticized (Wheeller, 1991). Equally, many alternative tourism experiences, such as eco-holidays, are promoted on the basis of their appeal to 'better' tourists, or those seeking 'real' travel experiences.

Of course, it has long been recognized that the criticism directed at the mass tourist is as much a social construct as it is a valid or accurate observation of a particular form of tourist behaviour, that the mass tourist is typically considered the lowest common denominator of tourism. As Buzard (1993: 5) notes, criticism of the mass tourist is 'more to do with the society and culture that produce the tourist than it does with the encounter any given tourist or "traveller" may have with a foreign society and culture', whilst looking down on other tourists is all part of the game of tourism: 'tourists can always find someone more touristy than themselves to sneer at' (Culler, 1981). At the same time, however, the spectacular growth of contemporary international mass tourism during the latter half of the last century coincided with the emergence, at least in the principal tourism generating countries, of the consumer society. In other words, the growth in the consumption of tourism experiences in particular reflected not only the equally rapid growth in the production and consumption of goods and services more generally, but also the increasing significance of consumption as a defining characteristic of social life. Thus, for many, consumption has become a key feature of (post)modern culture (Pretes, 1995); it has become 'the active ideology that the meaning of life is to be found in buying things and pre-packed experiences' (Bocock, 1993: 50). Tourism and, in particular, mass tourism – if it is, in fact, possible to identify mass tourism as a distinctive form of tourist consumption for, as argued elsewhere (Sharpley, 2008), is not all tourism simply a mass social phenomenon? – is quite evidently one of these 'pre-packed experiences' that is increasingly sought and consumed. Therefore, it would appear logical to ask the question, as does this short paper, whether consumerism, in encouraging the consumption of (mass) tourism, promotes what has long been considered to be 'bad' tourism.

The short answer, as I shall go on to suggest, is 'no'; a causal relationship between consumerism and 'bad' tourism does not necessarily exist. However, the question in the title of this piece also raises a number of other important questions. For example, what *is* consumerism? Is there a distinction between consumerism, with its perhaps negative connotations, and understandings of consumer culture? Is tourism an inevitable, reactive outcome of consumerism, or are tourists more proactive or aware in their consumption of tourism? Are there in fact forms of tourism (or tourist behaviour) that may be described as 'bad'? Should the focus be on the production, as opposed to the consumption, of tourism? And, is tourism simply an easy target for those who are critical of consumerism more generally?

It is not possible here to address these in any depth nor, indeed, to give the principal question highlighted in the title the attention it undoubtedly deserves. Nevertheless, this paper offers the opportunity to explore a number of relevant issues that may point towards some answers and that may, it is hoped, stimulate further debate. First and foremost amongst these issues is the distinction between 'consumerism' and 'consumer culture' and their relationship with the consumption of tourism.

Consumerism or Consumer Culture?

There can be no doubt that consumption, or the act of purchasing goods and services, has come to occupy a dominant position within contemporary Western societies. To put it another way, there has been a fundamental transformation in the relationship between production and consumption. As Featherstone (1990) has suggested, consumption was traditionally production-led. In an era of mass, homogenized production and consumption (as was certainly manifested in the early days of mass international tourism), producers were able to dictate what, when and how people consumed goods and services; consumption was 'the result of. . .the development of manufacturing and other forms of production' (Miller, 1987: 134). However, production has now become consumption-led; there has been a reversal in the production–consumption relationship as producers have been obliged to respond to increasing demands for variety and choice in goods and services. According to Featherstone (1990), this has been driven by the pursuit of the 'emotional pleasure of consumption', or consumption as the fulfilment of dreams. However, this increasing significance of consumption in social life may be conceptualized from two inter-related perspectives:

Consumerism

The practice of consumption is often described in a negative sense as consumerism, or the socio-economic process whereby consumption (or, more precisely, the desire to consume goods and services) is promoted both as the basis of economic growth and the pursuit of personal happiness. In this sense and from a demand perspective, consumerism can be equated with the concept of materialism, or the placing of a high value on the possession of wealth and material goods: that is, the purchasing of goods and services for the sake of having or experiencing them. In other words, consumerism or materialism may be seen as the devotion to possessions and acquisitions or, according to Pearce *et al.* (2011: 107), 'a shallow cultural value that impoverishes the human spirit and fuels narcissistic self-absorption'. They go on to suggest that not only is tourism a vehicle for 'fulfilling this devotion' (Pearce *et al.*, 2011: 106), but that the materialistic consumption of tourism may be a destructive force for both tourists themselves and for destinations. For example, materialist tourists obsess over the acquisition of goods or souvenirs, yet remain frustrated because their material goals cannot be met – the appetite for spending is insatiable (Pearce *et al.*, 2011: 107). As a consequence, the more that people pursue materialistic satisfaction through tourism, the less likely they are to achieve what Pearce *et al.* (2011) refer to as the 'good life'.

Moreover, according to these authors, materialistic tourism or consumerist-driven tourism consumption may also have negative consequences for destination societies and environments. They suggest that materialists, in placing greater importance on the personal values of possession and acquisition rather than social values, such as community, equity and quality of life, tend to consume unethically. That is, their personal need for material fulfilment or satisfaction supersedes any sensitivity or responsibility towards the needs or well-being of destination communities. Equally, and repeating well-rehearsed arguments, they also generalize that the materialism inevitably results in ecological costs: 'materialism in capitalist, market economies is a major cause of environmental decline as it fuels overconsumption and beliefs that are not environmentally friendly' (Pearce *et al.*, 2011: 111).

Such arguments cannot be divorced, of course, from broader debates surrounding sustainable development in general and sustainable consumption in particular. Though these are beyond the scope of this paper, it is generally accepted that a prerequisite to the achievement of sustainable development is sustainable consumption, or the adoption of a new social paradigm appropriate to sustainable living; indeed, advocates of alternative forms of tourism or 'responsible tourism' more generally base their arguments on the belief/assumption that tourists are adopting such a new paradigm, that tourists are becoming more environmentally aware and that the 'good tourist' is becoming a reality. Conversely, even the most ardent supporters of sustainable development and 'post-materialism' accept that such a fundamental transformation in consumption values remains the single greatest obstacle to sustainability (Porritt, 2007), whilst there is in fact little evidence in practice that tourists are becoming responsive or adhering to the principles of responsible tourism consumption (Sharpley, 2009; forthcoming). Either way, however, the consumerism perspective described here implies that all forms of consumerist-driven behaviour, including tourism, are 'bad'. That is, tourists acquire and consume tourism-related experiences and products for the sake of it, unaware of or ignoring the consequences of such behaviour for themselves or others.

The culture of consumption

The consumer culture thesis similarly recognizes consumption as a defining characteristic of contemporary (post)modern societies. Indeed, a variety of socio-economic transformations in post-industrial societies, such as the increasing availability of an ever-growing range of goods and services, pervasive advertising, the advent of 'leisure shopping' and widely available credit facilities have enabled the practice of consumption to assume a leading role in social life (Lury, 1996). However, rather than being negatively equated with the implicitly mindless pursuit of material happiness, consumption is seen to have a more positive significance. That is, it has long been recognized that commodities, whether goods or services, possess meaning beyond their economic exchange or utilitarian value (Appadurai, 1986): 'the utility of goods is always framed by a cultural context. . .material goods are not only used to do things, but they also have a meaning, and act as meaningful markers of social relationships' (Lury, 1996: 11). Thus, the way in which goods and services, including tourism, are consumed will vary according to the cultural context of individual consumers. Typically, consumption is related to postmodern identity construction (Bocock, 1993). In anomic, de-differentiated postmodern societies, consumption has become a key process in identity creation and classifying the self in relation to others and whilst tourism has in particular 'remained an expression of taste since the eighteenth century, it has never been so widely used as at present' (Munt, 1994: 109).

However, it is not only in the realm of identity creation that consumption is of significance (Holt, 1995). The manner in which particular goods or services are culturally embedded in the individual consumer's social world, for example, may determine how or why they are consumed, and tourism is no exception. Once considered the preserve of the wealthy or leisured classes, it has become 'democratized' (Urry, 2002), an accepted and, perhaps, expected element of contemporary social life. Moreover, tourism possesses different meanings to different consumers in relation to their personal cultural context; to some, for example, it may represent spiritual refreshment, to others the fulfilment of dreams or fantasies. Equally, tourism may be consumed as a means of experiencing a (temporary) social world with other tourists; it may be purposefully consumed in expectation of shared experiences, or what might be referred to as *communitas*.

Many other examples exist of the different modes of consumption of tourism (Sharpley, 2008). However, the important point is that, from this perspective, the consumption of tourism is, at least for tourists themselves, a positive experience. In other words, rather than displaying 'negative' consumerist behaviour as defined above, tourists seek out either instrumental or autotelic (that is, ends in themselves) experiences from which they expect to receive positive benefits. Therefore, in the context of this paper, it suggests that the extent to which they engage or participate in 'bad tourism' is determined only by the relationship or interaction that tourists have with destination environments and communities. The question then to be addressed is: under what circumstances can this interaction or relationship be considered 'good'/'bad' or, indeed, what constitutes 'good' or 'bad' tourism?

Tourism: The Good and the Bad (and the Ugly?)

As noted in the introduction of this paper, early criticism of tourism or, more precisely, tourists coincided with the advent of mass transport systems, although those responding to the opportunity to participate in tours offered by the first tour operators, such as Thomas Cook, were also not immune to the attentions of 19th century social commentators. Typically, their criticism was directed at either the perceived inability of tourists to appreciate or to engage with the cultural benefits of travel in general or at specific aspects of their behaviour in particular and, some 150 years later, it is probably true to say that little has changed. Mass package ('sunlust' or 'institutionalized') tourists continue to be looked down upon, implicitly or otherwise, for participating in such forms of tourism whilst, often quite legitimately, displays of specific types of behaviour draw criticism both in academic circles and in the media. Conversely, those who consume 'responsible' holidays are perhaps seen to be 'good' tourists and, hence, avoid criticism.

Since the 1970s, of course, the mass tourism 'product', epitomized perhaps by rapid, extensive and seemingly unplanned development of the Spanish 'costas' but repeated in numerous other destinations around the world, has also been widely criticized. From early works, such as Young's (1973) *Tourism: Blessing or Blight?*, to more recent books, such as Hickman's (2007) powerful, ethics-focused critique of tourism development across the globe, attention has been drawn to the well-known, long-debated and much-researched negative consequences of tourism development, and there is no doubt that numerous examples of 'bad' tourism development can be identified. At the same time, however, blanket condemnation of mass tourism development, as opposed to small-scale developments proposed under various guises by the alternative tourism school, overlooks the significant role played by tourism in socio-economic development. From an economic perspective, consumerism is a driver of economic growth – witness the recent attempts of governments around the world to stimulate consumer spending following the global economic crisis – and, for many countries, tourism (or tourist-consumer spending) is vital to economic development.

As others have more than adequately pointed out (for example, Butcher, 2008), the alternative tourism school is of little relevance to the many destinations that have benefited from mass tourism development: such development might be 'ugly', but it is not 'bad'.

Here, however, we are primarily concerned with the behaviour of tourists as consumers, the question being: does consumerism promote 'bad' tourism (or tourists)? What constitutes 'good' or 'bad' behaviour in the context of tourism is, inevitably, a value judgement, and very much depends upon conceptualizations of the relationship or interaction between tourists, the destination and local communities. Unfortunately, perhaps, there has long been the tendency to refer to this as a 'host–guest' relationship. That is, tourists are seen as the guests of destination communities and, as a consequence, the parameters of 'good' and 'bad' tourist behaviour are established according to how different stakeholders, within their own cultural context, believe guests should behave. It is not surprising, therefore, that, in principle, 'good' tourists (guests) adapt their behaviour, demonstrating respect for local communities, seeking to minimize any costs they impose on the destination and attempting to give as much as they receive from the tourist experience. 'Bad' tourists (guests), conversely, are perhaps inflexible, making little or no attempt to adapt their behaviour to reflect local customs and sensitivities.

In reality, however, tourists are not 'guests' in the strict sense of the word, nor are local communities 'hosts'. As I have argued elsewhere (Sharpley, 2009), tourism is a form of capitalistic endeavour; destinations seek to exploit their social, human, cultural and environmental capitals to produce products (services and experiences) which are then sold to tourists for a 'profit' (income, foreign exchange earnings, employment creation, etc). Tourists are, therefore, customers and destinations are service providers. Moreover, the nature of the products/services that destinations choose to provide inevitably influences the 'type' of tourist purchasing those products and services and their behaviour once they have done so. Thus, providing high volume, lower cost summer-sun holidays with associated facilities, amenities and experiences (as innumerable destinations do) will, of course, attract a particular type of customer seeking that type of holiday. It is, therefore, both illogical and unfair to label such tourism or tourists as 'bad', particularly when consumers of other, alternative forms of tourism, such as eco-tourism or responsible tourism, are labelled 'good' tourists simply by virtue of their purchasing a specific product. An 'eco-tourist' may behave irresponsibly (see Duffy, 2002); conversely, a mass tourist may behave entirely responsibly within the parameters of the experience being consumed.

From this perspective, it is then logical to conclude that it is not consumerism, however defined, that promotes bad tourism; indeed, the very term 'bad tourism' demands to be challenged. There are undoubtedly numerous examples of environmentally or socially damaging tourism developments, the responsibility for which lies with poor planning or mis-management at the level of the destination; equally, there are undoubtedly numerous instances of 'bad' or inappropriate behaviour on the part of tourists, irrespective of the context. However, to not only generalize a particular form of tourism (specifically, mass tourism) as 'bad', but also to propose a causal link with consumerism/consumer culture is, perhaps, to return to the elitism expressed by the social commentators of the 19th century. Contemporary tourism is a mass social phenomenon that should be celebrated for the benefits it brings to destinations, businesses and tourists themselves and, whilst the need for its sustainable development and management is inarguable, it should be recognised not as 'bad' but as 'good'.

2.2

Consumerism, Tourism and Voluntary Simplicity: We All Have to Consume, but do we Really Have to Travel So Much to be Happy?[1]

C. Michael Hall

We all have to consume. Consumption is an ecological necessity and is inherent in biological systems. The issue really is the nature of that consumption, and from whose – or what – perspective we are taking with respect to its appropriateness. In the case of tourism we are fundamentally dealing with two different, though related, aspects of consumption. First, there is the socio-economic dimension in which tourism is part of economic, cultural and lifestyle concerns that centre on economic, social and mobility capital. Second, there is the extent to which tourism consumes the non-human environment, what may be referred to as natural or ecological capital. Both aspects of consumption are deeply embedded within contemporary capitalism (Hall, 2010b).

Tourism, as we would recognize it and, as is well noted in the literature, existed well before the onset of the industrial age of tourism. However, the industrial revolution and the rapid growth of capitalist society clearly marked a radical change in the rate, nature and the promotion of consumption. Tourism, in the sense of travel for travel's sake, was intimately associated with this process as a new form of mass consumption and production that changed both people and places. In addition, tourism consumption came to be linked with identity (Baranowski & Furlough, 2001). By this we understand that consumers combined, adapted and personalized different travel and tourism discourses as a way of negotiating key existential tensions (Thompson & Haytko, 1997). Clearly, 'all societies, at all times and places, have prevailing sign systems. These systems are socially constructed by the participants and, over time, become social structures' (Murray, 2002: 428). Tourism, and the current taken-for-grantedness of leisure and business mobility, is a significant component of contemporary sign systems of consumerism, a mode of capitalism that has become so widespread over the past 30 years that it is the dominant sign system on the planet.

It has, of course, long been recognized that contemporary tourism activity must be understood within the context of contemporary capitalism (Britton, 1982, 1991; Hall, 1994), although this essential relationship is perhaps at times not to the forefront of thinking in tourism studies as it should be, particularly given its critical institutional role in the setting of economic, social and environmental relations. However, it may well be the case that capitalism, and especially its current neo-liberal form, is now so institutionalized in the academy that the capacity to think other, let alone do other, has been significantly reduced (Slaughter & Rhoades, 2004; Hall, 2010a). Nevertheless, there are significant lines of resistance to some of the dimensions of contemporary capitalism in tourism, especially with respect to alternative forms of tourism and leisure consumption and the desire by some to 'tread lightly' on our planet. Indeed, concern over anthropogenic global environmental change probably provides the most urgent driver for improving our understanding of consumption and consumptive practices (Gössling *et al.*, 2010).

Assessments of consumption can be subjective or objective in form. Subjective assessments are usually value judgements as to the appropriateness of tourism behaviour as well as to the extent of travel. This is often bound up in notions of cultural and local appropriateness, good form and concerns over deviant behaviour. A personal observation here would be that the academy has tended to focus on 'middle-class' tourism forms rather than much of what would be regarded as mass tourism. This is not to argue of course that such matters as heritage, cultural attractions and events, convention centres, national parks and wine and food are unimportant, but perhaps it does neglect the reality that for many people tourism really is about fun and sun, getting away and having a pleasurable time – and doing it cheaply. Of course, the academy may also be reflecting its own, predominantly white, highly mobile, middle-class concerns and that what it does is travel and what other people do is tourism. Furthermore, the vast majority of work in tourism studies is fundamentally about getting people to consume more. The tourism academy could be loosely described as a bunch of relatively time and money rich people trying to find ways of getting other relatively time and money rich people to travel and travel more, sometimes with a good cause in mind like conserving heritage or creating jobs for the poor, but it's still about encouraging consumption.

The encouragement of consumption has real effects. Some of these are tied up with identity, lifestyle and quality of life, more of which will be discussed below. But it also raises objective concerns with respect to the impact of tourism on natural capital (Hall, 2010b, 2011). Objective assessments of tourism consumption are indicating that tourism has a massive affect on the environment. This is in relation not only to climate change but also the introduction of invasive species, biodiversity loss, land use change, pollution and water consumption. And despite all the sustainable tourism policies, voluntary codes of conduct and admonitions to be a responsible tourist, the negative impacts of tourism are continuing to increase (Hall, 2011), while the global employment and contribution to GDP generated by tourism has decreased in relative terms since the early 1990s (World Travel and Tourism Council, 2011).

There are clear links between subjective and objective assessments of tourism consumption, particularly with respect to the relationship between behaviour and consumption. Indeed, there is substantial interest in tourism in ways of encouraging consumption behaviour that would have a smaller environmental impact. Much of this falls under the rubric of what is usually described as sustainable consumption. The Organization for Economic Cooperation and Development (OECD, 2002) uses a Norwegian Ministry of Environment (1994) definition of sustainable consumption as 'the use of goods and services that respond to basic needs and bring a better quality of life, while minimising the use of natural resources, toxic materials and emissions of waste and pollutants over the life-cycle, so as not to jeopardise the needs of future generations' (OECD, 2002: 16). A more expansive definition is provided by the UNEP (2001) which identified four 'strategic elements', the first of which is *dematerialization* (efficient consumption from increased resource productivity), which can also be described as a green growth or efficiency approach to sustainable consumption and is the approach most favoured by tourism industry groups. The three others are ways of optimizing consumption and include *different consumption* patterns arising from changes in choices and infrastructure, mainly on the part of governments and industry, but also consumers; *appropriate consumption*, where overall consumption levels and patterns are addressed by society at large, local communities and citizens; and *conscious consumption*, where consumers are primarily responsible for choosing and using more wisely. The three behavioural approaches can be broadly described as a slow or sufficiency approach. However, to be effective in reducing humanity's ecological footprint it is critical that all approaches are used (Hall, 2010b).

Of course, a desire for more sustainable or appropriate consumption patterns is not new and, in the West, has its roots in Quakerism, Transcendentalism and even elements of Puritanism (Shi, 1986), while in a more modern form it also finds expression in the counter cultures of the 1960s (Musgrove, 1974). More recently, it has also found expression in the notion of voluntary simplicity, which refers to 'the choice out of free will (rather than being coerced by poverty, government austerity programmes, or being imprisoned) to limit expenditures on consumer goods and to cultivate non-materialistic sources of satisfaction and meaning' (Etzioni, 2003: 7).

In tourism, the concept of voluntary simplicity has arguably had some impact with respect to the development of staycations and slow tourism, as well as recognition of the environmental necessity of appropriate consumption (Hall, 2010b, 2010c). One area of contribution has been part of the growing criticism of the inadequacy of GDP figures as a measure of sustainable development (Costanza & Daly, 1992; Czech, 2003; Daly, 2008). In tourism, high visitor numbers and spend per tourist are almost always regarded as good in policy terms even though there is an increasing realization of the inadequacy of many of the economic measures of consumption that fail to measure its environmental effects (Hall, 2008). Unfortunately, there is an overwhelming tendency to conflate growth with well-being and, by using GDP as a measure whether by political entity or per capita, there is an implicit assumption that all economic activity is good (Hall, 2010b). In the same way organizations such as the United Nations World Tourism Organization, the World Travel and Tourism Council and many national and regional tourism organizations also continue to present figures on growth in international tourism arrivals and the economic contribution of travel and tourism without providing a broader appreciation of their socio-cultural costs and benefits, the contribution to equity and their environmental effects. In other words, they do not provide the details of the extent to which tourism contributes to sustainability or not (Hall, 2010b, 2011). The tacit assumption seems to be that the more we travel, the better it is for both individual and collective well-being.

The concerns of voluntary simplicity and sustainable consumption suggest that we should be interested as much in the quality of the tourism experience as the quantity. Despite the best efforts of neo-liberal economists, marketers, corporations and governments to persuade others, the promise of consumerism, that the more goods and services – including travel and tourism – a person uses, the more satisfied that person will be, is not true. Money and materialism does not buy happiness. Research on income and subjective well-being shows that among the non-poor, increased income has little or no lasting impact on happiness (Myers, 1993, 2003; Frank, 1999; Ahuvia, 2008). There is a clear necessity to ensure that basic material needs as well as health and education are met, both between and within countries. But the transfer of intensive consumerism to the newly and less developed countries only appears to be creating new sets of problems rather than providing solutions. As Myers (2003) noted in the American context:

> We have bigger houses and broken homes, higher income and lower morale, more mental health professionals and less well-being. We excel at making a living but often fail at making a life... The evidence leads to a startling conclusion: our becoming much better off over the past four decades has not been accompanied by one iota of increased psychological well-being. Economic growth has provided no boost to our collective morale. (Myers, 2003: 50)

Such a situation should be setting off alarm-bells in the academy. Not just in terms of issues of equity in relation to tourism and leisure mobility (Hall, 2010c), but more profoundly with respect to the environmental and social effects of its consumption (Gössling *et al.*, 2009, 2010). Do we really need to travel so often and so far to be happy?

This is not to suggest that tourism cannot be meaningful and pleasurable. It clearly can be. Moreover, as has been suggested elsewhere, 'the most authentic tourists of all may be those wanting to visit friends and relations because of the connectedness it provides' (Hall, 2007: 1140). Authenticity is born from everyday experiences and connections which are often serendipitous, not from things 'out there'. They cannot be manufactured through promotional and advertising deceit or the 'experience economy' (Pine & Gilmore, 1999) which is inherently grounded in fakery (Boyle, 2004). So why is there so much attention being given to encouraging people to consume more by travelling more often and usually further and then pretending that it will make them happier or more fulfilled? Indeed, the transition from consumption tied to satisfaction of basic needs to consumerism (the preoccupation with gaining ever higher levels of consumption, including a considerable measure of conspicuous consumption of status goods and cultural capital, including travel and tourism) appears to become even more pronounced as GDP increases and societies becomes 'wealthier' (Etzioni, 2003).

It has long been recognized that consumers have a major role in changing consumption behaviour (Leonard-Barton, 1981; Ebreo *et al.*, 1999). But there remain real institutional barriers at different scales for this to be the case, whether they be political, industrial and/or cultural, as well as the dominant mode of economic thinking itself. From this position there is a need to reconcile the two dominant perspectives on symbolic consumption. The first perspective, 'sign experimentation', assumes that consumption is an 'expressive movement' (Levy, 1981: 51), by which consumers distinguish themselves from alternative values and meanings by expressing desired symbolic statements (Murray, 2002). In this interpretation, symbolic consumption is associated with identity politics, with agency often being expressed in 'new' social movements (Best & Kellner, 1997) such as environmental activism, the Slow Food movement, Fair Trade and voluntary simplicity, all of which have implications for tourism. In this context 'new' demotes the fragmentation of labour and class structural inequalities that accompanies the growth of contemporary consumer culture, and emphasizes the development of social movements around fashion, style, identity and 'emotional communities' (Murray, 2002).

The second perspective, what Murray (2002) refers to as 'sign domination', emphasizes the elimination of agency in favour of structural processes, and combines a post-Marxist semiology with a critical sociology of consumer society (Kellner, 1989). The persistent demand by consumers to adopt the appropriate images and signs of the everyday reinforces Gramsci's notion of domination (Forgacs, 2000), whereby 'without critical reflection, consent to hegemonic social structures is more likely than resistance' (Murray, 2002: 428). From this perspective, socialization within a consumer culture creates a mass of 'good consumers', all struggling for the signs that fuel corporate capitalism (Harvey, 1990). 'Sign value' is thus an institutional practice that sits at the very foundation of values and social integration (Baudrillard, 1981). Nevertheless, as Murray (2002: 428) suggests, 'The use of signs and radical imagery to resist the system only creates a feeling of resistance. As a way of managing crisis and change, radical identities are also fashioned by the system'. Such a view is significant as it suggests that much consumer research in tourism has not noted the political dimensions of consumption, and symbolic consumption in particular. The assumption of agency has emphasized the creative role of the consumer, which has only been reinforced by the current marketing fashion of reference to co-creation, while simultaneously turning away from the political and oppressive potential of the symbolic (Murray, 2002).

Indeed, there is a real need to question the way in which concern with the present economic system is bound up by many in the tourism industry as a form of anti-consumption consumption, such as the promotion of green travel experiences in some foreign land which take no account of the emissions in getting to and from the latest fashionable ecotourism destination. This is not to suggest that green consumption is a fad or a fashion statement. There is clear evidence to suggest that some consumers are making decisions based on environmental and social concerns and are interested in transferring these to a tourism context (Miller, 2003). However, there are often substantial systemic barriers to tourism products being as environmentally and socially friendly as they could be. This includes not only the lack of independent capacity to monitor green claims but also the relatively weak regulation of such businesses. Moreover, a belief that green growth via greater efficiency is the most appropriate way forward to reduce tourism's contribution to climate emissions or that encouraging corporations to treat green consumers as an attractive market (Kleanhous & Peck, 2006), without dealing with the fundamental implications of consumerism and consumption means that tourism related global environmental change will continue to grow and that quality of life will continue to decline (Hall, 2010b, 2011).

Nevertheless, it is also important to recognize the link between consumption and identity. Particularly, as identity is forged as much by the meanings the consumer 'feels impelled to resist as by those that are tacitly embraced' (Thompson & Haytko, 1997: 38). For those who are embracing alternative modes of tourism and leisure consumption, this is not a denial of consumption but a move towards greater equity in terms of the benefits that consumption can bring – including access to leisure and recreation time and travel. It is, perhaps, no accident that many of those who are seeking to adopt a voluntary simplicity lifestyle also tend to be engaged in active leisure activities (Iwata, 2006). Furthermore, such active leisure can often be engaged in locally. Again, this is not to deny the possibility of long-distance travel but at least those who do engage in it may also be seeking to be fully environmentally responsible consumers and pay the full environmental costs of their travel. A responsible tourist is still a tourist after all.

As noted above, there are attempts by consumers to escape the 'totalizing logic of the market' by attempting to construct localized 'emancipated spaces' that are constructed by 'engaging in improbable behaviors, contingencies, and discontinuities' (Firat & Venkatesh, 1995: 255). Such improbable behaviours include distinguishing acts that appear to be outside the logic of commercialization, such as voluntary simplicity and anti-consumption. These are all part of the new politics of consumption (Schor, 2003). Nevertheless, focusing on lifestyle alone, without challenge to the role of structure and the cultural forces of production may well mean that alternative consumption paths become 'appropriated by experts, packaged, and sold, [and] loses its distinctive character. When this happens, even a lifestyle based on anti-consumption becomes defined in terms of commodities, possessions, sign value, and commercial success' (Murray, 2002: 439). The breakage of the false nexus between consumerism and happiness and the potential realization of improving quality of life through leisure, tourism and travel lies not just in the advocacy of the benefits of sustainable consumption and voluntary simplicity. Instead, it will require a much more fundamental and overt critique of the tourism industry's and tourism academy's role in reinforcing and supporting contemporary consumerism and neo-liberal forms of capitalism and its institutions than what has hitherto been the case. Agency is important. But understanding it in the context of structure – and how structures continue to be replicated – is critical.

Consideration of the interplay between agency and structure, between power and interests, and asking basic questions such as who benefits, how and why in tourism and tourism development forces one to consider issues of 'good' and 'bad'. If it could be treated in isolation, travel and tourism would be neither inherently bad or good. But things should not be treated in isolation. If tourism did bring benefits to destinations, individuals, and the planet without also incurring substantial costs then it should be celebrated. Yet it does not. Instead, we need to recognize that the vast majority of commercial travel and tourism is inseparable from contemporary consumerism and neo-liberal capitalism. More often than not tourism is serving a system with a narrow range of political and economic interests in which wealth is ever concentrated, in which the gap between rich and poor is greater than ever, in which happiness is conflated with materialism and quality with quantity, and in which the rights of other species to exist are being denied. This is bad.

Tourism is not value free. There are no problems in recognizing that tourism has values. The problem is in failing to see the implications of those values and how they are linked to interests and power. The fundamental question is not why we want to engage in leisure and travel. The question is why so many people have increasingly come to believe that consuming such mobility will somehow make them happier and improve their life.

2.3

Not All Consumerism has a Shady Side!

Joan C. Henderson

Consumerism and tourism on mass scales are two defining characteristics of modern society which have both inspired much criticism. They are also closely connected and certain examples of tourism are seen as undesirable manifestations of consumerism. Questions of whether the censure of the phenomena is justified and the nature of their relationship are addressed in this brief commentary written in response to the research probe which raises enduring and still pertinent issues about tourism as a human and commercial activity, underlying dynamics and impacts.

Reactions to the proposition that consumerism encourages bad tourism partly depend on the meanings ascribed to the terms. The notion of bad tourism is frequently associated with selected instances of modern mass tourism and its consequences, exemplified by the urbanization of coastlines around the world deemed by detractors to have been 'ruined'. Large resorts and developed seashores are not, however, intrinsically bad and are a core visitor attraction with the capacity to perform valuable functions for tourists and destination residents. Their popularity may alleviate pressures elsewhere and protect more sensitive and vulnerable socio-cultural and physical environments which are under threat from the shift away from standardized tourism to a series of markets of vacationers looking for novelty of setting and experience. Smaller scale tourism of a sort typified by ecotourism and cultural tourism is lauded by suppliers and participants. It is hailed as the antithesis of and superior to modern mass tourism because of the rectitude of motives and positive effects, yet it is not necessarily better or to be welcomed. Tourists who search out pristine territories and remote communities could cause disruption to fragile eco-systems and everyday life, despite protestations about responsible travel (Butcher, 2003).

What constitutes bad and good tourism is clearly subject to interpretation and perceptions reflect the position of the individual and group alongside circumstances prevailing in countries of departure and arrival. Academic opinion is also divided within and between disciplines, although social scientists as a whole have been very fierce critics of tourism which is perhaps something of an easy target for them. The long standing dispute about whether tourism is a blessing or blight (Young, 1973) would seem to be one that will never be fully resolved; for almost every argument, there is a counter-argument which has resulted in a circularity of discourse. Tourism as a field of enquiry and action is full of complexities and contradictions, encompassing an immense range of individual behaviour and commercial products as well as spaces where it occurs. Generalizations become difficult and opinions about badness or goodness are determined by personal, economic, political, socio-cultural, geographical and historical contexts. Interestingly, the debate is not confined to instances from contemporary society and Baranowski (2004) writes about consumerism and mass tourism in the *Third Reich* in her study of the Nazi 'Strength through Joy' programme. It claimed to be one of the largest leisure and travel organizers in the world and a force for good, integrating work and leisure, a view which soon became untenable.

Despite ambiguities about its composition, there is a consensus that good tourism is equated with sustainability which is at the forefront of discussions of tourism. While employment of the phrase is ubiquitous, there are no universally accepted standards of sustainable tourism and accurate assessment and measurement of progress are problematic. Descriptions of sustainable tourism as a journey or beacon to strive towards, rather than a final state, acknowledge that full realization of the goal of protecting the assets on which tourism relies so that they are available for the use and enjoyment of future generations (UNEP/WTO, 2005) may be over-ambitious. There are obvious tensions, especially in the shorter term, between pursuing profits and the practice of sustainable tourism which gives due regard to socio-cultural and environmental landscapes (Mowforth & Munt, 2003). However, tourism is an industry in which commercial objectives and the social and environmental welfare prerequisites at the heart of sustainability can be mutually reinforcing. It is reliant on healthy natural resources and well-preserved cultural heritages, together with hospitable destination residents who ideally are involved in decision- making, if it is to survive and prosper. Businesses and tourists, alongside citizens and national and local governments, thus have a vested interest in nurturing sustainable tourism; this is eminently preferable to unsustainable tourism which is ultimately self-destructive as well as harmful to people and places. Nevertheless, there are conflicts between the demands of conservation and commerce which leads to consideration of the merits of consumerism and how it shapes tourism.

Consumerism has been attacked by many thinkers who decry the manner in which consumers and consumer societies are compelled to produce and acquire goods for which the buyers have no real need. Materialism dominates and individuals and communities are spiritually diminished as a consequence (Barber, 2008). Such assertions are not new in the developed world, but still resonate and are being heard in emerging economies where the burgeoning middle class are demonstrating a keen appetite for branded merchandise and luxury labels. Rampant consumption is hard to condone and reflective of greed and self-centred lifestyles dedicated to meeting hedonistic wants with little thought for the implications (Jackson, 2009). Within the theatre of tourism, it has been argued that consumerism is predicated on the assumption of a right to travel. The right is, however, only accessible to the privileged elite of wealthy nations who are often served by the poor in developing countries, giving rise to exploitation and commodification. In this reading, tourism is an expression of neo-imperialism (Higgins-Desbiolles, 2010) and capitalism at its most pernicious. It is 'securely rooted in the real world of gross political and economic inequalities between nations and classes' and is 'doubly imperialistic; not only does it make a spectacle of the Other, making cultures into consumer items, tourism is also an opiate of the masses in the affluent countries themselves' (Crick, 1989: 334). Such critics may well take comfort from the idea that tourism, like capitalism, could contain the seeds of its own destruction. The end of capitalism and tourism might seem to be unlikely, and doomsday scenarios to their advocates, but the global financial crisis has tested the old economic order and found it wanting (Harvey, 2010) with ongoing uncertainty and some fear about what the future holds.

Defenders of consumerism, and by implication of capitalism, might reply that it propels economies. Businesses have customers and jobs are created, fuelling growth and enhancing standards of living for all citizens. Consumer spending is fundamental to capitalism which, whatever its drawbacks and the deficiencies exposed by recent upheavals, has proved to be a resilient system that is more efficacious than alternatives. Avid tourist shoppers may be emblematic of a consumer culture, yet their presence supports a varied retail mix which is an amenity for residents and injects money into urban economies. Guests expect pampering at expensive hotels, but hospitality companies are increasingly integrating sustainable principles and practices into their business models in alignment with policies of Corporate Social Responsibility (Porter & Kramer, 2006) which can make a real difference to local communities. Consumer movements such as those backing fair trade and action on environmental improvements can also influence corporate and official strategies, favouring the less privileged and affording better protection for scarce resources. In addition, the contention that tourism is an arena of neo-colonial struggle between domineering tourists from the developed world and passive disadvantaged Third World inhabitants is no longer as persuasive as perhaps it once was. Pro-poor tourism initiatives in impoverished nations endeavour to cultivate a more equitable distribution of power and rewards (Ashley, 2006) while tourism in some developing regions is being driven by mounting domestic and inter-regional demand, evidenced within ASEAN (Association of South East Asian Nations). Governments and populations of South East Asian countries are not inevitably dupes at the mercy of outside agencies, but informed about the workings of the international tourism industry and active in securing a fair share of its benefits.

It must also be remembered that consumerism is not always synonymous with mindless acquisitiveness as its detractors complain. Not everyone in established or emerging economies is a slavish devotee of consumerism, yet they still undertake a degree of consumption. Many are seeking a deeper meaning and purpose to their existence, recognizing that quality of life is derived from more than just possessions. Such people are not immune from the imperatives and pleasures of shopping and travel might be an expenditure item of great appeal to those who see journeys as 'new age' pilgrimages which yield non-material fulfilment (Attix, 2002). Tourism inspired by a spiritual quest can thus be viewed as an illustration of consumerism broadly defined and whether it constitutes bad or good tourism is a question which brings the review back to where it began, highlighting the diversity of experiences and interpretations. There is an assumption that participants will be more enlightened and sensitive to native cultures and environments than their mass market counterparts, but they could be focused on selfish ends and fail to display the respect for people and places encountered which are usually at the heart of conceptions of good tourism.

In conclusion, any discussion about consumerism and tourism must acknowledge what might be unpalatable to some observers. Tourism is an important, albeit imperfect, industry and consumerism is essential to modern societies and economies, although dangerous if unchecked. It would seem premature to herald the arrival of an era of either post-consumerism or post-tourism in light of their current entrenchment and scale. Neither consumerism nor tourism is intrinsically bad or good, yet some of the many forms they take may be more or less desirable for society at large and individual members in countries of origin and destination. The two phenomena are also locked into a close relationship which appears to be both potentially constructive and destructive. Tourism can be conceived of as a beneficial expression of consumerism and consumers have the capacity to direct the tourism industry towards greater sustainability. At the same time, consumerism can engender tourist activity of a sort which leads to the perpetuation of social inequities and environmental degradation or tourism which is unsustainable. The formidable challenge for the future is to encourage a kind of ethical consumerism which is aware of and gives appropriate attention to the wider costs and consequences of consumption. When applied to tourism, this would translate into knowledgeable consumers with a sense of responsibility who purchase services which are the product of a genuine commitment to more sustainable tourism on the part of public and private sectors.

Concluding Remarks

In addressing the broad question 'Consumerism and tourism: Are they cousins?', the contributions to this chapter have highlighted a number of themes relevant to the relationship between consumerism and tourism in particular, and between consumerism and global capitalism/development more generally. Indeed, if tourism can in fact be thought of as a metaphor for the social world (Ryan, 2002) or as a lens through which our contemporary world may be studied and understood (Sharpley, 2011), then the debate in this chapter is as much about global capitalism and the role (and potential threat) of rampant consumerism in the global economy as a whole as is it about the consumption of tourism in particular. And, perhaps the conclusions are also of such wider relevance.

Three key points arise from the preceding contributions. Firstly, tourism, as an economic sector that stimulates ever-increasing levels of consumption of touristic services and experiences in ever more numerous and diverse locations, is, in an economic development sense, a 'good thing'. Just as the global economy as a whole and, indeed, the success of capitalism is dependent upon the growth in consumption (referring here specifically to consumer spending on goods and services rather than the value-laden concept of consumerism), so too do innumerable destinations and societies rely on tourism as a valuable source of income, foreign exchange and investment, government revenues, employment and, if managed appropriately, environmental protection and enhancement, social improvements, and so on.

Secondly, however, tourism is not always managed appropriately relevant to the needs and constraints of particular places where it occurs. There is not necessarily a 'bad' tourism; nor are there 'bad' tourists. However, the negative consequences of tourism are all too visible while, as Michael Hall observes in his contribution, the relative benefits in terms of economic contribution have been decreasing for the last two decades. Thus, the need undoubtedly exists to manage tourism development (and, implicitly, how it is consumed) in order to ensure its sustainability – both tourism's survival as an economic phenomenon and its contribution to socio-economic development. This may well require a shift in emphasis from quantity to quality. In other words, just as national and global economic policy focuses on growth – if an economy is not growing it is implicitly failing – so too do organizations such as the UNWTO promote the continual growth of tourism. However, common sense suggests that unfettered growth is unsustainable; sooner or later, there will be insufficient resources to support continued growth in tourism or, indeed, other economic sectors.

Yet, thirdly, continuing growth is driven by consumer demand. Nowhere is this more evident than in global tourism where, despite the contemporary challenges facing the world economy, the demand for tourism continues to grow at a remarkable rate. This raises the question, again noted by Michael Hall, that if consumerism is a manifestation of the pursuit of happiness (albeit an unsuccessful pursuit), then why is tourism in particular so widely consumed as an assumed route to happiness? Or is it? Perhaps the consumption of tourism is, as Henning (forthcoming) suggests, no more than a habit or, ultimately, an addiction. And could not the same be said for any form of discretionary consumption?

In conclusion, then, understanding the relationship between consumerism and tourism is important not because it may reinforce atavistic, elitist and, ultimately, unsupportable arguments about different forms of tourism consumption. Rather, it is important because it focuses attention on the much more fundamental debate surrounding consumer driven capitalism and economic growth in general. Sooner or later, not only will consumerism (including the consumption of tourism) as the basis for material and existential satisfaction or happiness need to be challenged; so too will contemporary economic growth-based models of development of which tourism remains a significant element.

Discussion Questions

(1) Can the negative consequences of tourism simply be blamed on tourists?
(2) Has tourism become a defining element of contemporary consumer society?
(3) Is the continuing growth in tourism a good thing?
(4) What drives the growth in tourism: demand or supply?
(5) Tourism: luxury, necessity or habit?
(6) Can the world afford more tourism?
(7) Is Capitalism/Consumerism sustainable?

Notes

(1) ©C.M. Hall

References

Ahuvia, A. (2008). If money doesn't make us happy, why do we act as if it does? *Journal of Economic Psychology* 29(4): 491–507.

Appadurai, A. (Ed) (1986). *The Social Life of Things: Commodities in Cultural Perspective.* Cambridge: Cambridge University Press.

Ashley, C. (2006). *How Can Governments Boost the Local Economic Impacts of Tourism: Options and Tools*. London: Overseas Development Institute.

Attix, S.A. (2002). New age-oriented special interest travel: An exploratory study. *Tourism Recreation Research* 27(2): 51–58.

Baranowski, S. (2004). *Strength Through Joy: Consumerism and Mass Tourism in the Third Reich*. New York: Cambridge University Press.

Baranowski, S. and Furlough, E. (eds) (2001). *Being Elsewhere: Tourism, Consumer Culture, and Identity In Modern Europe and North America*. Ann Arbor: University of Michigan.

Barber, B.R. (2008). *Consumed: How Markets Corrupt Children, Infantilise Adults and Swallow Citizens Whole*. New York: Norton.

Baudrillard, J. (1981). *For a Critique of the Political Economy of the Sign*. St. Louis: Telos.

Best, S. and Kellner, D. (1997). *The Postmodern Turn*. New York: Guilford.

Bocock, R. (1993). *Consumption*. London: Routledge.

Boorstin, D. (1964). *The Image: A Guide to Pseudo-Events in America*. New York: Harper & Row.

Boyle, D. (2004). *Authenticity: Brands, Fakes, Spin and the Lust for Real Life*. London: Harper Perennial.

Britton, S.G. (1982). The political economy of tourism in the third world. *Annals of Tourism Research* 9(3): 331–358.

Britton, S.G. (1991). Tourism, capital and place: Towards a critical geography of tourism. *Environment and Planning D: Society and Space* 9: 451–478.

Butcher, J. (2003). *The Moralisation of Tourism: Sun, Sand and Saving the World?* London: Routledge.

Butcher, J. (2008) Ethical travel and well-being: Reposing the issue. *Tourism Recreation Research* 33(2): 219–222.

Buzard, J. (1993). *The Beaten Track*. Oxford: Oxford University Press.

Costanza, R. and Daly, H. (1992). Natural capital and sustainable development. *Conservation Biology* 6(1): 37–46.

Crick, B. (1989). Representations of international tourism in the social sciences: Sun, sex, sights, savings and servility. *Annual Review of Anthropology* 18: 307–344.

Croall, J. (1995). *Preserve or Destroy: Tourism and the Environment*. London: Calouste Gulbenkian Foundation.

Culler. J. (1981) Semiotics of tourism. *American Journal of Semiotics* 1(1&2): 127–140.

Czech, B. (2003). Technological progress and biodiversity conservation: A dollar spent, a dollar burned. *Conservation Biology* 17(5): 1455–1457.

Daly, H.E. (2008). *A Steady-State Economy*. London: Sustainable Development Commission.

Duffy, R. (2002). *A Trip Too Far*. London: Earthscan.

Ebreo, A., Hershey, J. and Vining, J. (1999). Reducing solid waste: Linking recycling to environmentally responsible consumerism. *Environment and Behavior* 31: 107–135.

Etzioni, A. (2003). Introduction: Voluntary simplicity-psychological implications, societal consequences. In D. Doherty and A. Etzioni (eds) *Voluntary Simplicity: Responding to Consumer Culture* (pp. 29–41). Oxford: Rowan and Littlefield Publishers.

Featherstone, M. (1990). Perspectives on consumer culture. *Sociology* 24(1): 5–22.

Firat, A.F. and Venkatesh, A. (1995). Liberatory postmodernism and the reenchantment of consumption. *Journal of Consumer Research* 22: 239–267.

Forgacs, D. (2000). *The Antonio Gramsci Reader: Selected Writings, 1916–1935*. New York: New York University Press.

Frank, R.H. (1999). *Luxury Fever: Why Money Fails to Satisfy in an Era of Excess*. New York: Free Press.

Goodwin, H. and Francis, J. (2003). Ethical and responsible tourism: Consumer trends in the UK. *Journal of Vacation Marketing* 9(3): 271–284.

Gössling, S., Ceron, J-P., Dubios, G. and Hall, C.M. (2009). Hypermobile travellers. In S. Gössling and P. Upham, (eds) *Climate Change and Aviation* (pp. 131–149). London: Earthscan.

Gössling, S., Hall, C.M., Peeters, P. and Scott, D. (2010). The future of tourism: A climate change mitigation perspective. *Tourism Recreation Research* 35(2): 119–130.

Hall, C.M. (1994). *Tourism and Politics: Policy, Power and Place*. London: John Wiley.

Hall, C.M. (2007). Response to Yeoman et al.: The fakery of 'the authentic tourist'. *Tourism Management* 28: 1139–1140.

Hall, C.M. (2008). *Tourism Planning*. Harlow: Prentice-Hall.

Hall, C.M. (2010a). Academic capitalism, academic responsibility and tourism academics: Or, the silence of the lambs? *Tourism Recreation Research* 35(3): 298–301.

Hall, C.M. (2010b). Changing paradigms and global change: From sustainable to steady-state tourism. *Tourism Recreation Research* 35(2): 131–145.

Hall, C.M. (2010c). Equal access for all? Regulative mechanisms, inequality and tourism mobility. In S. Cole and N. Morgan (eds) *Tourism and Inequality: Problems and Prospects* (pp. 34–48). Wallingford: CABI.

Hall, C.M. (2011). Policy learning and policy failure in sustainable tourism governance: From first and second to third order change? *Journal of Sustainable Tourism* 19(4&5): 649–671.

Harvey, D. (1990). *The Condition of Postmodernity: An Enquiry into the Origins of Cultural Change*. Oxford: Blackwell.

Harvey, D. (2010). *The Enigma of Capital and the Crisis of Capitalism*. London: Profile Books.

Henning, G. (2012). The habit of tourism: Experiences and their ontological meaning. In R. Sharpley and P. Stone (eds) *The Contemporary Tourist Experience: Concepts and Consequences* (pp. 25–37). Abingdon: Routledge.

Hickman, L. (2007). *The Final Call: The Search for the True Cost of Our Holidays*. London: Transworld Publishers.

Higgins-Desbiolles, F. (2010). The elusiveness of sustainability in tourism: The culture-ideology of consumerism and its implications. *Tourism and Hospitality Research* 10(2): 116–129.

Holt, D. (1995). How consumers consume: A typology of consumption practices. *Journal of Consumer Research* 22(1): 1–16.

Iwata, O. (2006). An evaluation of consumerism and lifestyle as correlates of a voluntary simplicity lifestyle. *Social Behavior and Personality* 34(5): 557–568.

Jackson, T. (2009). *Prosperity without Growth: Economics for a Finite Planet*. London: Taylor and Francis.

Kellner, D. (1989). *Jean Baudrillard: From Marxism to Postmodernism and Beyond*. Stanford: Stanford University Press.

Kleanhous, A. and Peck, J. (2006). *Let Them Eat Cake: Satisfying the New Consumer Appetite for Responsible Brands*. Godalming: World Wildlife Fund UK.

Leonard-Barton, D. (1981). Voluntary simplicity lifestyles and energy conservation. *Journal of Consumer Research* 8: 243–252.

Levy, S.J. (1981). Interpreting consumer mythology: A structural approach to consumer behavior. *Journal of Marketing* 45: 49–61.

Lury, C. (1996). *Consumer Culture*. Cambridge: Polity Press.

Miller, D. (1987). *Material Culture and Mass Consumption*. Oxford: Blackwell.

Miller, G.A. (2003). Consumerism in sustainable tourism: A survey of UK consumers. *Journal of Sustainable Tourism* 11: 17–39.

Ministry of the Environment (1994) *Oslo Roundtable on Sustainable Production and Consumption*. Oslo: Ministry of the Environment.

Mowforth, M. and Munt, I. (2003). *Tourism and Sustainability: New Tourism in the Third World* (2nd edn). London: Routledge.

Munt, I. (1994). The 'other' postmodern tourism: Culture, travel and the new middle classes. *Theory, Culture and Society* 11(3): 101–123.

Murray, J.B. (2002). The politics of consumption: A re-inquiry on Thompson and Haytko's (1997) Speaking of Fashion. *Journal of Consumer Research* 29: 427–440.

Musgrove, F. (1974). *Ecstasy and Holiness: Counterculture and the Open Society*. Bloomington: Indiana University Press.

Myers, D.G. (1993). *The Pursuit of Happiness: Who is Happy and Why?* New York: Avon.
Myers, D.G. (2003). Wealth and happiness: A limited relationship. In D. Doherty and A. Etzioni (eds) *Voluntary Simplicity: Responding to Consumer Culture* (pp. 41–52). Oxford: Rowan and Littlefield Publishers.
Organization for Economic Cooperation and Development (OECD) (2002). *Towards Sustainable Household Consumption? Trends and Policies in OECD Countries*. Paris: OECD.
Pearce, P., Filep, S. and Ross, G. (2011) *Tourists, Tourism and the Good Life*. London: Routledge.
Pine, J.B. II, and Gilmore, J.H. (1999). *The Experience Economy: Work is Theatre and Every Business a Stage.* Boston: Harvard Business School Press.
Poon, A. (1993). *Tourism, Technologies and Competitive Strategies*. Wallingford: CAB International.
Porritt, J. (2007). *Capitalism as if the World Matters*. London: Earthscan.
Porter, M. and Kramer, M.R. (2006). Strategy and society: The link between competitive advantage and corporate social responsibility. *Harvard Business Review* (December): 1–16.
Pretes, M. (1995). Postmodern tourism: The Santa Claus industry. *Annals of Tourism Research* 22(1): 1–15.
Ryan, C. (2002). *The Tourist Experience* (2nd edn). London: Continuum.
Schor, J. (2003). The problem of over-consumption – Why economists don't get it. In D. Doherty and A. Etzioni (eds) *Voluntary Simplicity: Responding to Consumer Culture* (pp. 65–82). Oxford: Rowan and Littlefield Publishers.
Sharpley, R. (2008). *Tourism, Tourists and Society* (4th edn). Huntingdon: Elm Publications.
Sharpley, R. (2009). *Tourism, Development and the Environment: Beyond Sustainability?* London: Earthscan.
Sharpley, R. (2011). *The Study of Tourism: Past Trends, Future Directions*. Abingdon: Routledge.
Sharpley, R. (forthcoming). Responsible tourism: Whose responsibility? In A. Holden and D. Fennell (eds) *A Handbook of Tourism and the Environment*. London: Routledge.
Shi, D.E. (1986). *The Simple Life: Plain Living and High Thinking in American Culture*. New York: Oxford University Press.
Slaughter, S. and Rhoades, G. (2004). *Academic Capitalism and the New Economy: Markets, State and Higher Education*. Baltimore: Johns Hopkins University Press.
Thompson, C. J. and Haytko, D.L. (1997). Speaking of Fashion: Consumer's uses of fashion discourses and the appropriation of countervailing cultural meanings. *Journal of Consumer Research* 24: 15–42.
UNEP (2001) *Consumption Opportunities: Strategies for Change*. Geneva: UNEP.
UNEP/WTO (2005). *Making Tourism More Sustainable: A Guide for Policy Makers*. Paris/Madrid: UNEP/WTO.
Urry, J. (2002). *The Tourist Gaze* (2nd edn). London: Sage Publications.
Wheeller, B. (1991). Tourism's troubled times: Responsible tourism is not the answer. *Tourism Management* 12(3): 91–96.
Wood, K. and House, S. (1991). *The Good Tourist: A Worldwide Guide for the Green Traveller.* London: Mandarin.
World Travel and Tourism Council (2011). Economic Data Search Tool. Online at: http://www.wttc.org/eng/Tourism_Research/Economic_Data_Search_Tool/index.php. Accessed 11 May 2011.
Young, G. (1973). *Tourism: Blessing or Blight?* Harmondsworth: Penguin.

Further Reading

Brown, F. (1998) *Tourism Reassessed: Blight or Blessing?* Oxford: Butterworth-Heinemann.

Durning, (1992) *How Much Is Enough?: The Consumer Society and the Future of the Earth*. London: W. W. Norton & Company.

Graham, C. (2009) *Happiness Around the World: The Paradox of Happy Peasants and Miserable Millionaires*. Oxford: OUP.

Jackson, T. (1997) *Prosperity without Growth: Economics for a Finite Planet*. London: Routledge.

Porritt, J. (2009) *Capitalism as if the World Matters*. London: Earthscan.

Sharpley, R. (2009) *Tourism, Development and the Environment: Beyond Sustainability?* London: Earthscan.

Chapter 3
Is Small Tourism Beautiful?

David Harrison, David Weaver and Richard Butler

Context

In this chapter, David Harrison sketches the historical origins of the 'small is beautiful' debate and suggests that, in regard to tourism, what is 'small' is difficult to specify and covers a wide variety of operations, that there is little evidence that small operations are more efficient, more locally-orientated or more 'beautiful' than larger tourism enterprises, and that tourists themselves opt for a variety of accommodation types, often preferring large-scale to small-scale resorts. David Weaver concurs, arguing that small-scale tourism normally relies on, and frequently *leads* to, mass tourism, that evidence exists that larger-scale enterprises are often adopting more sustainable practices, and that there are increasing opportunities for synergies across alternative tourism and mass tourism. Similarly, Richard Butler links Schumacherian approaches to the craft movement in the UK, and suggests we differentiate the characteristics and apparent benefits of smallness across different various components of tourism. Not all tourists value smallness; it does not necessarily correlate with economic, social or cultural sustainability, and the 'smallness' of past tourism was often elitist, ethically questionable and unsustainable. All agree that, in tourism, one size does not fit all.

3.1

Tourism: Is Small Beautiful?

David Harrison

The term 'small is beautiful' is closely associated with E.F. Schumacher, who went on to publish *Small is Beautiful: A Study of Economics as if People Mattered*, a series of essays that became a milestone in development thinking (Schumacher, 1974). In 1996, he established the Intermediate Technology Development Group, now known as Practical Action (Practical Action, 2012). Much influenced by Buddhism, and holding a holistic view of development, he confronted the prevailing belief that the main aim of economic 'development' was the growth of gross national product (GNP), to be achieved through large-scale production and modern technology. Rather, he suggested, economic production (preferably in smaller units) should serve humanity's needs, with sparing use of fossil fuels and other non-renewable capital assets, and should, as far as possible, use intermediate technology, especially in developing countries. In general, production should be people-centred and facilitate human creativity:

> Man is small, and, therefore, small man is beautiful. To go for gigantism is to go for self-destruction. And what is the cost of a reorientation? We might remind ourselves that to calculate the cost of survival is perverse. No doubt, a price has to be paid for anything worthwhile: to redirect technology so that it serves man instead of destroying him requires primarily an effort of the imagination and an abandonment of fear. (Schumacher, 1974: 133)

Along with the work of such other humanitarian economists as Seers (1969, 1977), Schumacher's contribution has had a lasting impact on development theory, especially approaches to 'alternative' development (Burkey, 1993: 196; Reid, 1995: 69).

Since the 1970s, tourism has emerged as a tool for development, in both developed and developing societies (Harrison, 2001b: 1–22; Sharpley, 2009: 1–27). Initially, links with development theory were tenuous, but in recent years there have been several attempts to situate tourism within the wider context of 'development' (Sharpley, 2002; Telfer, 2002; Mowforth & Munt,

2009), a focus given added point by the fact that while most tourism enterprises are likely to be small in scale (Gartner, 2004), large-scale tourism has taken an increasing share of the global tourist market. This development has been almost universally condemned and mass tourism's many critics have generally incorporated the notion of small being beautiful as a feature of some kind of 'alternative' or 'sustainable' tourism, even though Schumacher's influence has rarely been acknowledged, and then only obliquely (Singh, 2010: 211). Despite the general preference for small-scale tourism, however, some academics have defended mass tourism (Sharpley, 2000, 2009; Butcher, 2003; Aramberri, 2010).

Examples abound of support for small-scale tourism development, often linked to greater community participation, itself the topic of a Research Probe in *Tourism Recreation Research* (Singh, 2010). An early advocate was Rodenburg, who concluded that, in Bali, development objectives were best met by 'craft and small industrial tourism' rather than 'large industrial tourism,' as profits were more likely to go to local people and there was a better 'fit' with traditional Balinese culture (1980: 194). Later, Brohman (1995, cited in Telfer, 2002: 59) was to argue that alternative development strategies stress 'small-scale, locally-owned developments, community participation, and cultural and environmental sustainability' and a similar emphasis is found in Telfer (2002: 67–75); Fennell (1999: 9), Scheyvens (2002: 13), Mowforth and Munt (2009: 98–119) and Honey (1999: 25), whose definition of ecotourism is 'travel to fragile, pristine, and usually protected areas that strives to be low impact and (usually) small scale'.

Particularly strong support for small-scale tourism enterprises comes from Dahles, who notes:

> These forms of tourism depend on ownership patterns that are in favour of local, often family-owned, relatively small-scale business rather than foreign-owned transnationals and other outside capital. By stressing smaller scale, local ownership, it is anticipated that tourism will increase multiplier and spread effects within the host community and avoid problems of excessive foreign exchange leakages. (1999: 2)

Furthermore, following Echtner (1995) she *assumes* (my emphasis) that 'small-scale tourism developments and active resident involvement in the ownership and operation of facilities' will not produce the negative economic and socio-cultural effects associated with foreign ownership and will also enable enterprises to respond more quickly to changes in the tourism market (Dahles, 1999: 2).

Non-academics are similarly enthusiastic. Non-government organizations have long been involved in small-scale tourism initiatives, though not always successfully (Simpson, 2008: 7–9) and the United Nations Economic and Social Commission for Asia and the Pacific (UNESCAP), for instance, considers 'smaller-scale business operators are more appropriate for activities related to ecotourism' (UNESCAP, 2001: 6). Similarly, the UNWTO (2000: 11), when compiling a list of examples of good practices in sustainable tourism development, notes that 80% of them are 'small or medium sized projects,' though it also admits that 'sustainability in tourism is not necessarily reserved for small-scale operations'.

There are some who buck the above trend. They include critics of small-scale tourism, who suggest it is prone to many of the problems of mass tourism (Duffy, 2002: 155–160) and, by contrast, those who recognize that truly sustainable tourism development must *include* mass tourism (Weaver, 2001; Aramberri, 2010). However, most academics writing on tourism, along with practitioners in aid agencies and non-government organizations, are inclined to dismiss mass tourism, especially when it involves transnational corporations in developing countries, as politically, economically and environmentally unsustainable (and therefore ugly?) and to prioritize small-scale tourism enterprises, *particularly* those that are locally-owned (and which, being small, must therefore be seen as beautiful)! And if these enterprises are community-owned, even non-capitalistic, so much the better. Put crudely, political correctness here is on the side of the small!

What are we to make of all this? First, even allowing for the fact that a small industry in one place might be considered big in another, it is deceptively simplistic (even for Schumacher) to talk of 'small' being beautiful. For him, the crux of the matter was that enterprises should be worker-friendly, give workers a sense of belonging, have a substantial element of public ownership and accountability, and should be run on humanistic principles. The chemicals company he provides as a role model – the Scott Bader Commonwealth – was then run as a Trust, on Quaker lines, and its 379 employees in 1971 were all co-owners, with guaranteed shares in the profits, a high percentage of which was donated to charity (Schumacher, 1974: 230).

Scott Bader still exists, with a similar structure of co-ownership, as 'a multinational chemical company employing 600 people worldwide with manufacturing sites in Europe, Middle East, South Africa, and has a turnover of 220 million Euros' (Scott Bader, 2010). For many 'small is beautiful' advocates, this large, impressively-organized international company, operating and *succeeding* in a capitalist environment, might not be a very apt example of 'small'. What is 'small', then, depends on the lens through which we are viewing the enterprise.

Second, is small 'beautiful' because it is better for the economy, or in some way more efficient? The evidence for such a view is decidedly shaky. As one extensive review of the literature for the World Bank puts it:

> [I]t is questionable whether SMEs 'deliver the goods' as advertised. The claim that SME promotion will improve the income distribution is based on two presumptions: (i) that SMEs are particularly effective vehicles for expanding employment and (ii) that growth of SMEs and the

> employment they create disproportionately benefit the poor. As it happens, both of these conclusions are questionable. . . . [T]he evidence indicates that SMEs do not appear to be any more effective at job creation than large firms. (Biggs, 2002: 29)

Where the focus is specifically on *tourism* SMEs, the story is much the same. True, there are cases when small-scale tourism seems to be highly successful, especially at the initial stages of tourism development (Harrison & Schipani, 2007), but in developing societies, especially, individual owners of SMEs frequently lack economic, social and cultural capital (Dieke, 2001; Harrison, 2001a: 253) and donor-funded projects, intended to increase community participation and, in some cases, conservation, generally prove to be economically unsustainable (Goodwin, 2006: 1; Mitchell & Ashley, 2010: 54–58).

Third, smallness covers a variety of categories. 'Small' tourism enterprises operate across a wide range of price and facilities, and may be locally- or foreign-owned. In Fiji, for example, there are many small, backpacker 'budget' resorts, especially in the Yasawa island chain, often with less than 20 rooms, but there are also (equally) small, upmarket 'boutique' resorts, catering for wealthy visitors, including some whose rooms cost well in excess of US $1500 a night (FijiMe, 2010). None of these categories of small establishments, however, conform to Schumacher's criteria, as all operate in a capitalist environment and generally have individual owners. Furthermore, even where landowners may hold some kind of equity, this is no guarantee of strife-free operations (Harrison, 2004a: 10–11)!

Fourth, how one should assess the 'beauty' of these different types of 'small' firms is problematic. As indicated earlier, leakages may be less in the cheaper, locally-owned establishments (less leakages = more beauty?) but so are *total* receipts, and wage rates are likely to be higher, and training better, in the foreign-owned resorts. Furthermore, most of the staff in both are likely to be indigenous Fijian – no less contented in their jobs, and no less hospitable (but possibly more professional) in the upmarket resorts than in those that are locally-owned. Indeed, even where (foreign) ownership and management is heavily (arguably oppressively) 'top-down', employees reported they liked their work and the resort had demonstrable positive economic and social effects in the region (Harrison, 2004b).

Fifth, it might be argued that the impacts of 'small' tourism enterprises are 'beautiful' because they have fewer negative impacts on the physical environment. Such an argument, though possibly sustainable for some *individual* establishments, is deeply flawed. Indeed, it is commonly accepted, even by critics of mass tourism, that 'small-scale' tourism cannot and will not *replace* mass tourism. The economies of scale that characterize large tourism establishments *include* environmental conservation: a 300-room hotel, operating to international standards, is likely to accommodate, feed, cool, entertain and collectively cosset its many guests with far less damage to the environment than the equivalent of fifteen 20-rooms hotels, spread out over a wider area. That some large-scale hotel developments have had negative environmental impacts is undeniable but there is *also* evidence that – as in parts of Spain – with proper planning and legislation, mass tourism can be made more sustainable (Batle, 2000; Aguilo *et al.*, 2005).

Sixth, just as some small-scale resorts may be characterized by heavy-handed, even alienating managements (circumstances in which workers have no place to hide), such negative characteristics are not necessarily the norm in large-scale international hotels. Just as small need not be beautiful, big need not be ugly. I know of no evidence that workers are more contented in small organizations, though there is a growing literature on the importance of empowerment in large-organizations, including large-scale hotels, which suggests that – in *some* cultures – empowered employees suffer less stress and offer a better service, and this is increasingly being recognized by the hospitality industry (Gill *et al.*, 2010; Klidas, 2002; Edwards, 2010; Mohsin, 2008; Mohsin & Kumar, forthcoming). As Klidas (2002: 5) notes, though, how far employers are prepared to delegate authority, and how far employees accept or reject it, will both be crucially affected by cultural factors.

Finally, if it is problematic to argue that 'small is beautiful' at the operational (supply) side of the hotel industry, do *consumers* value small accommodation providers over larger ones? Not necessarily. Whether or not the tourists are wealthy, on business or holiday, and in developed or developing societies, the answer is the same: some prefer small-scale accommodation and some do not! With some exceptions, though, the more 'developed' the tourist destination, the more rooms will be supplied by large-scale accommodation providers, and the more tourists will visit. Why? Because, for convenience, comfort and cost, 'many people seem to enjoy being a mass tourist' (Butler, 1990: 40).

I fondly remember the cosy British pubs, that small hotel in the French Alps, the amazingly cheap but truly authentic guest house by the River Mekong and the Caribbean home-stay. On holiday, I value peace and quiet and loathe deafening music in nightclubs and bars. I also attend soccer games, though, and concerts and theatres, all packed with other enthusiasts. I like cities and prefer them to beaches! And if I am on business, I need internet facilities, good food and (probably) access to the central business district. Clearly, my requirements are very mixed. By contrast, I have friends who love the Sheraton or Hilton, or even Butlins, and especially favour places where their children can play safely and where other families – holidaymakers – become friends. They also spend days on the beach and enthuse over the nightclubs I hate. It is surely similar for the supply side: some may prefer to cater for tourists in large establishments, while others favour some version of 'small'.

This range of preferences is not atypical, and any blanket statement to the effect that, in tourism (and elsewhere?) 'small (or big) is beautiful' simply makes no sense. One size, large or small, simply does not fit all.

3.2

Small can be Beautiful, but Big can be Beautiful Too – And Complementary: Towards Mass/ Alternative Tourism Synergy

David Weaver

The 'small is beautiful' school of tourism, inspired at least indirectly by Schumacher, has a long pedigree rooted in the Grand Tour and some forms of pilgrimage, but found compelling contemporary expression since the early 1980s under the guise of 'alternative tourism' (AT) (Dernoi, 1981; Gonsalves, 1987; Holden, 1984). Perhaps it was the rapidity with which conventional mass tourism appeared to conquer large swaths of the global coastline after World War Two, and the shock of realizing that the promised economic benefits were often accompanied or superseded by a bevy of unwelcomed costs, but the supporters of AT were for the most part subsequently and unfortunately inclined to view the constructs of AT and mass tourism in sharply conflicted black and white terms (Weaver, 2006). As an ideal type contrasted with mass tourism, AT emphasized small scales of engagement presumably more appropriate for the small and marginalized communities that were beginning to be incorporated into the global pleasure periphery.

This 'good' tourism, unlike 'bad' mass tourism, would empower local people, celebrate their authentic local culture, and attract small numbers of sensitive visitors whose expenditures were expected to foster linkages rather than leakages, especially in Third World countries or peripheries of developed countries inhabited by indigenous people. An especially popular and resilient manifestation that continues to resonate with government development agencies, NGOs such as Tourism Concern and many academics is 'community-based tourism' (CBT) (Murphy, 1985), where local residents are invested with control over the planning and management of the sector, and the (presumably equitable) disbursement of benefits, with concomitant benefits for the environment (Sebele, 2010).

Despite strong levels of continued support in some quarters for CBT and other forms of AT, the actual track record is more ambivalent, as discussed by Harrison. Small can be nasty to the extent that members of small and insular communities are acutely aware from their own perspective of who does and does not benefit 'fairly' from such activity. Intersecting family and clan affiliations exacerbate the potential for internal and external conflict, and for elites to consolidate or reinforce their existing power through tourism (Ranck, 1987; Reed, 1997). Small scales of operation also imply limited bargaining power and exposure, and low levels of tourism-related experience, skill and expertise that increase the probability of poor service standards and visitor dissatisfaction. The desire of some 'alternative tourists' to intrude into the private spaces and times of residents is also cited as a salient cost among segments such as backpackers (Ooi & Laing, 2010). Highly germane as well is Harrison's observation that small-scale products are not all created equally, and may include ultra-luxury boutique hotels and ecolodges that grotesquely exaggerate host–guest inequalities of wealth and power.

The dynamics of CBT are especially illustrative given its continued broad support by academics, development agencies and others. Butcher (2007) describes how these outside agencies exercise enormous power over participating communities and exhibit almost neo-imperialistic behaviour that makes support conditional upon adherence to a 'small is beautiful' mindset favoured by those agencies that effectively frames the community as an antediluvian cultural museum, regardless of what community members really want. That these communities develop dependency on these agencies and their ideologies is demonstrated by the tendency of CBT initiatives to collapse once agency support is withdrawn (Salafsky *et al.*, 2001), given their susceptibility to the same diseconomies of scale that plague the performance of tourism SMEs (small and medium enterprises) more generally (McKercher, 1998). It is perhaps partly in recognition of such scale-related limitations that various forms of neo-alternative tourism, such as 'slow tourism' (Conway & Timms, 2010) and de-growthed 'steady state' tourism (Hall, 2009), have not yet gained traction as widely embraced options, even though they appear, in large part, as responses to the potentially apocalyptic effects of climate change, which did not figure prominently in the critiques that attended mass tourism starting in the late 1960s.

All this criticism of AT per se, however, distracts from the more fundamental point that the latter, at least since the 1950s, has rarely, if ever, been unadulterated by the influence of mass tourism and its broader economic context, as the ideal type would suggest. If AT is basically a counter-reaction to the dependency that is fostered in the pleasure periphery by conventional mass tourism, then how does one reconcile the purported empowerment and relative autonomy of participants with the obvious point that most alternative tourists must include a flight with a major carrier as part of their transit journey? Visitors and stakeholders are likewise closely invested with and implicated in the modern globalized mass market economy through the use of credit cards and the internet, fossil fuel consumption, television viewing, importation of food and other basic transactions. It is disingenuous to suggest otherwise. Tourist behaviour, in addition, is largely conditioned by expectations of clean bedding, professional services, safe and tasty food, and reliable scheduling that are shaped by patterns of exposure to conventional large-scale tourism and other sectors, often perhaps experienced as part of the same overall trip. It would be interesting to analyze the extent to which the problematic aspects of AT are respectively exacerbated and ameliorated by this close affinity with mass tourism and other conventional market dynamics.

The listed dynamics are mostly inevitable and 'taken for granted', if also mostly denied. Of great interest, therefore, is the apparent *deliberate* contemporary movement of many AT products and destinations to a larger scale of engagement. Notably, both Bhutan and Dominica, long regarded as AT paragons (Weaver, 2006), are moving in the direction of controlled, smart or sustainable growth. Bhutanese tourism, long constrained by an annual Western international visitor quota of a few thousand, raised this quota to 35,000 in 2010 with approval for further increases to 65,000 and 100,000 in 2011 and 2012 respectively (personal communication, Tourism Council of Bhutan, 2010). The sustainable growth model of Dominica, articulated in 2005 and refined in 2010, calls for expansion of the main airport to accommodate far more stayover visitors, and encourages continued exponential growth of the cruise sector (Joseph, 2010).

The gauging of the motivations for moving to the new sustainable mass tourism paradigm would make for fascinating research. Presumably a larger scale model of tourism is deemed by Bhutanese authorities as compatible with the goal of increasing the country's 'gross national happiness' through added revenue and other outcomes. A paradox of AT has always been that the exclusiveness, 'authenticity' and desirability conferred by limited access serves to increase demand, thereby triggering actions to accommodate this demand in a presumably sustainable way. For Bhutan, this entails the expansion of approved tourist activity to new areas. Other destinations may reject this diffusionist approach in favour of concentrating more visitors into basically the same areas through additional site hardening. The task is more difficult for settlements or districts within countries, which lack the ability to control incoming numbers through quotas or other formal restrictions and thus are more likely to experience the unhappy succession of stages towards unsustainable mass tourism as hypothesized by Butler

(1980) in his tourism area life-cycle model. In all cases however, one discerns confidence that the 'sustainability threshold' can be lifted proactively to accommodate high levels of tourism demand. At a deeper level, it is perhaps ingrained in all tourism managers that growth is still ultimately good and preferable as long as it is pursued sustainably, as per the fundamental premise of the greatly influential Brundtland Report (WCED, 1987).

Going one step further, it can also be argued that mass tourism is moving in some ways towards AT, suggesting a trend towards convergence. This has long been manifested, for example, in encounters between conventional tourists and informal sector street vendors, and in impromptu drives from large coastal resorts to hinterland villages. However, destination managers are now more involved in creating experiences that amplify 'sense of place' and increase product diversity. Exposure of mass tourists to these AT areas, though conditioned by concerns about safety, image and revenue diversion, has rarely been characterized by the same animosity that might attend the exposure of alternative tourists to the corrupting mass tourism interface, perhaps because of the differences in power. Nevertheless, it may mean that large-scale tourism is more open to innovation that engages AT-type characteristics than AT proponents are to engagement with mass tourism.

Nowhere is this openness more evident than in the industry's embracing of triple bottom line sustainability, long regarded as the exclusive domain of AT. Accusations of greenwashing, some of them valid, have led many to regard this trend with cynicism or at least skepticism (Font & Harris, 2004). Beyond the universally supportive rhetoric, sustainability is most evident on the operational side of the industry, where increasingly ubiquitous practices such as energy conservation and recycling indicate a pragmatic environmentalism that contributes significantly and directly to the financial bottom line while cultivating credibility with the public through green branding (Bohdanowicz, 2009). This pragmatic element, which complements the basic market ethos, prompts me to describe this trend as an example of 'paradigm nudge' rather than paradigm shift (Weaver, 2007) – that is, it represents an opportunistic response to recent trends and technological developments that does not threaten basic values of capitalism and growth.

Continuing technological innovation (as for example with solar panels and LED lighting), escalating natural resource prices, and increasing environmental sensitivities among the public, suggest that this trend will continue even if it does not evolve into a more transformational paradigm shift. Economies of scale, in addition, confer various advantages to large-scale tourism that are absent from most SMEs, such as the negotiation of bulk purchase LED lights and solar panels at favourable prices. Less progress, overall, is evident in the realm of social or cultural (as opposed to environmental) sustainability, though increased engagement with communities through charitable activity, sponsorship and other initiatives is evident, often pursued under the banner of corporate social responsibility (CSR) (Bohdanowicz & Zientara, 2009; McGehee *et al.*, 2009).

These examples of the deep reciprocal linkages between AT and mass tourism suggest that synergy between the two 'opposing' ideal types is a longstanding although mostly unrecognized reality. Surely it is appropriate to formally recognize AT and mass tourism as two at least *potentially* synergistic components of a single tourism system and incorporate this principle into the fundamentals of tourism planning and management for any destination. This recognizes that all destinations intersect with both types of tourism, directly and indirectly, and that these intersections can be positive. For rural destinations oriented more towards AT, this could entail constructive partnerships with appropriate intermediaries to enhance destination marketing (or demarketing), facilitate visitor education and participate in donation and sponsorship schemes (e.g. a large airline sponsoring a tree-planting initiative in Third World destinations). Personnel from larger-scale tourism businesses could also help local communities to improve service quality standards and operational/management skills, and assist with the implementation of energy saving and waste reduction practices. Reciprocally, formal consultation with local residents could help to avoid socially or culturally insensitive actions on the part of large companies whilst cultivating an understanding of the factors that contribute to the local sense of place, and encouraging the sale of local artefacts and food in, for example, large beachside hotels.

Higher order protected areas such as National Parks (category II under the IUCN [International Union for Conservation of Nature] classification system) have long implicitly recognized this synergy through internal zoning structures that essentially distinguish between large areas deliberately maintained in a relatively undeveloped condition (as per their mandate to protect biodiversity and ecosystem essentials) and small areas where mass tourism is concentrated (as per their mandate to provide complementary educational and recreational opportunities) (Lawton, 2001). Formal entry requirements, as with Bhutan or Dominica, allow visitation flows to be carefully monitored and controlled. It will be more difficult for villages, towns or districts to effect a similar amalgamation, although zoning can be similarly employed to designate a mass tourism-focused frontstage and an AT-focused backstage. Destinations so diversified benefit by offering the range of tourism choices described by Harrison in his closing paragraphs.

Most tourists would perhaps agree that small and big can both be beautiful if they exhibit their respective inherent strengths through sustainable practice and complementary mutual engagement, and it is on this goal that a new conception of destination planning should be focused.

3.3

Small is Beautiful, but Size can be Important

Richard Butler

The enduring concept of *Small is Beautiful* (Schumacher, 1974) appeared at a time of reaction to many of the post-war elements of economic growth, and was to some degree a contemporary of other provocative and enlightening publications such as Rachel Carson's *Silent Spring* (Carson, 1962) and the Club of Rome's *Limits to Growth* (Meadows *et al.*, 1972), all expressing concern and doubt about the validity of perpetual and rapid economic growth. Thus, Schumacher's ideas have become intermingled with those of environmentalists and to a degree also with those of anti-capitalists and perhaps represent a throw-back to the ideas of William Morris and the craft movement of the early 20th century in Britain (Cumming & Kaplan, 1991). This was a movement predicated on the greater value of hand-crafted items over those that were mass-produced, which inevitably failed to gain widespread support because of the greater cost of hand-produced items, as well as the variation in quality which could ensue from this method of production. For those who could afford to utilize the best craftspeople, cost and quality were not of concern, but for the populace at large, economic considerations were always a prime consideration. The brief discussion which follows is essentially a personal commentary on the general concept and the two preceding viewpoints. It is mostly in support of those viewpoints but to some extent disputes the connection suggested between the 'small is beautiful' argument and the concept of alternative or sustainable tourism.

In applying Schumacher's concept to tourism the issues are less clear-cut than perhaps in other areas. Cost does not always equate with size in the context of tourism, particularly as far as hospitality and food and beverages are concerned. In terms of transportation, however, the basic law of economies of scale generally does apply very much as far as costs are concerned. To reach a destination in a small plane is normally more expensive on a per capita basis than to fly in a 'jumbo' jet, and to use mass transportation is generally far more economical, and often less damaging to the environment on the same basis. We need, therefore, when applying Schumacher's proposal to tourism, to differentiate between the different components of tourism, at least with respect to the levels of travel, accommodation and subsistence, and on-site services and facilities. We should also consider the specific site (resort/development/community), as one could well argue that the same considerations apply in the context of a destination itself.

The concept is inextricably tied into other issues already familiar to tourism researchers, one of which is carrying capacity (Butler, 2010). It can be argued that size (volume of tourism) itself need not be a problem if demand does not exceed the respective carrying capacities (environmental, physical, social and economic) of the destination. A destination operating beyond any of its capacities can be 'ugly' or unattractive irrespective of its absolute size and damage to any or all of its elements can result. Thus, issues such as accessibility, environmental vulnerability, cultural strength, political coherence and sympathetic governance (expressed in appropriate policy implementation) are all relevant to how size is interpreted in terms of reactions by both tourists and residents to tourism development.

As well, as the cliché suggests, 'Beauty lies in the eye of the beholder' and a crowded noisy beach resort may be the ideal destination to one segment of the tourist market and anathema to many other segments. Lucas's (1964a, 1964b) and Wagar's (1964) studies of several decades ago showed clearly that users of an area engaging in different forms of recreation had different tolerance levels for the presence of other people in that area, and that these varied with the activities those others were engaged in. Thus size, expressed in numbers of people, may be a key element for those seeking solitude and/or quietness, and low numbers (i.e. small) may indeed be not only beautiful but essential to an enjoyable experience. However, a near empty visitor attraction (small numbers of people) might make that destination unattractive to other potential users as well as being economically unsustainable to the operators.

Thus proponents of Schumacher's proposal should be careful in linking his concept to that of sustainability as the two ideas are very far apart when one examines them in detail. As has been argued before (Butcher, 2003; Butler, 1999; Wheeller, 1993) smallness does not automatically equate with sustainability, certainly not in terms of economics, one of the three 'bottom lines', and not necessarily in terms of social-cultural or environmental aspects either (WCED, 1987). One person in an inappropriate location, depending on their activities, may have a greater negative impact on both the environment and the resident population than a much larger number in an appropriate location. Numbers alone are not the simple answer to sustainability difficulties, although they are certainly a part of the problem. A fundamental problem with the 'small is beautiful' concept is that the 'genie is out of the bottle' in the sense that tourism is massive (not necessarily the same as mass), and if UNWTO (2009) figures are even remotely accurate (which is questionable), there are many millions, perhaps close to a billion, international trips made each year, with many more made domestically annually, some of these being what most researchers would regard as tourist trips. As the official figures include almost everyone travelling for whatever reason as long as they are alive and not part of a military force, they are all counted as tourists. In reality, the number of people on holiday and recreational trips, which is what most researchers deal with in their work, is probably considerably smaller, but still must constitute many millions each year.

These numbers are not likely to decline appreciably short of global near-destruction. Wars, plagues/epidemics, catastrophes, recessions/depressions and increased fuel costs have all failed to significantly decrease tourist numbers at the global scale, although they may have had generally short-term impacts on specific locations and regions. Thus, tourism is not likely to become 'small' in the way that it was in the pre-Second World War era in many locations that have since been developed as tourism destinations. That 'smallness' was only beautiful, if it really was that, because it was designed for and populated by the affluent and offered levels of services and facilities that were high (in cost as well as standards) while environmental and social impacts were relatively low because of small numbers and isolation from local residents. When one considers some of the activities engaged in by the privileged few participants such as big game hunting, collection of local species and artefacts, use of locals as amusement and entertainment, the scenario is hardly one of 'beauty'.

The enlargement of tourism to encompass 'the great unwashed' (at least until they reach the sea) has been one of the great social benefits of the 20th century, whatever problems it may have introduced (Sharpley, 2000). To attempt to reduce tourism to 'small' levels would be anti-social, discriminatory and generally unacceptable, even though some places might become a little more 'beautiful' through lower levels of use (Hall, 2009). Such a process would have to involve increasing costs per capita, almost inevitably pricing many people out of the market and returning tourism to the levels and markets of the 1920s, 1930s and 1950s. Why then might the argument in support of Schumacher's (1974) proposal emerge in the tourism literature in 2011? Undoubtedly tourism, its proponents and participants have 'ruined' (changed significantly) many destinations, although to the puzzlement of some researchers many residents of such areas still support tourism in their communities (Harrison, 2001b) and it is often tourists rather than tourism which they find less desirable. If one reads the travel supplements of national newspapers, or travel magazines aimed at the affluent or travel guides, there is a common thread, namely, that only places not yet 'discovered' by the masses are worth visiting. Inevitably such places have small numbers of tourists and thus the facilities they offer are generally small in scale. Boutique hotels and other forms of accommodation, in particular, are praised to the heavens, as are safari camps with capacities of under 20 people, where prices may reach several hundred dollars a night and where expeditions, tours and trips are personal and cater to only a handful of people at a time. One can imagine that such experiences are exceptional and could not be obtained by being in a large group or staying in a large hotel. Thus, in these situations, small is indeed beautiful, but it does not mean that the impacts of such forms of tourism are necessarily small or less negative (or more positive) simply by the numbers involved being small. The few can be as irresponsible (inappropriate, tactless and unsympathetic) as the mass (Butler, 1990).

Where then does this leave us? Perhaps almost inevitably one has sympathy and even support for the concept of 'small is beautiful' and what is implied through that phrase. Terms such as benign, supportive, appropriate, responsible, sympathetic, come to mind, terms that one does not often apply to large-scale tourism at least, if to tourism at all, except to identify very specific niche forms of that activity. It would be incorrect and inappropriate to deny that the concept has meaning or value. Some forms of small tourism are truly beautiful, however we might define that term. The familiar advertising scene of a couple romantically linked on a sunset beach does not have the same impact when the shot shows several hundred people on the same beach. As a lapsed bird-watcher this author has no doubt that finding a rarity alone or with one or two friends is vastly superior in satisfaction to seeing a rarity while standing in a crowd of several hundred (I state that from personal experience). On the other hand, watching one's favourite sports team win an important game with a few fellow supporters is not as stimulating or satisfying as seeing that team win in a full national stadium before a hundred thousand spectators (especially when many of those are supporters of the other team).

I perhaps have to end with the conclusion that it is a case of 'horses for courses'. In some contexts of tourism, small is appropriate and desirable, for example in sensitive environmental and/or heritage locations, such as the Galapagos or Machu Picchu, not so much because the small number of tourists makes it more enjoyable for participants but because the site is only capable of withstanding small numbers if it is to remain viable. In other places, with appropriate design and facilities, large numbers of tourists can be accommodated with high levels of satisfaction, and low (at least per capita) environmental and social negative impacts with high economic benefits to the community. Tourism is long past the stage of one size fitting all; it is already too big to try to reduce it to return to a supposedly golden past when tourist numbers were small – that would be impossible and socially and economically (and almost certainly politically) undesirable. We need to learn to make mass tourism beautiful, or at least less ugly (Weaver, 2007), if we are to try to reach what is implicit in Schumacher's statement, namely, that relative scale is important, both to the beholder and to the participant and recipient of tourism.

Concluding Remarks

Several conclusions emerge from the preceding pages, and might be used to direct further discussion and research. First, we need to move away from blanket condemnation or commendation of specific types of tourism. Simply to dig in our heels and insist that any form of tourism – 'small' or 'mass' in all their variations – is inherently superior to any other form is to allow personal preference to influence academic judgement. This is not to argue that academics' personal preferences are irrelevant. Far from it. They inevitably influence our selection of research topics and, through our preferred methodology, our choice of research techniques. Nevertheless, to contemptuously argue that all forms of mass tourism (and all mass *tourists*) are problematic is to allow personal preference to cloud academic integrity.

Secondly, it is necessary to conceptualize small and mass tourism enterprises as part of an overall tourism *system* of demand. Frequently considered to cater for separate markets, in fact they are often interdependent, with small-scale tourism supplementing attractions offered by larger-scale enterprises. Tourists staying at large hotels might go on treks organised by small-scale tour operators, visit plantations and/or villages, handicraft producers, and so on. Similarly, backpackers may prefer guest houses with relatively few rooms, but they also seek out large establishments where they can meet other backpackers and share notes about where to find the best value accommodation and most satisfying travel experiences. There are large as well as small hotels and hostels for backpackers.

Thirdly, one has to beware of evaluations based on criteria grounded in Western culture. While North Americans or Europeans are likely to agree on the desirability of 'natural' landscapes and 'untouched' wilderness settings, such perceptions may not be shared by visitors from East or South-East Asia, who will perhaps prefer muddy tracks to be concreted over, their walking tours to be accompanied by piped music, and trees and shrubs to be uniformly planted. For such visitors, too, an empty beach might be considered unacceptable, while one that is crowded an attraction.

Finally, there is a need to be circumspect when discussing 'sustainability', a term which (arguably) over-use has rendered almost meaningless. As contributors to this debate have noted, it is simply inaccurate to equate small-scale with sustainability (and pro-poor benefits) and large-scale with unsustainability (and exploitation). Rather, as Weaver notes in his discussion piece, we should consider the extent to which 'sustainability' is (or can be) increasingly a feature of mass tourism, and develop a research agenda which includes a primary focus on the operation of large units (including those of transnational corporations), their contribution to 'development', and ways in which they can be made increasingly sustainable over the long term.

To repeat the key conclusion agreed by all three contributors to this discussion: in the demand for and supply of tourism, one size really does *not* fit all! Tourists and those who cater to their various demands, be it accommodation, tours or attractions, are not all of a piece. Rather, they epitomize the individual and social complexity that makes the study of tourism in all its aspects so exciting and rewarding, and to try to encapsulate such variety into a binary scale of small = beautiful and mass = ugly is both mistaken and unsustainable. It is also to make highly questionable assumptions about the nature of beauty:

> A doubtful good
> A gloss, a glass, a flower,
> Lost, faded, broken,
> Dead within an hour.
> (Shakespeare, *A Passionate Pilgrim*: XIII)

Does this mean that beauty itself is unsustainable? That is the topic of another paper!

Discussion Questions

(1) Why is mass tourism so often treated as the unacceptable face of international tourism?
(2) Discuss the idea that small-scale or alternative tourism is beautiful or sustainable because it focuses on pristine and untouched nature.
(3) In what ways are there cultural differences in how tourists approach 'beauty' and 'sustainability'?
(4) What problems have arisen as a result of small-scale tourism?
(5) What environmental advantages might mass tourism have over small-scale tourism?
(6) Consider how small-scale tourism supplements and/or complements mass tourism.
(7) 'A tourist in his/her time plays many parts.' Discuss how the taste and requirements of tourists may vary in the course of a single stay at a destination.
(8) How might small-scale and mass tourism be converging and together becoming more sustainable?

References

Aguilo, E., Alegre, J. and Sard, M. (2005). The persistence of the sun and sand tourism model. *Tourism Management* 26(2): 219–232.

Aramberri, J. (2010). *Modern Mass Tourism*. Oxford: Elsevier.

Batle, J. (2000). Rethinking tourism in the Balearic Islands. *Annals of Tourism Research* 27(2): 524–526.

Biggs, T. (2002). Is small beautiful and worthy of subsidy? *Literature Review*. Washington. International Finance Corporation. Online at www.worldbank.org/PapersLinks/Open.aspx?id=2482. Accessed 5 December.

Bohdanowicz, P. (2009). Theory and practice of environmental management and monitoring in hotel chains. In S. Gössling, C.M. Hall and D. Weaver (eds) *Sustainable Tourism Futures: Perspectives on Systems, Restructuring and Innovations* (pp. 102–130). New York: Routledge.

Bohdanowicz, P. and Zientara, P. (2009). Hotel companies' contribution to improving the quality of life of local communities and the well-being of their employees. *Tourism and Hospitality Research* 9: 147–158.

Brohman, J. (1995). Economic and critical silences in development studies: A theoretical critique. *Third World Quarterly* 16(2): 297–318.

Burkey, S. (1993). *People First: A Guide to Self-Reliant, Participatory Rural Development*. London: Zed Books.

Butcher, J. (2003). *The Moralisation of Tourism: Sun, Sand – and Saving the World?* London: Routledge.

Butcher, J. (2007). *Ecotourism, NGOs and Development*. New York: Routledge.

Butler, R.W. (1980). The concept of a tourist area cycle of evolution: Implications for management of resources. *Canadian Geographer* 24: 5–12.

Butler, R.W. (1990). Alternative tourism: Pious hope or Trojan Horse? *Journal of Travel Research* 28(3): 379–391.

Butler, R.W. (1999). Sustainable tourism: The state of the art. *Tourism Geographies* 1(1): 7–25.

Butler, R.W. (2010). Carrying capacity in tourism: Paradox and hypocrisy? In D.G. Pearce and R.W. Butler (eds) *Tourism Research: A 20-20 Vision* (pp. 53–64). Oxford: Goodfellow.

Carson, R. (1962). *Silent Spring*. Boston: Houghton Mifflin.

Conway, D. and Timms, B. (2010). Re-branding alternative tourism in the Caribbean: The case for 'slow tourism'. *Tourism and Hospitality Research* 10: 329–344.

Cumming, E. and Kaplan, W. (1991). *The Arts and Crafts Movement*. London: Thames and Hudson Ltd.

Dahles, H. (1999). Tourism and small entrepreneurs in developing countries: A theoretical perspective. In H. Dahles and K. Bras (eds) *Tourism and Small Entrepreneurs: Development, National Policy and Entrepreneurial Culture: Indonesian Cases* (pp. 1–19). New York: Cognizant.

Dernoi, L. (1981). Alternative tourism: Towards a new style in North-South relations. *Tourism Management* 2: 253–264.

Dieke, P. (2001). Human resources in tourism development: African perspectives. In D. Harrison (ed) *Tourism and the Less Developed World: Issues and Case Studies* (pp. 61–76). Wallingford: CABI.

Duffy, R. (2002). *A Trip too Far: Ecotourism, Politics and Exploitation*. London: Earthscan.

Echtner, C. (1995). Entrepreneurial training in developing countries. *Annals of Tourism Research* 22(1): 119–134.

Edwards, M. (2010). A look at empowerment: Essential tool for effective management. Online at: www.hotelinteractive.com/article.aspx. Accessed 4 December.

Fennell, D. (1999). *Ecotourism: An Introduction*. London: Routledge.

FijiMe (2010) Online at: http://www.fijime.com/ Accessed 30 November 2010.

Font, X. and Harris, C. (2004). Rethinking standards from green to sustainable. *Annals of Tourism Research* 31: 986–1007.

Gartner, W.C. (2004). Factors affecting small firms in tourism: A Ghanaian perspective. In R. Thomas (ed) *Small Firms in Tourism: International Perspectives* (pp. 35–70). Oxford: Elsevier.

Gill, A., Flaschner, A.B. and Bhutani, S. (2010). The impact of transformational leadership and empowerment on employee job stress. *Business and Economics Journal*. Online at: http://astonjournals.com.bej. Accessed 5 December.

Gonsalves, P. (1987). Alternative tourism: The evolution of a concept and establishment of a network. *Tourism Recreation Research* 12(2): 9–12.

Goodwin, H. (2006). *Community-Based Tourism: Failing To Deliver?' id21insights* 62. Brighton: Institute of Development Studies.

Hall, C.M. (2009). Degrowing tourism: Décroissance, sustainable consumption and steady state tourism. *Anatolia* 20: 46–61.

Harrison, D. (2001a). 'Afterword'. In D. Harrison (ed) *Tourism and the Less Developed World: Issues and Case Studies* (pp. 251–263). Wallingford: CABI.

Harrison, D. (2001b). Less developed countries and tourism: The overall pattern. In D. Harrison (ed) *Tourism and the Less Developed World: Issues and Case Studies* (pp. 1–22). Wallingford: CABI.

Harrison, D. (2004a). Tourism in Pacific islands. *The Journal of Pacific Studies* 26(1&2): 10–11.

Harrison, D. (2004b). Working with the tourism industry: A case study from Fiji. *Social Responsibility* 1(1&2): 249–270.

Harrison, D. and Schipani, S. (2007). Lao tourism and poverty alleviation: Community-based tourism and the private sector. In C.M. Hall (ed) *Pro-Poor Tourism: Who Benefits? Perspectives on Tourism and Poverty Reduction* (pp. 84–120). Clevedon: Channel View Publications.

Holden, P. (ed) (1984). *Alternative Tourism: Report of the Workshop on Alternative Tourism with a Focus on Asia*. Bangkok: Ecumenical Coalition on Third World Tourism.

Honey, M. (1999). *Ecotourism and Sustainable Development: Who Owns Paradise?* Washington: Island Press.

Joseph, E. (2010). Director of tourism presents five-year tourism action plan to government. Government Information Service. Online at: http://www.gis.dominica.gov.dm/news/feb2010/dofp5ysptg.php. Accessed 22 December 2010.

Klidas, A. (2002). Employee empowerment in the european cultural context: Findings from the hotel industry. Paper presented at Athens University of Economics and Business, CRANET 2nd International Conference on 'Human Resource Management in Europe: Trends and Challenges'.

Lawton, L. (2001). Public protected areas. In D. Weaver (ed) *The Encyclopedia of Ecotourism* (pp. 287–302). Wallingford: CABI.

Lucas, R.C. (1964a). The recreation carrying capacity of the quetico-superior area. *US Forest Service Research Paper* LS-15, St Paul. USDA.

Lucas, R.C. (1964b). Wilderness perception and use: The example of the Boundary Waters Canoe Area. *Natural Resources Journal* 3(3): 394–411.

McGehee, N., Wattanakamolchai, S., Perdue, R., and Calvert, E. (2009). Corporate social responsibility within the U.S. lodging industry: An exploratory study. *Journal of Hospitality and Tourism Research* 33: 417–437.

McKercher, B. (1998). *The Business of Nature-Based Tourism*. Melbourne: Hospitality Press.
Meadows, D.H., Meadows, D.L., Randers, J. and Behrens, W.W. (1972). *Limits to Growth*. New York: Universal Books.
Mitchell, J. and Ashley, C. (2010). *Tourism and Poverty Reduction: Pathways to Prosperity.* London: Earthscan.
Mohsin, A. (2008). How empowerment influences revenue management and service quality? A case of New Zealand Hotel. *International Journal of Revenue Management* 2(1): 92–106.
Mohsin, A. and Kumar, B.C. (2010). Empowerment education and practice in luxury hotels of New Delhi, India. *Journal of Hospitality and Tourism Education* 22(4): 43–50.
Mowforth, M and Munt, I (2009). *Tourism and Sustainability: Development and New Tourism in the Third World.* London and New York: Routledge.
Murphy, P. (1985). *Tourism: A Community Approach*. New York: Methuen.
Ooi, N. and Laing, J. (2010). Backpacker tourism: Sustainable and purposeful? Investigating the overlap between backpacker tourism and volunteer tourism motivations. *Journal of Sustainable Tourism* 18: 191–206.
Practical Action (2012). Online at http://practicalaction.org. Accessed 12 May 2012.
Ranck, S. (1987). An attempt at autonomous development: The case of the Tufi Guest Houses, Papua New Guinea. In S. Britton and W. Clarke (eds) *Ambiguous Alternative: Tourism in Small Developing Countries* (pp. 154–165). Suva: University of the South Pacific.
Reed, M. (1997). Power relations and community-based tourism planning. *Annals of Tourism Research* 24: 566–591.
Reid, D. (1995). *Sustainable Development: An Introductory Guide.* London: Earthscan.
Rodenburg, E.E. (1980). The effects of scale in economic development: Tourism in Bali. *Annals of Tourism Research* 7(2): 177–196.
Salafsky, N., Cauley, H., Balachander, G., Cordes, B., Parks, J., Margoluis, C., Bhatt, S., Encarnacion, C., Russell, D., and Margoluis, R. (2001). A systematic test of an enterprise strategy for community-based biodiversity conservation. *Conservation Biology* 17: 1585–1595.
Scheyvens, R. (2002). *Tourism for Development: Empowering Communities.* Harlow: Pearson Education.
Schumacher, E.F. (1974). *Small is Beautiful: A Study of Economics as if People Mattered.* London: Sphere Books.
Scott Bader (2010) Online at: http://www.scottbader.com/ Accessed 28 November 2010.
Sebele, L. (2010). Community-based tourism ventures, benefits and challenges: Khama Rhino Sanctuary Trust, Central District, Botswana. *Tourism Management* 31: 136–146.
Seers, D. (1969). The meaning of development. *International Development Review* 11(4): 2–6.
Seers, D. (1977). The new meaning of development. *International Development Review* 19(3): 2–7.
Sharpley, R. (2000). In defence of (mass) tourism. In M. Robinson, J. Swarbrooke, N. Evans, P. Long and R. Sharpley (eds) *Environmental Management and Pathways to Sustainable Tourism* (pp. 269–284). Sunderland: The Centre for Travel and Tourism in Association with Business Education Publishers.
Sharpley, R. (2002). The evolution of tourism and development theory. In R. Sharpley and D.J. Telfer (eds) *Tourism and Development: Concepts and Issues* (pp. 35–78). Clevedon: Channel View Publications.

Sharpley, R. (2009). *Tourism Development and the Environment: Beyond Sustainability?* London: Earthscan.

Simpson, M. (2008). Community benefit tourism initiatives – A conceptual oxymoron? *Tourism Management* 29(1): 1–18.

Singh, S. (2010). Community participation – In need of a fresh perspective. *Tourism Recreation Research* 35(2): 209–211.

Telfer, D. (2002). The evolution of tourism and development theory. In R. Sharpley and D. Telfer (eds) *Tourism and Development: Concepts and Issues* (pp. 35–78). Clevedon: Channel View Press.

UNESCAP (2001). Managing sustainable tourism development. *ESCAP Tourism Review* No. 22. Bangkok: United Nations.

UNWTO (2000). *Sustainable Development of Tourism: A Compilation of Good Practices.* Madrid: WTO.

UNWTO (2009). *Tourism Highlights.* Madrid: UNWTO.

Wagar, J.A. (1964). The carrying capacity of wild lands for recreation. *Forest Science Monograph* 7. Washington D.C.: Society of American Foresters.

WCED (1987). *Our Common Future*. Oxford: Oxford University Press.

Weaver, D. (2001). Mass and alternative tourism in the Caribbean. In D. Harrison (ed) *Tourism and the Less Developed World: Issues and Cases* (pp. 161–174). Wallingford: CABI.

Weaver, D. (2006). *Sustainable Tourism: Theory and Practice*. London: Butterworth-Heinemann.

Weaver, D. (2007). Towards sustainable mass tourism: Paradigm shift or paradigm nudge? *Tourism Recreation Research* 32(3): 65–69.

Wheeller, B. (1993). Sustaining the ego. *Journal of Sustainable Tourism* 1(2): 121–129.

Further Reading

Butcher, J. (2003). *The Moralisation of Tourism: Sun, Sand – and Saving the World?* London: Routledge.

Butler, R.W. (1999). Sustainable tourism: The state of the art. *Tourism Geographies* 1(1): 7–25.

Font, X. and Harris, C. (2004). Rethinking standards from green to sustainable. *Annals of Tourism Research* 31: 986–1007.

Harrison, D. (1996) Sustainability and tourism: Reflections from a muddy pool. In L. Briguglio, B. Archer, J. Jafari and G. Wall (eds) *Sustainable Tourism in Islands and Small States: Issues and Policies* (pp. 69–89). London: Cassell.

Honey, M. (1999). *Ecotourism and Sustainable Development: Who Owns Paradise?* Washington: Island Press.

Mitchell, J. and Ashley, C. (2010). *Tourism and Poverty Reduction: Pathways to Prosperity.* London: Earthscan.

Rodenburg, E.E. (1980). The effects of scale in economic development: Tourism in Bali. *Annals of Tourism Research* 7(2): 177–196.

Simpson, M. (2008). Community benefit tourism initiatives – A conceptual oxymoron? *Tourism Management* 29(1): 1–18.

Weaver, D. (2007). Towards sustainable mass tourism: Paradigm shift or paradigm nudge? *Tourism Recreation Research* 32(3): 65–69.

Chapter 4

Does Community Participation Empower Local People?

Jim Butcher, David Weaver and Shalini Singh

Context

Community participation has become a major theme in development over the last few decades. In fact, the language of 'community' has increasingly been invoked over this period in many aspects of social and political life. Tourism is no different, and in fact the 'person to person' character of the industry accentuates the emphasis on community.

In this chapter, the opening probe draws on exemplary literature to argue that community participation has become a mantra that, far from being an ethical development in development and tourism industry practice, masks some rather less than progressive assumptions about the destinations in question. Butcher argues that whilst the talking up of community sounds democratic, especially when accompanied by terms such as 'empowerment', it all too often flatters to deceive. The localism that invariably accompanies community participation in tourism is limiting in itself *vis a vis* national and large scale development, and the reality is that the direction of development is rarely if ever a part of the process of community participation. Butcher makes some hard hitting points about the limits and problems with the mantra of community participation.

In response, two eminent authors in the field make a number of important points. David Weaver argues that community participation has a very questionable record – successes are few and far between. He argues for a more pragmatic rationale for community participation, one that is clear about its aims and about its intended beneficiaries.

Shalini Singh argues that community participation is so central to the life of a community that it must, despite the myriad versions and criticism, remain a priority. She focuses on two examples, arguing that local communities should not be at the mercy of the global tourism industry, and that community participation should be viewed in light of this imperative.

This chapter covers a theme that is very prominent in discussions about contemporary tourism.

4.1
The Mantra of 'Community Participation' in Context

Jim Butcher

Introduction

The following probe introduces a critique of the contemporary emphasis on community participation in literature on tourism planning, and tourism generally. At the outset it is important to note that this is not a critique of the extent to which community participation operates well in practice, or is genuine and not tokenistic (of which there are plenty). Rather, it argues that community participation – as currently constituted through academic literature and practitioner opinion – does not warrant its generally-held status as part of a progressive shift in development politics, specifically with regard to tourism.

The emphasis here is on rural tourism in the developing world, but perhaps some of the points made have wider application too.

Community Participation as Orthodoxy

Community participation has become a popular cause in many areas of society – planning, architecture, the arts and, of course, tourism development. There are many reasons for the resonance of 'community' and 'participation' in modern politics, especially when placed together, but it is fair to say that they receive a favourable hearing.

Community participation has become an important point of reference in development studies, especially with regard to rural development. The neopopulist maxim that development should be what people do for themselves rather than what is done to them by others, has influenced attempts in theory and practice to make development assistance, and development generally, responsive to the local communities most directly affected by it (Brohman, 1996). Chambers' (1983) advocacy of 'Putting the Last First' is a good example of its influence.

The literature on tourism is no exception to this. In fact, I would say that this literature generally veers on the side of being overly deferential and uncritical of the community participation agenda, as I shall argue.

Empowerment

One of the key concepts underlying the advocacy of community participation is 'empowerment'. Community participation can, it is held, empower local people to take greater control over development of tourism-related projects. This is often presented as an antidote to a lack of sensitivity to the needs of the local community, and a lack of any serious consultation with them, on the part of big government and big business.

The most commonly invoked example of a scale for gauging this is Pretty's typology (Pretty, 1995). Pretty's seven levels of participation feature 'manipulation' at one end and 'self mobilization' at the other. Pretty's analysis presents a greater level of participation as 'good', with the ideal being this 'self mobilization'. Here, communities instigate, as well as plan and see through, conservation and development projects within their community.

Pretty's typology emphasizes the question of the *distribution of power within a community*. Its focus is on inter-personal and inter-group power. It has nothing to say about the prior limits placed on the community from without. If living a subsistence existence, closely reliant on the immediate natural environment, is considered a limitation on the community's ability to develop economically, then such limits cannot be challenged by any amount of community participation.

Scheyvens provides a nuanced definition of empowerment (1999: 247–249; see also Scheyvens, 2002), but one that also sees power as a micro-political category, *within* the community. Even her category of 'political empowerment' envisages this at the micro-level, within the community.

Many of the projects for rural development in the developing world that emphasize community participation also tie development possibilities to the conservation of the immediate natural environment – for example, ecotourism is a popular vehicle for Integrated Conservation and Development Projects (ICDPs). Yet, through the language of 'empowerment', 'participation' and 'control', such limits are presented as reflecting the agency of the community – *their* culture and *their* aspirations (Butcher, 2007). The key issue arising here is whether the lauding of empowerment on a micro-political level rationalizes, or makes acceptable, a lack of power, or unequal power relations between the developed and developing worlds. If so, empowerment and community participation, in context and in content, may be less than progressive.

Broader issues of power between nations, between the developed and developing world, between social classes, and notions of social power beyond the immediate experience of individuals, are either deprioritized or non-existent in most accounts that focus on empowerment. Reed (1997), for example, looks directly at the issue of power in her article entitled 'Power relations and community based tourism planning', yet the conception of power adopted is restricted to interpersonal and inter-group power within the community. This is problematic when inflated claims are being made for community participation, and where it is fêted as part of a new, progressive and even radical politics of development.

Democracy

Community participation also suggests itself as part of a democratic agenda – greater choice, empowerment and control all evoke a greater degree of democracy in development. It has at its heart a promotion of the agency of the popular majority, usually within a locality.

But the issue of *what is being participated in* is decided prior to the community's involvement – and this is no less true in the case of small-scale, 'sustainable' initiatives than it is with mainstream tourism. Even NGO-backed projects that seek to promote sustainable, community-led development, that are often seen as part of a 'new' development perspective, are dependent on the funding priorities of milieus of NGO professionals and environmental activists. These priorities are prior to any substantial community input.

A few writers have noted the obvious dilemma that can occasionally arise in community participation. What happens when communities opt for alternatives – mass tourism perhaps – that are not in keeping with the aims of funding authorities such as NGOs or Western development agencies? Weaver articulates this as follows: 'If [these] experts attempt to impose an AT (alternative tourism) model or to re-educate the local people so that they change their preferences, the entire issue of local decision making, control and community based tourism is called into question' (Weaver, 1998: 15). However, this dilemma may rarely surface, as whilst communities may have opportunities to engage with how a project is implemented, and how its benefits are distributed, the broader issue of choosing development priorities is usually foreclosed – there simply is not a mechanism through which communities can play a part in this.

Pragmatism is more characteristic of community participation than is democracy. Faced with the possibility of assistance or investment tied to a particular type of project, or no assistance at all, the pragmatic choice is to accept assistance regardless of any unfavourable terms attached (White, 2000). Hence, participation does not involve real choice at all, as there is likely to be an absence of alternatives on offer. Rather, participation by the community is likely to be instrumental to the prospect of some limited financial assistance, the terms of which they have little if any control over.

Also, the language through which this funding is rationalized and presented by donors ('participation', 'ownership', 'empowerment', etc.) is likely to be adopted by recipients based on a recognition of its instrumental value rather than a deep-seated commitment to the development ideas it expresses (see Hann & Dunn, 1996). Hence the democratic credentials of participation, seen in a slightly wider context, are illusory.

It is through national democratic political activity that people can have some control over bigger political questions such as the trajectory of development. Localism situates control at the local level, and assumes that the big question of the type of development has been resolved beyond debate in favour of 'sustainable development', which is often interpreted as being rooted in the pre-existing relationship between people and their local environment.

Certainly, in the case of the advocacy of ecotourism, national government and large-scale development is often ignored or denigrated. Agency is talked up at the local level, but at the same time limited to that level. Scheyvens articulates this sentiment well in her popular book *Tourism for Empowerment* which she asserts '. . .is not a book about how governments can extract the greatest economic benefits from encouraging foreign investment in tourism. [. . .] Rather, the interests of local communities in tourism development are placed foremost' (2002: 8). These interests are assumed to reside in tourism which supports 'sustainable livelihoods' (Scheyvens, 2002), which in turn is often interpreted as small-scale eco-development.

The Critics

It would be wrong to argue that community participation, and its associated armoury of terms such as empowerment and control, are accepted uncritically. Mowforth and Munt recognize that 'the push for local participation comes from a position of power, the first world' (Mowforth & Munt, 1998: 242). In similar vein, Midgeley writes with insight that 'the notion of community participation is deeply ideological in that it reflects beliefs derived from social and political theories about how societies should be organised' (Midgeley, 1986: 4). Midgeley is referring here to the notion that the rhetoric of community participation could be a cover for Western style 'modernization', an argument also prominent in Mowforth and Munt's *Tourism and Sustainability: New Tourism in the Third World* (1998).

Many critics have questioned the efficacy of community participation along these lines, regarding it as either tokenistic, or a cover for commercial or preservationist schemes. For example, Woodwood's research argued that the norm in South African ecotourism projects was to adopt a participatory approach primarily in terms of its public relations value (Woodwood, 1997: 166). Similarly, Scheyvens cites the work of the Conservation Corporation of Africa (CCA) as an example of an organization that she believes works with local communities only out of a sense of economic pragmatism rather than a commitment to the communities themselves (Scheyvens, 2002: 192–193).

Yet, is this approach, roundly criticized by Scheyvens, so different from the alternative examples she and others cite as being progressive? Private companies may introduce participation on an instrumental basis, for its public relations value. In the case of the conservation organizations and rural development agencies, there seems to be a similar instrumental approach to participation – it is participation *for a specific end*, an end no more the product of the community's unfettered desires than is the case in the above example featuring the CCA. And Midgeley's statement, referred to above, that community participation is 'deeply ideological' (Midgeley, 1986: 4) holds true with regard to the neopopulist alternatives – these alternatives have been developed in, and are funded from, a particular milieu in the developed world. They, too, are ideological – that they emanate from civil society, rather than government or commerce, does not preclude this.

Local and National Perspectives

The emphasis on the community as a local, as opposed to national, phenomenon, as the appropriate spatial unit for development, is a feature of neopopulist thinking. There is relatively little justification for this within the tourism-specific literature – writing on national tourism trends on one hand, and on rural, participatory tourism on the other, seldom meet in the middle.

Yet, the idea of community can equally be applied to the nation – the national community – and inevitably what takes place, or does not take place, in the forests of a developing world country, also affects people living in the cities. It is not clear where this leaves local participation in relation to national participation through elections – should the local be privileged over the national, as it seems to be the case in much of the analysis?

One author who does consider critically the relationship between community participation in tourism and national priorities, Scheyvens (2002), simply sees the issue in terms of central government's role in facilitating community level development. Scheyvens' view reduces national priorities to acting as an enabling state in rural areas, enabling the functioning of locally based sustainable development. She is ultimately interested in development through 'local agency' (Scheyvens, 2002: 56). This is a common theme, developed by Chambers and other neopopulist writers. Yet, it is not at all clear where this leaves national agency and the imperative of national development, in a global world economy based upon trade between nation states.

In similar vein, Parnwell argues that community participation is desirable in order to compensate for a lack of democracy or good governance at a national level. He argues that the ability of NGOs and communities themselves to shape tourism in a fashion that is positive for the community, depends on the 'prevailing socio-economic context' (Parnwell, 1998: 217), and goes on to contend that developing world governments may encourage international capital to benefit the elites rather than to benefit the majority of the people. Once again, the state is posed as a limiting factor upon the community, and its ability, in conjunction with NGOs, to yield development and conservation through tourism.

This privileging of the local over the national does have merit. For example, Scheyvens (2002: 33) is partly right to claim that modernist discourse has been preoccupied with macro-level improvements, rather than a broader concern for well-being. However, some seem to morally elevate the local, community level above macro level indicators, and indeed quite a lot of the discussion about development through ecotourism fails to mention national perspectives at all. As such, they may replace a bias towards macro indicators – a national bias which neopopulists claim is symptomatic of modernization as development – with an inability to envisage development as anything other than a locally based phenomenon.

For example, Scheyvens (2002: 54), adopting this localism, articulates the case for development through tourism. She argues that 'A concern for livelihoods should be integral to development efforts, based on the recognition that local people need to benefit from the existence of natural resources in their area. . .'. To argue that a concern for livelihoods should be central to development is uncontentious but vague. However, to suggest that *local* people need to benefit from the existence of natural resources *in their area* is more difficult to accept. In most contexts in developed countries, with an international division of labour and global trade, people do not benefit from the natural resources in their area. They tend to benefit from resources in the widest sense – every time they switch on a light, light the gas oven, drive their car, read a book or visit a museum they are benefiting from resources produced far from their own communities. Resources that communities in the developed world have at their disposal, and the efficiency with which they can transform them into goods, is shaped by modern development, which itself is premised on an international division of labour. Scheyvens' localized, eco-development approach eschews this legacy in favour of self-sufficiency and smallness of scale, not as a stepping-stone to something else, but as a point of principle.

Also, concerns about the lack of a 'trickle down effect' from nationally based development can be well justified. However, some of these concerns seem to dismiss rather readily that in a world of nation states, development of any substantial scale has to have a strong national perspective if it is to contribute to the transformation of national economies away from developing world status towards a more developed one.

A Progressive Alternative?

It is notable that community participation is often viewed by its advocates as 'alternative' or as radical, as a counter to overbearing governments and the rhetorical free market agenda associated with the big global financial institutions the World Bank and IMF. This is the tenor of Scheyvens' *Tourism for Development: Empowering Communities* (2002) and of Tourism Concern's *Community Tourism Guide* (Mann, 2000). Activist Anita Pleumaron even argues that true 'grassroots' participation is necessary as part of the construction of 'an alternative "new world order" in which people themselves, rather than outside interests, determine and control their lives' (Pleumaron, 1994: 147).

Ironically, the radical rhetoric of 'empowerment' and 'community control' masks a shared outlook with the proponents of the free market rhetoric that the advocates of community participation often claim to oppose, that shared outlook being a diminished view of the importance and efficacy of the state in development.

What the mantra of local participation does is to represent political agency as a local phenomenon affecting local people, premised upon their local environment. This is viewed favourably *vis-a-vis* the nation state and the potential for states to organize economic development in the interests of the populace.

Its radical credentials and the rhetoric of empowerment and democracy mask the fact that local participation is normally about involving people (and incentivizing people) in playing a role in seeing through a predetermined project. In rural parts of the developing world, tourism projects are typically organized around creating a symbiosis between the community and nature – ecotourism brings revenue, revenue premised upon conserving the environment and changing very little. In such circumstances, community participation arguably becomes about participating in modifying the terms of one's poverty.

Local participation – in the abstract never a bad thing – is in fact a product of political developments which are far from progressive, or at least which should be contested much more vigorously. Development thinking, and the potential of tourism in development, is ill served by the deference to local participation.

4.2

Community-Based Tourism as Strategic Dead-End

David Weaver

Jim Butcher rightly points out how community-based tourism (CBT) has maintained its status as orthodoxy in some academic and non-profit quarters despite its numerous flaws and contradictions which, as also correctly noted, have not received the recognition or attention from tourism scholars that they deserve. This blindness suggests that proponents are still locked into the ideologies and assumptions that framed its introduction in the early 1980s as the *sine qua non* of alternative tourism. Corporate mass tourism, as is well known, was regarded by many academics of the left as inherently evil, and locally controlled small-scale tourism served as a diametrically opposed and morally correct counterpoint (Dernoi, 1981). Although long discredited as simplistic and naïve (see for example Butler, 1990), this dogma unfortunately is still reflected and advocated in most of the CBT discourses that have proliferated in subsequent years.

A Questionable Record of Success

If replication alone is regarded as a parameter of success, then CBT has been highly successful, given the hundreds of examples that have been established throughout the world during the past 30 years and partially inventoried by scholars such as Buckley (2003) and Zeppel (2006). CBT, undoubtedly, remains the option of choice for most non-governmental organizations (NGOs) and government agencies that include tourism in their developmental portfolio. Honey (1999) describes how just one agency, USAID, was involved with 105 CBT-focused ecotourism projects in the mid-1990s. More recent initiatives, such as the trans-boundary Heart of Borneo proposal, continue to advocate CBT as a core developmental principle (Hitchner *et al.*, 2009).

With other more visceral criteria, however, the evidence is far less certain. A very important aspect that is not addressed in the opening probe is individual-level performance. No rigorous collective evaluation of CBTs has yet been undertaken, but anecdotal information indicates a less than stellar track-record of financial sustainability. Buckley (2003), for example, includes many CBT operations in his collection of ecotourism case studies and argues that the number of unqualified successes is 'quite limited'. More specifically, an analysis of 37 operations funded in Asia by the Biodiversity Conservation Network revealed that 10 covered both their variable and fixed costs, 13 covered only their variable costs, three produced 'minimal' revenue and four produced no revenue at all (Salafsky *et al.*, 2001). It is additionally apparent from the descriptions of these cases and follow-up internet searches that a large proportion no longer exists, which is further consistent with findings about tourism-related small and medium enterprise (SME) constraints more generally (Getz & Carlsen, 2005; Hall *et al.*, 2005).

Opportunities to Fail

The impressive inventory of indigenous CBT products compiled by Zeppel (2006) is especially effective in emphasizing the degree to which they, unlike private SMEs, are dependent on continuing financial and other resources provided by NGOs and government agencies. Indeed, they are situated by Zeppel (2006: 14, 17) as a key feature of indigenous CBT, especially within developing countries. Such assistance, normally, is time-limited and intended to foster circumstances that will allow the operation to be self-sustaining and fully or largely community-operated by a given time. In practice, it appears that most of these operations fail soon after the aid is cut, rendering the concept of 'empowerment' as farcical and reflecting diverse opportunities for failure in the incubation phase. One potential problem not discussed by Butcher is contestation over the boundaries of the local 'community' that is privileged to make core decisions about the operation and reap most of the anticipated benefits. Though those benefits may well be confined to the disbursement of financial and other aid, they will generate ongoing conflicts over community membership and the designation of power to decide the latter (Medina, 2005).

This issue of power reminds us that communities, however defined, are not homogeneous, but collections of individuals, families, clans and factions continually competing for power and resources. While the ideal of consensus is admirable, it is seldom encountered in the real world and even then may be superficial, indicating a series of compromises or a North Korean-type of conformity to the local socio-political status quo where certain individuals or factions, through fear or convention, are acknowledged as the power-brokers, and stability is arguably the only real benefit for the disempowered. Just as the ideal of community control is questioned by Butcher in light of the reality of systemic global and national core–periphery relationships, so can it be further challenged by the presence of 'micro' cores and peripheries within every community (Beeton, 2006).

When these inequities are combined with the complex relationships that exist with external and other stakeholders (e.g. locally born individuals now residing in urban areas, their descendents, recent arrivals into the local area from other ethnic groups, absentee landowners, etc.), it may be that the contemporary construct of 'community' is better depicted as a series of oscillating concentric circles with a small and more stable core of real power at the centre. Resulting inter-group dynamics can inhibit the cultivation of factors that increase the likelihood of successful resident-focused rural tourism, which include responsive and genuinely supported leadership, broad-based participation, access to ecotourism venues, formation of effective partnerships, and skill and capacity acquisition (Weaver, 2008).

Partnership-Based Tourism: A Practical Alternative?

The vagaries of human nature, as reflected in the above critique, are one fundamental but usually ignored influence on CBT performance that needs to be taken into account when attempting to implement sustainable tourism strategies. A concurrent influence is failure to appreciate the implications associated with becoming involved in tourism. The latter is a uniquely globalized and globalizing activity, involving as it does varying degrees of association with an array of self-interested intermediaries within the distribution chain, and face-to-face contact with an equally complex and self-interested array of visitors at the receiving end. We must bear in mind that relatively remote settlements, aside from the internal dynamics described above, are inherently disadvantaged in their negotiations with these other interests by scale, lack of tourism-related business and management skills and other deeply entrenched constraints. Intermediaries and tourists have plenty of alternative destinations to deal with if a particular place does not provide satisfactory terms of engagement. Attempts to implement a model of tourism that challenges the inequitable fundamentals of the global core/periphery tourism system, accordingly, are quixotic.

How can proponents of CBT imagine that such realities, inherent in 'alternative' as well as mass tourism, will *not* increase instability, internal competition and external dependency, perhaps setting the stage for a more intensive (though perhaps more desired) level of engagement with tourism as per Butler's (1990) Trojan horse? The idealistic autarkic goals of CBT, and high levels of local initiative such as 'self-mobilization', 'citizen control' and 'spontaneous participation' (Tosun, 2006), are arguably more likely to be achieved by avoiding tourism altogether.

Butcher does not actually propose any genuinely 'progressive alternative' to current models of CBT despite the title of the final section, perhaps because it is too difficult to imagine what this might look like. The search for a more effective *practical* alternative should perhaps begin by asking what benefits are being sought, and for whom. Maximum well-being is an obvious and attractive theoretical goal, but one that still begs logistical questions of definition and measurement. However, since tourism depends on a complex web of interdependencies, and since none of the aforementioned intermediaries or tourists are likely to participate in any tourism that does not attend to their own well-being, it is reasonable to argue that any model of tourism responsive to providing maximum benefits at the local scale must strategically also meet the needs of the other stakeholders, each of which will have their own perception of what these needs are.

From the local perspective, information as to what benefits residents in less developed regions really want from tourism is lacking, expressions of same tending to be made by patronizing aid agencies rather than the people themselves, as Butcher points out. It is probable, however, that there would be widespread agreement with those dimensions of the empowerment model of Scheyvens (1999) such as universal self-esteem and confidence, provision of opportunity, respectful cooperation among residents and a representative and responsive political system, that both result in and from an improvement in material well-being. While it might logically appear that the challenge is to facilitate these objectives while allowing other stakeholders to achieve *their* own objectives, it may be that the pursuit of collective self-interest *is* the best way for local residents to achieve this empowerment, through strategies that evoke the principle of reciprocal altruism/enlightened self-interest. If so, then the establishment of effective partnerships and relationship networks among residents, intermediaries, government and tourists is perhaps the most critical dimension of successful and sustainable tourism (Boyd & Singh, 2003; Fuller *et al.*, 2005).

It is unlikely that any tourism situation where a clearly defined community exercises real control, takes the initiative, makes all decisions on the basis of consensus, promotes a sustainable local natural environment and indigenous economy, supports traditional culture, retains most economic benefits internally, allows these benefits to be distributed equitably, and is financially self-sustaining, actually exists anywhere or is likely to emerge. Yet, the proselytization of a 'pure' form of CBT, which offers the fool's hope of such outcomes, continues unabated. Contradictions multiply, and a new paradigm is called for to replace this dead-end approach. The idea of a pragmatic rather than 'progressive' partnership-based tourism model is mooted here as a more effective alternative that promotes a realistic empowerment of local residents. However, unlike CBT, it does so by embracing integration with the outside world, recognizing collective self-interest and multiple nodes of power and decision-making as assets, and unabashedly situating material well-being – within reasonable limits – as a primary desired outcome of their engagement with tourism. All these characteristics, I would argue, better reflect the realities of human nature and the realities of becoming involved in tourism, and increase the likelihood of successful outcomes even from the perspective of local residents.

4.3

Community Participation – In Need of a Fresh Perspective

Shalini Singh

Jim Butcher's exposition on community participation is fairly substantive in terms of the scope of the subject as well as its practice. That community participation is a delightful oxymoron seems to be the clear message. Based on the multi-pronged arguments, the author, then, appears to conclude that it is generally an unavailing pursuit and can conveniently be relinquished in theory and practice by scholars and communities alike. Such conclusions tend to arise from a despairing conflict of utopian perspective with realities. Given that engagement, interaction and action (collectively referred to as 'communitarian') are so intrinsic to humankind, would it really be possible to abandon a concept that is ingrained in human culture and dynamism? If collective human action is representative of communities then is it possible for scholars to deny the existence of community participation? This probe intends to present two distinct approaches to community participation through two case examples. These different approaches not only convey distinct meanings of community participation but also tease out the challenges peculiar to each. My contention would be that, more than communities, it is probably the advocates whose short-sighted utopianism may have inflicted disrepute on its relevance by simply reducing communitarian symbiosis to 'goal-driven community participation'.

The shrinking of distances and barriers to access resources and markets, through the processes of globalization, has spawned much confusion between 'what is' (realities) and 'what ought to be' (intentionalities). With the latter often taking precedence, the pursuit of scientific and utilitarian rationalization compels people to reconstruct and align their cognition and action in compliance with the prevailing dictates of global restructuring (see for example Murphy & Pauleen, 2007). Apparently the global market has been raised to such levels of importance so as to convey definitive meaning and intent to society, communities and individuals (component parts). In the process, community participation becomes less about culturally relevant conscious choices of a local collective and instead a sure way for locals to reap the fruits of economic globalization merely by buying into the market-driven new world order (e.g. Murphy, 1988; Tosun, 2000).

Such thinking now claims almost every aspect of community being, existence and mutuality. Our relationship with the self, each other, the elements and the world are increasingly emphasized in terms of expectability, stability, order and linearity as opposed to interdependency, non-linearity and complex adaptive systems (Farrell & Twining-Ward, 2004). In some cases this chaos and complexity is derided as a major constraint (for example see Tosun, 2000; Li, 2006) (e.g. the labelling of tradition as 'backward') which should make way for the rational linearity of modern day thinking. There are, however, some instances when the non-linearity is acknowledged (such as Murphy & Pauleen, 2007; Reid *et al.*, 2004). Clearly, the two previous discussants are lamenting and questioning the circumstantial constraints that are imposing reductionism in the well intentioned community systems approach.

However, as true as this may seem, there is also a need to acknowledge that the human-scape is subject to historical processes of 'becoming'. Community participation, like most other human processes, is emergent. Because of this it is imperative to understand it from within, '*in-situ*', and also in historical context. These assertions can be briefly considered through the two case studies below. The cases, and the contrasts between them, support the 'reductionist' claim and the propriety of a longitudinal community approach.

The first case comes from Ladakh – a small region in the higher reaches of the Indian Himalaya. Until the early 1980s the way of life in this Himalayan region may be termed as primitive, where subsistence economy not only integrated with the community's way of life and natural environment but apparently reflected folk ethos, aspirations and value systems (see the film edited by Norberg-Hodge & Page, 1993). Farming, weaving, construction, religion and other aspects of life have been pursued with aspirations for prosperity, joy, peace and survival. All actions and thoughts were directed by the spirit of sustaining human life. Community traditions had developed around and through human thought for, and contemplation of, people's linkage with their land. Local people participated in communal life within their respective villages and among themselves, thereby nurturing supportive relationships through generations. Communitarian reciprocity thrived despite the heterogeneous religious, demographic and economic orientations of community members. This diversity notwithstanding, Buddhist ideology permeated all aspects of life to the extent that it became a cohesive force within the community. Interdependencies and relationships spanned local scales for the simple reason that this was a human scale of functioning. Ladakhi sense of place interprets living in the context of valuing life, resources and local-scale interdependencies as opposed to merely increasing production. In essence, the practice of community participation may be considered as a way of thinking and action. The means and ends of co-operative living were steeped in a sense of timelessness of routines of life, and living made possible through a shared and collective sense of being and belonging to land and culture, and where local symbiosis safe-guarded the self-determined in(ter)dependence and self-reliance.

Recent changes, prompted by development initiatives, have cast afflictions on the communitarian fabric, causing upheavals that challenge the community's timeless culture. Pressures to modernize, globalize and integrate into the Indian poitico-nomic system have been undeniable. In the past two decades or more, the Ladakhis have experimented with the dialectic of these processes, for the purpose of adoption or adaptation (Michaud, 1996). The experience was problematic, although the community's values endured. That the community has survived the ravages of global processes so far is a testimony to their tenacity and the strong essence of their identity.

In the second case, the focus shifts to Uganda (Africa) where community members have resolved to join hands to 'improve' their lot, faced with political instability, economic backwardness and the AIDS epidemic. A set of communities receive succour through a partnered project (a local Foundation; a public sector Tourism Association and a British charity) to 'jump-start' their life. A community tourism project is initiated with an explicit objective to meet 'market needs' so that the locals may be financially equipped to afford vital services and provisions (schools, hospitals and daily sustenance). Local participation is deemed a conduit for 'harnessing' the windfalls of global market economy. To this end, the communities were called upon for tourism product formulation (as also formalization) such as establishing nature trails, crafting souvenirs, instituting organizational structures and infrastructural development, so as to enable them to participate in the mainstream global economy of goods and services. Project patrons, namely partners, local government and community leaders, worked relentlessly to build capacity for destination development, training personnel for management, marketing, book-keeping, administration and fund raising, guiding services and craft production. The seed-grant was devoted to 'breaking new ground' for an impoverished community. Now that the community has 'come to realize' that their cultural and natural assets have market value, their 'spirits are high' with thoughts of a future. The commentators (in the film by Sue Hurdle, 2004) assert that the 'project' is 'good' for – communities, government, business and conservation. This is combined with an imperative to undertake an 'enormous amount of work' for the future of the project. So far, the Kabaka Trail project has been successful in providing schools for children, livestock for impoverished households, food for many and care for the sick. For all the intervention, idea development and 'hope for the future', the locals are utterly grateful to the 'outside help' for 'showing' them the way of the world – in guiding the development of local cultural heritage sites. Without such stewardship, the local custodians (referred to as cultural guardians) admitted to 'confusion' about their 'responsibility' and options for survival.

The two illustrations are not exhaustive albeit the cases convey contrasting scenarios of community participation. Despite having a lot in common, both cases can be seen as constituting the extremities of a continuum of community participation concept and practice. In the case of Ladakh, the locals were critical of the reductionist forces that threatened their place-culture synergies and their timelessness. Over several years of communitarian living, the members' beliefs and actions precipitated and crystallized into a cohesive value system for the community as a whole. Their deep-seated sense of communitarian well-being, in all its socio-cultural, environmental and economic dimensions, served as a point of reference for them. With this wholesomeness still in place, the Ladakhis are able to question the realities of the new world order before adopting or adapting to it. The synthesis of the sense of place, meaning and selfhood as a product of human 'collective memory' (Todd, 2004) had been developed and enriched over several years of their existence and is characterized by its non-linearity and complex adaptive systems. The communes of Uganda, on the other hand, sought survival by aligning themselves with the new opportunities brought to them by outside agents whose interest was focused on 'connecting' the locals to global markets. Such an approach to community participation relies heavily on commodification of the local alongside engendering dependencies on external global structures. Each community differs in their perspective on participation – while the Ladakhis maintain an 'inside-to-outward' perspective, the second case illustrates an 'outside-to-inward' perspective. The points of difference are in (a) the source of impetus and information for adaptation; (b) the extent to which community participation is subsumed under the rubric of capacity building and social/intellectual capital (Murphy & Pauleen, 2007; Reid *et al.*, 2004) (c) the extent to which traditional modes of community are undermined or undercut.

Also, two aspects of significance to the theme emanate from these examples. Firstly, that community participation exists in a spectrum of myriad hues. In the first case, the nature of participation is a manifestation of a community's inherent cultural elements that are fortified by the resilience and resolve of its members who, in turn, thrive on the abstract idealism and esoteric abstraction of their particular collective identity. In contrast, the second example projects community participation as a requisite for any change in an *'in-situ'* circumstance. Innumerable permutations of community participation modes/formats thus emerge in response to variations of the internal and external influences over time. Secondly, that community participation (engagement and exchange) is indeed the only way for the realization of people's aspirations. As a corollary, one must also acknowledge that the level, intensity and dynamics of collaboration will vary with time, context and cause. Furthermore, these observations reflect that community participation is as much a communitarian reality as it is an ideal.

Social thinkers do not necessarily consider reductionism as particularly wrong or damaging. Some admit reductionism as a way of understanding what actually constitutes the larger world in its totality (see Todd, 2003, 2004). This approach to comprehending the world, however, cannot be mistaken as the goal. A common error usually made by the advocates of community participation is that community participation has become an intended goal within the big picture of globalization. This is certainly questionable as community participation is actually a historical (rather a continuing) process of becoming a community in and of itself, where participation is naturally a social and dynamic landscape of evolving human values. With this in mind, then, it may be quite difficult to propose 'progressive alternatives' to current models (see Weaver in this chapter), since it may be a repetition of the same mistake that the scholars are attempting to remedy!

Admittedly, however, the academic angst and frustration vented by Butcher is a very genuine concern too! Unfortunately, most of the present-day advocacy in favour of community participation, perpetuates the systemic transition of small-scale and/or subsistence economies to larger global cooperation – which is the very bane of scholarly idealism. The dictates of globalization propel the participation of small (and mid-sized) communities towards a mega-organization for global-(re)ordering. Those who encourage such ordering are seen as performing 'noblesse oblige' while those engaging in the mega-merger consider it a 'privilege of the proletariat'! At best, community participation has fallen upon difficult times where choices are expected to be made from the limited few. Attempts to define, interpret, model and 'induce' community participation in the present realities of this 'our' globalized world could well be its anti-thesis.

At this point, then, some questions that seem relevant here may be posed – why must 'local' (community) refigure itself to configure as the global (world-at-large)? Would it be advisable for local people to depend on the global systems for their 'Schumacherian' existence? Is it possible for local inherent values and visions of communities to be substituted by, if not reflected in, global market forces? Simply put, does/will local wisdom have a place and role in global rationalization? If the answer to this question is in the negative, then, certainly community participation invokes a false sense of the future and is a mere buzzword. But, if there is a genuine role of community participation in global ordering, then its advocates must rise to the challenge. After all, community participation, like sustainability, is fundamentally about addressing valued elements and dynamics that are the lifeblood of a community.

Concluding Remarks

It is notable that studies of community participation tend to focus on technique and positivist inspired assessments of what worked and what did not. All three pieces question a reliance on this – the character of community participation is influenced by the political and social context in which it operates.

There is a case for stripping away some of the rhetoric around project based assessments of community participation, and accepting a more prosaic approach to the issue. Weaver's response suggests a pragmatic and balanced approach to community participation – one that embraces integration rather than being insular, and that recognizes that increasing material well-being is desired by the communities. This seems a sensible approach for operationalizing community participation.

Yet Singh's response calls into question integration with global markets and problematizes material well-being itself. As such the two responses provide a useful contrast, arguably one relatively pro-development and one influenced by post-development ideas.

Singh points out that community participation is surely no more or less than an extension of the basic human instinct to interact and try to shape one's fate. Taking issue with Butcher, she argues against 'despair' – failures of the past should lead to a redoubling of effort to establish community participation as a part of our humanity.

However, maybe we should distinguish between community participation as a basic aspect of everyday life, and community participation as a distinctive political strategy, most often derived externally to the community in question. Indeed, one argument implicit in Butcher's lead piece is that community participation, as constituted through milieus of academics and NGOs, *undermines* community through treating it as a stage army, brought out (and incentivized) to give moral back up to schemes that may have little if anything to do with the agency of the community.

In another sense, Singh's piece is very much at odds with the lead piece, and also to an extent with Weaver's response. Singh paints a picture of the Ladakhi community as one that embodies a harmony between people and nature, a 'human scale of functioning' and a culture directed by the spirit of sustaining human life. This relativist defense of culture sits uneasily alongside the desire for material betterment, a desire that must play some part in the exodus of young people from Ladakh to the city.

Any assessment of community participation rests on assumptions, and these should be explicitly stated. Otherwise we have ideology masquerading as technique. Equally it is legitimate to assess community participation as a technique, but perhaps not to then make grand claims of empowerment and progress on that basis. This would constitute technique masquerading as ideology. Both are to be avoided in the evolving debates on community participation.

Discussion Questions

(1) What are the potential advantages and disadvantages of the localism implicit in the community participation agenda?
(2) In what sense could community participation be seen as intrinsic to the tourism product itself?
(3) What role should community participation play in wider, national development?
Consider community participation in the light of the debates on improving governance in the developing world.
(4) Critically consider the importance of 'empowerment' in community participation. 'Community' can be suggestive of a group of people united by culture or common interest. How should we view 'community' as a concept?
(5) Research the CAMPFIRE project in Zimbabwe. Some have argued it is an example of good practice in community participation, whilst others have strongly criticized it. Identify and discuss the perspectives and assumptions of the critics and the advocates respectively.

References

Beeton, S. (2006). *Community Development through Tourism*. Collingwood, Australia: CSIRO.
Boyd, S. and Singh, S. (2003). Destination communities: Structures, resources and types. In S. Singh, D. Timothy and R. Dowling (eds) *Tourism in Destination Communities* (pp. 19–33). Wallingford, UK: CABI.
Brohman, J. (1996). *Popular Development: Rethinking the Theory and Practice of Development*. Oxford: Blackwell.
Buckley, R. (2003). *Case Studies in Ecotourism*. Wallingford, UK: CABI.
Butcher, J. (2007). *Ecotourism, NGOs and Development: A Critical Analysis*. London: Routledge.
Butler, R. (1990). Alternative tourism: Pious hope or Trojan Horse? *Journal of Travel Research* 28: 40–45.
Chambers, R. (1983). *Rural Development: Putting the Last First*. London: Longman.
Dernoi, L. (1981). Alternative tourism: Towards a new style in North-South relations. *International Journal of Tourism Management* 2: 253–264.
Farrell, B.H. and Twining-Ward, L. (2004). Reconceptualizing tourism. *Annals of Tourism Research* 31(2): 274–295.
Fuller, D., Buultjens, J. and Cummings, E. (2005). Ecotourism and indigenous micro-enterprise formation in northern Australia: Opportunities and constraints. *Tourism Management* 26: 891–904.
Getz, D. and Carlsen, J. (2005). Family business in tourism: State of the art. *Annals of Tourism Research* 32: 237–258.

Hall, D., Kirkpatrick, I. and Mitchell, M. (eds) (2005). *Rural Tourism and Sustainable Business*. Clevedon: Channel View Publications.
Hann, C. and Dunn, E. (eds) (1996). *Civil Society: Challenging Western Models*. London: Routledge.
Hitchner, S., Apu, F., Tarawe, L., Galih, S. and Yesaye, E. (2009). Community-based transboundary ecotourism in the heart of Borneo: A case study of the Kelabit Highlands of Malaysia and the Kerayan Highlands of Indonesia. *Journal of Ecotourism* 8: 193–213.
Honey, M. (1999). *Ecotourism and Sustainable Development: Who Owns Paradise?* Washington: Island Press.
Hurdle, S. (2004). *A Force for Change: Tourism as a Positive Tool for Development* (Film). Bristol, UK: The CREATE Centre.
Li, W. (2006). Community decision making participation in development. *Annals of Tourism Research* 33(1): 132–143.
Mann, M. (2000). *The Community Tourism Guide: Exciting Holidays for Responsible Travelers*. London: Earthscan.
Medina, L. (2005). Ecotourism and certification: Confronting the principles and pragmatics of socially responsible tourism. *Journal of Sustainable Tourism* 13: 281–295.
Michaud J. (1996). A historical account of modern social change in Ladakh (Indian Kashmir) with special attention paid to tourism. *International Journal of Comparative Sociology* 37: 286–301.
Midgeley, J. (1986). Introduction: Social development, the state and participation. In J. Midgeley, A. Hall, M. Hardiman and D. Narine (eds) *Community Participation, Social Development and the State* (pp. 1–11). New York: Methuen.
Mowforth, M. and Munt, I. (1998). *Tourism and Sustainability: New Tourism in the Third World*. London: Routledge.
Murphy, P. and Pauleen, D. (2007). Managing paradox in a world of knowledge. *Management Decision* 45(6): 1008–1022.
Murphy, P.E. (1988). Community driven tourism planning. *Tourism Management* 9(2): 96–104.
Norberg-Hodge, H. and Page, J. (1993). *Ancient Futures: Learning From Ladakh* (Film). London: ISEC.
Parnwell, M. (1998). Tourism, globalisation and critical security in Myanmar and Thailand. *Singapore Journal of Tropical Geography* 19(2): 212–231.
Pleumaron, A. (1994). The political economy of tourism. *The Ecologist* 24(4): 142–148.
Pretty, J. (1995). The many interpretations of participation. *In Focus* 16: 4–5.
Reed, M. (1997). Power relations and community based tourism planning. *Annals of Tourism Research* 24(3): 566–591.
Reid, D.G., Mair, H. and George, W. (2004). Community tourism planning – A self-assessment instrument. *Annals of Tourism Research* 31(3): 623–639.
Salafsky, N., Cauley, H., Balachander, G., Cordes, B., Parks, J., Margoluis, C., Bhatt, S., Encarnacion, C., Russell, D. and Margoluis, R. (2001). A systematic test of an enterprise strategy for community-based biodiversity conservation. *Conservation Biology* 17: 1585–1595.
Scheyvens, R. (1999). Ecotourism and the empowerment of local communities. *Tourism Management* 20(2): 245–249.
Scheyvens, R. (2002). *Tourism for Development: Empowering Communities*. Harlow: Prentice Hall.
Todd, J. (2003). The virtues of nonreduction, even when reduction is a virtue. *Philosophical Forum* 34(2): 121–140.

Todd, J. (2004). Reductionism and antireductionism: Rights and wrongs. *Metaphilosophy* 5(5): 614–647.
Tosun, C. (2000). Limits to community participation in the tourism development process in developing countries. *Tourism Management* 21(6): 613–633.
Tosun, C. (2006). Expected nature of community participation in tourism development. *Tourism Management* 27: 493–504.
Weaver, D. (1998). *Ecotourism in the Less Developed World*. Oxford: CABI.
Weaver, D. (2008). *Ecotourism* (2nd edn). Milton, Australia: Wiley.
White, S.C. (2000). Depoliticising development: The uses and abuses of participation. In D. Eade (ed) *Development, NGOs and Civil Society* (pp. 142–155). Oxford: Oxfam.
Woodwood, P. (1997). Cashing in on the Kruger: The potential of ecotourism to stimulate real economic growth in South Africa. *Journal of Sustainable Tourism* 5(2): 166–168.
Zeppel, H. (2006). *Indigenous Ecotourism: Sustainable Development and Management*. Wallingford, UK: CABI.

Further Reading

Butcher, J. (2007). *Ecotourism, NGOs and Development: A Critical Analysis*. London: Routledge (Chapter 4).
Cooke, B and Kothari, U. (2001). *Participation, the New Tyranny?* London: Zed Books.
Hall, D and Richards, G (eds) (2000). *Tourism and Sustainable Community Development*. London: Routledge.
Hickey, S. and Mohan, G. (eds) *Participation: From Tyranny to Transformation?* London: Zed Books.
Scheyvens, R. (2002). *Tourism for Development: Empowering Communities*. London: Prentice Hall.
Stone, I and Stone, T. (2011). Community-based tourism enterprises: Challenges and prospects for community participation: The Karma Rhino Sanctuary Trust, Botswana. *Journal of Sustainable Tourism* 19(1): 97–114.
Taylor, G. (1995). Current issue: The community approach: Does it really work? *Tourism Management* 16(7): 487–489.
Tosun, C. (2000). Limits to community participation in the tourism development process in developing countries. *Tourism Management* 21(6): 613–633.
Tosun, C. (2005). Stages in the emergence of a participatory tourism development approach in the developing world. *Geoforum* 36(3): 333–352.

Chapter 5
Does Tourism Reduce Poverty?

Regina Scheyvens, Dorothea Meyer, David Harrison and Paul Peeters

Context

This chapter explores the relationship between tourism and poverty reduction, examining particularly the way the term 'pro-poor tourism' (PPT) is being used, and promises made under the guise of PPT. Is there real meaning in the term PPT and, perhaps more importantly, does the term have the potential to transform the way in which governments, the private sector and other agencies are working to ensure that tourism is actually delivering a greater proportion of tangible benefits to the poor?

The four authors contributing to this chapter take different views on the subject. Scheyvens begins with a provocative piece noting that despite all of the statistics circulated about the growth of tourism in developing countries, its potential to provide jobs and contribute to development through skills training and infrastructural improvements, a number of concerns remain. Scheyvens suggests such 'harsh realities' include the continued inequalities associated with international tourism, the use of PPT as a way of providing good publicity for the tourism sector to deflect criticism for its negative environmental and social impacts, and the fact that only small net benefits may be received by the poor. Dorothea Meyer, while agreeing with some of Scheyvens' arguments, supports the very pragmatic approach to PPT which has been adopted by several key research organizations and technical assistance organizations. They have focused on how tourism businesses can make a more significant contribution to the poor, and what other stakeholders can do to support them in this work. Harrison's piece, while recognizing some weaknesses of PPT, feels that it has been something of a 'soft target', thus he challenges Scheyvens on some of the criticisms she has put forward. Finally, the environmental is brought centre stage in the writing by Paul Peeters on the 'elephant in the room', that is, the fact that travel to developing countries usually involves medium to long-haul flights which generate significant carbon emissions, thus increasing tourism to poorer countries which will directly contribute to climate change.

5.1

Pro-Poor Tourism: Is There Value Beyond the Rhetoric?

Regina Scheyvens

Introduction

The concept of Pro-Poor Tourism (PPT), which was first used in a British report in 1999 (Bennett *et al.*, 1999), has since received widespread support from development agencies, donors, governments and various tourism organizations. While some have adapted the term – the UNWTO, for example, preferring ST-EP (Sustainable Tourism – Eliminating Poverty), and Zhao and Ritchie (2007) favouring APT (Anti-Poverty Tourism) – there has been relatively little critical assessment of the concept itself by academics or others.[1] On the contrary, most reports have preached enthusiastically about the potential of PPT to contribute to poverty-reduction in a wide range of countries and contexts. While many of us are attracted to the prospect that what is claimed to be the world's largest industry, an industry which continues to grow in many of the world's poorest countries, could deliver substantially greater benefits to the poor, it is important to be realistic about whether or not this is likely given the divergent agendas of various stakeholders currently endorsing the concept.

This research probe thus holds the concept of PPT up for direct scrutiny. It is important to critically appraise PPT in order to ensure that we are not being drawn into either another neoliberal agenda (Scheyvens, 2007a), a publicity campaign for agencies wishing to promote continued growth of tourism or a half-hearted attempt by aid agencies to reframe existing development programmes under a poverty-alleviation agenda. The article explores whether PPT as an approach has the potential to transform the way in which governments, the private sector and other agencies are working to ensure that tourism is actually delivering a greater proportion of tangible benefits to the poor. In doing so, it identifies several 'harsh realities' which may stand in the way of PPT achieving its purported promises.

The Promise of PPT

Tourism has been identified as a strategy to overcome poverty partly because tourism is a significant or growing economic sector in most developing countries with high levels of poverty. Developing countries now have a market share of 40% of worldwide international tourism arrivals, up from 34% in 2000 (UNWTO, cited in PATA, 2008). Tourism contributes up to 40% of GDP in developing countries compared with 10% of GDP in Western countries (Sofield *et al.*, 2004: 2). For over 50 of the world's poorest countries tourism is one of the top three contributors to economic development (World Tourism Organization, 2000, cited in Sofield, 2003: 350). While the least developed countries (LDCs) receive only 2.6% of international tourist arrivals (UNWTO, 2005: 1), there has been 110% growth in these arrivals between 2000 and 2007 (compared with an overall 32% increase in worldwide international arrivals for this period) (UNWTO, cited in PATA, 2008).

Such statistics have led to grand claims from the president of Counterpart International, Lelei LeLaulu, who asserts that tourism represents 'the largest voluntary transfer of resources from the rich to the poor in history, and for those of us in the development community, tourism is the most potent anti-poverty tool ever' (eTurboNews, 2007).

His stance is somewhat supported by UNCTAD (1998), which refers to tourism as the 'only major sector in international trade in services in which developing countries have consistently had surpluses'. Many countries are being forced to look beyond their traditional agricultural exports because of the declining value of these products (e.g. bananas, cocoa, coffee and sugar). Comparing the value of such crops with tourism receipts for South Pacific countries over a 20-year period, it has been found that 'in every case the value of these primary products in real terms has declined and the only sector to demonstrate a continuous upward trend has been tourism' (Sofield *et al.*, 2004: 25–26).

It is argued that tourism as a sector 'fits' nicely with pro-poor growth because:

> it can be labour-intensive, inclusive of women and the informal sector; based on natural and cultural assets of the poor; and suitable for poor rural areas with few other growth options. (Ashley & Roe, 2002: 61)

Tourism can purportedly contribute to the well-being of the poor directly through generation of jobs (the tourism industry already employs over 200 million people worldwide: IIED, 2001) and income-earning opportunities and, indirectly, by providing revenue for infrastructure development (such as roads and water supplies), opportunities to interact with 'outsiders' and gain access to markets and encouraging conservation of natural and cultural assets. By enhancing local livelihoods, tourism can enable some rural communities to survive rather than seeing the out-migration of their youngest and brightest members.

Since 1999 'pro-poor tourism' has been promoted by an increasingly wide range of agencies, including bilateral donors such as DFID (the UK's Department for International Development), SNV (the Netherlands' bilateral aid agency) and GTZ (the German bilateral aid agency), tourism organizations such as WTTC (the World Travel and Tourism Council) and PATA (the Pacific and Asia Travel Association) and multilateral agencies including the ADB (Asian Development Bank) and the UNWTO (United Nations World Tourism Organization).

The Harsh Realities

Supposedly, there is great merit then to the concept of PPT; however, one must realize that most endorsement of PPT to date has come either from development agencies who see PPT as fitting in nicely with a broader pro-poor growth agenda (to be discussed below), or tourism organizations which have a clear self-interest in continuing to promote the tourism industry. Thus, the discussion below focuses on a number of harsh realities that stand in the way of realizing the fervent dreams of advocates of PPT. It is important to face up to these realities, otherwise there is no way that they can be overcome.

International tourism is founded on inequalities

The notion of PPT immediately raises contradictions. How can PPT help to overcome the vast inequalities between tourists and local people when international tourism is in many ways predicated upon inequalities between wealthy tourists and impoverished locals (Harrison, 2001a: 252)? For example, even budget travellers such as backpackers can somehow raise the funds to set out on extensive sojourns through 'exotic', 'adventurous' locations around the world, where they may seek out people living in 'traditional villages', go on a tour of a shanty town or favela, and have opportunities to mix with 'the locals'. These same 'locals', of course, are unlikely to have the means to travel outside their own region, let alone internationally.

At the other end of the scale are luxury tourists who have heavy resource demands for both local resources that might be in short supply (e.g. energy, fresh water) and for imported goods. This just compounds situations of inequality. As the United Nations Development Programme (UNDP) notes, 'Golf courses and enormous pools are an insult to more than 1.3 billion people denied access to clean water' (1999, cited in Richter, 2001: 50).

The figures do not add up

The fact that the vast majority of international tourists still visit Western countries, particularly Europe, leads Hall to conclude that 'the potential of tourism to contribute to the economic development of the developing countries. . .would appear to be questionable unless there are massive shifts in flows of international arrivals' (2007: 112).

For many years now, tourism researchers have critiqued the political economy of the tourism industry, showing how relationships between Western and Third World interests often lead to poverty being entrenched within Third World destinations (see for example Britton, 1982; Brohman, 1996). This is reinforced by evidence of leakages from tourism, which are highest in countries with small and/or weak economies. In the early 1990s, for example, leakages from charter operations to the Gambia were around 77% (Dieke, 1993). Rather than suggesting that tourism might 'cure' poverty then, the available data could in some cases be used to show that tourism has impacted negatively on the lives of the poor. This leads Hall to seriously refute some suggestions about the benefits of PPT:

> The notion espoused by the UNWTO that 'tourism exchanges benefit primarily the countries of the South' *is a ridiculous one* and hides the reality that not only is the consumption of tourism the domain of the wealthy, but in many ways so is its production. (Hall, 2007: 116 – emphasis added)

Plüss and Backes (2002: 10) further claim that in 10 of the 13 countries which are home to 80% of the world's people who live in extreme poverty, tourism has not been able to reduce poverty. Thus it seems we must look very carefully at how the statistics are portrayed by those who advocate PPT.

PPT as 'window dressing'

There is a real danger that, like a number of trends before it (e.g. 'ecotourism' in the 1990s), PPT will remain something of a fad, a new way of dressing up the tourism industry to reclaim its credibility not just as an engine of growth but also as a 'soft' industry that is both socially beneficial and environmentally benign. Tourism industry players put on 'green lenses' in the 1990s, and along with a revival of interest in the environment due to the rising profile of climate change issues, a commitment to poverty reduction seems to be a key focus for the industry in the first decade of the new millennium.

It can be argued then that the majority of tourism businesses and organizations are attracted mainly to the rhetoric of PPT only – whatever will work to enhance the image of the global tourism industry and continue to encourage increases in travel and tourism. This leads Higgins-Desbiolles (2006: 1201) to claim that PPT is not a transformative type of tourism but, rather, 'a program of minor reforms for a marketized tourism sector to deflect criticisms and prevent unwanted regulation'. Cynics may argue that this is the only reason that the UNWTO launched its ST-EP initiative in 2002.

Are small net benefits sufficient and sustainable?

PPT is commonly defined as tourism that generates net benefits for the poor. This definition has recently come under fire, however, as it still allows wealthier people to benefit *more* than the poor (Chok *et al.*, 2007; Schilcher, 2007). Even if PPT advocates find ways to ensure that the poor receive a larger slice of the tourism pie, their proportion may still be only a fraction of what middle- to high-income earners receive.

Zhao and Ritchie also identify a problem with the conventional definition of PPT, as this does not preclude the poor from benefiting largely as passive beneficiaries of development, rather than active agents of change. This leads these authors to formulate their own term, anti-poverty tourism, which they hasten to explain 'is not consistent with charity and philanthropy in that it focuses on the capacity building of the poor and income generation rather than directly giving to them' (Zhao & Ritchie, 2007: 132).

The PPT literature also suggests that the poor can benefit from growing the overall size of the pie. It is difficult to find evidence, however, of PPT advocates seriously grappling with the implications of a growth-oriented strategy, particularly in terms of environmental sustainability. Tourists can place a heavy demand on limited resources, and the issue of carbon emissions from air travel is now absorbing a great deal of attention (see Gössling *et al.*, 2008). Similarly, the social and cultural impacts of growth of tourism on local communities need to be carefully monitored and managed.

Businesses are in existence to make profits, not to serve the poor

Major players in the tourism industry, as in any industry, are centrally concerned with profit maximization (Ashley & Haysom, 2006; Zhao & Ritchie, 2007). Why then should we assume that they might have some ethical commitment to ensuring that their businesses contribute to poverty-alleviation?

Certainly, when one examines what real changes PPT has brought about in the practice of the tourism industry, the examples appear tokenistic rather than transformational. Even in South Africa, where the government has a strong commitment to promoting fair trade in tourism, it is struggling to get the industry to commit to and implement changes (Briedenham, 2004).

Thus, for example, when the resort of Sun City took part in the PPT Pilot Programme, through which it received assistance to implement pro-poor approaches, its major initiative was to support development of two small enterprises from which it could procure products, one producing glasses from recycled bottles, the other manufacturing greeting cards. As noted by Ashley and Haysom (2006: 274), 'The scale of these enterprises is tiny compared to Sun City's budget'.

Reflection on the PPT Pilot Programme led to the conclusion that while happy to engage in philanthropy, such as asking for donations from guests for a community development fund, few companies were willing to make more enduring changes to their business strategies and practices (Ashley & Haysom, 2006). One such strategy could be implementation of comprehensive policies that support labour rights (including training, health and retirement schemes). If we cannot expect such wide-ranging changes as a result of a PPT approach, then surely it can only hope to benefit a minority of people.

While larger businesses might be able to afford to devote a proportion of their funds to Corporate Social Responsibility (CSR) goals (e.g. Sun City's CSR department receives 1.5% of pre-tax profits – Ashley & Haysom, 2006), such as resources to help upskill nearby farmers so that the hotel can source more produce locally, it can be particularly difficult for small-scale businesses to reorientate their activities to meet the interests of the poor because their margins are slim. Smaller enterprises often work in an extremely competitive environment where daily efforts to keep afloat are the reality.

The neoliberal poverty agenda

A neoliberal agenda dictates that governments should take a 'hands-off' approach to the economy, instead implementing an 'enabling environment' in which market-led growth can occur. The market, however, bows to the interests of consumers, not to the interests of the poor.

Nevertheless, it is neoliberal logic that was behind the move to make poverty-alleviation the leading development agenda in the 1990s. To donors, the tourism sector fits in perfectly with a pro-poor growth agenda as it is typically a growing industry which is focused on the service sector and is thus labour intensive. SNV's recent growth of commitment to PPT is testament to this: from eight tourism advisors in six countries in 2000, to over 40 advisors in 26 countries in 2005 – a more than 500% increase in just five years (Leijzer, 2006).

Six countries ——————► 26 countries
Eight advisors ——————► Over 40 advisors

While it certainly sounds honourable that donors are seeking to make poverty-alleviation central to every aspect of their work, when approached from a neoliberal perspective, poverty-alleviation '...is considerably circumscribed in its premise of economic growth as the foundation of development' (Mowforth & Munt, 2003: 34). Furthermore, the 'poverty consensus' has been criticized because it tends to overlook important environmental, social and political issues. Such criticisms lead Hall and Brown (2006: 13) to seriously question the potential of PPT: 'does PPT simply offer another route by which economic imperialism, through tourism, may extend its tentacles, or is it an appropriately liberating and remunerative option?'.

Setting aside a neoliberal agenda, there are other ways of viewing poverty-alleviation that are based on deeper understandings of poverty itself. Poverty is not necessarily, or only, concerned with a lack of access to the basic means of living (e.g. food, shelter, clothing). For example, Goulet articulates that development should not simply focus on basic needs, rather, it includes freedom from servitude and dignity/self-esteem (cited in Todaro & Smith, 2003: 52–54). Goulet's ideas would then lead us to question whether employment in the tourism industry necessarily equates with pro-poor development. Thus, is the bonded employment of a 10-year-old in an Indian hotel – where s/he must work at least 16 hours per day every day of the week, is allowed to visit home once a year and is faced with regular verbal and physical abuse – really an example of tourism employment contributing to the alleviation of poverty? While this example may seem extreme, there are many others from around the world of a lack of labour rights in the tourism industry, as seen in evidence from Tourism Concern's campaign on 'Sun, sand, sea and sweatshops' (Tourism Concern, 2012).

A neoliberal lens leads researchers of PPT to focus largely on economic outcomes as well. In a similar bent, a recent report on tourism 'value chains' in Danang, Vietnam, counted 'massage' as another sub-sector which brought in money from tourists (Mitchell & Phuc, 2007). No comment was made on the fact that this term is commonly used in Vietnam as a euphemism for the sex trade – nor on the ethics of this practice. Sex tourism in Vietnam has been associated with forced prostitution of children (Duong, 2002).

Even when tourism initiatives are successful in bringing economic development to poorer countries and regions, there is no guarantee that this will actually impact on the extent and severity of poverty. As Ghimire and Li have noted (2001: 102), tourism has brought economic benefits to rural communities in China, as evidenced by a proliferation of televisions and satellite dishes. However, living conditions have not improved on the whole. Ghimire and Li (2001: 102) thus question whether poverty can be seen to have been alleviated in this context where there is still a lack of potable water, unreliable access to energy sources, poor sanitation and inadequate health care facilities.

Local realities: Corruption, cronyism and elite capture

Numerous research articles on tourism have pointed to the challenge of delivering tangible benefits to the poorest members of society (see, for example, Akama, 1996; Mansperger, 1995; Mowforth & Munt, 2003). In the main, involvement of the poorer classes has been ignored in tourism policy except in cases where it is asserted they will benefit from job creation or the assumed 'trickle-down' of any economic benefits which occur. Even within poorer communities it is those with capital, connections, confidence and foreign language skills – that is, those who definitely do not belong to the category 'poorest of the poor' – who have gained the greatest benefits from tourism development (Zhao & Ritchie, 2007: 130).

At higher levels, development practitioners have noted that efforts to spread the benefits of tourism more widely have been constrained due to cronyism and corruption.[2] For example, a local government authority which asks for tenders regarding development of new tourism infrastructure (e.g. the tarsealing of a road which leads to a renowned temple or waterfall) may feel pressured to award that contract to a particular influential family from a nearby town, even though the family is widely known for its maltreatment of workers and use of inferior materials in its road work.

Conclusion

This research probe into pro-poor tourism has raised some serious concerns about the legitimacy of claims that tourism is a good strategy for alleviating poverty. Can the interests of the poorest members of a society really be served by promoting expansion of a global industry that is founded on inequalities, where individual businesses strive to meet the interests of the market, not the poor, and where elites often capture the majority of benefits of any development which does occur? Until we have adequate responses to such concerns, it will be difficult to proclaim that tourism is really making a significant contribution to eliminating poverty around the globe.

While the tone of this article has been critical, this author genuinely welcomes the pro-poor approach to tourism, but also calls for caution. Critical reflection on this concept, rather than unadulterated support, is needed if PPT is to have any chance of achieving the outcome that so many of us are hoping for: better lives for impoverished people struggling to survive at the margins of society. Their well-being should be central to PPT efforts.

Tokenistic efforts to employ a few people to weave mats for a resort, on the one hand, or to encourage guests to contribute to a community development fund on the other, do not provide sufficient evidence that an industry which is focused on making profits by catering for the hedonistic pursuits of the world's middle and upper classes, can actually be transformed.

Perhaps if PPT is to work, it will require what Chok *et al.* (2007) call a fundamental shift in ideology, from relying on supposed altruism, to a more solid foundation of ethics. As Ashley and Haysom (2006) noted in the case of the PPT Pilots in southern Africa, there is a clear need for tourism enterprises to move beyond philanthropy if they wish to show a solid commitment to enhancing the well-being of the poor. Furthermore, we may also need a shift in policy, from a focus on growth to equity (Schilcher, 2007). This, however, seemingly contradicts the interests of a number of current advocates of PPT, including both donors who are driven by neoliberal poverty-alleviation agendas, and those agencies which act in the interest of the tourism and travel industry, such as the WTTC and PATA.

5.2

Pro-Poor Tourism: Is There Actually Much Rhetoric? And, if so, Whose?

Dorothea Meyer

Regina Scheyvens, the author of the research probe, deserves applause for critically raising the issues related to Pro-Poor Tourism (PPT) with the academic community engaged in tourism research. To respond to this probe was a challenging task, not least because the research probe by Scheyvens should have deserved a very critical and detailed reply. Being restricted to just a couple of pages, this response will focus primarily on the role of the tourism–private sector in the quest for poverty reduction. First, however, it is important to clarify what the original PPT approach initiated by Overseas Development Institute (ODI) and International Institute for Environment and Development (IIED) in 1999 was all about – and to a large extent still is – and maybe to question whether the probe has raised any new issues.

Is PPT Understood?

Somehow, it seems that discussions about PPT at the moment take place on two distinctly different levels, with fundamentally different agendas. For one, there are the newly emerging academic discussions related to PPT (see, in particular, *Current Issues in Tourism*, Volume 10, Number 2–3, 2007) that seem largely inspired by dependency paradigms *à la* Britton (1982), Brohman (1996) and Clancy (1999), and essentially focus on the need to 'revolutionize' the world system. On the other hand, there are the PPT practitioners' studies (ODI, SNV, etc.) being well aware of the ills of this world and attempting to find practical solutions. . .after all, it is their job to advise in the fight against poverty. The surprising bit is possibly that the development community only quite recently changed its attitude; gone are the days, it seems, when tourism was viewed as a hedonistic, northern-controlled, white-men's industry and now there is the pursuit of tourism as a potential tool for poverty reduction with gusto – and I must admit I welcome this.

More than anything the misalignment between these two camps illustrates the lack of communication and collaboration between academics and practitioners. This is not a surprise by any means. Academics need to publish conceptual and theoretical ideas – this is how funding for higher education is assessed in most countries – and how academics survive. Practitioners, on the other hand, are judged by their outputs/outcomes in the field, i.e. has a particular intervention worked? Academics generally receive a certain amount of research hours (the UK's Research Assessment Exercise (RAE) springs to mind) to theorize and publish these 'world-changing' ideas. Organizations such as ODI and IIED and, in particular, technical assistance agencies such as SNV and GTZ, on the other hand, are reliant on funding from donor organizations who are probably less interested in revolutionizing the global system but more focused on what happens on the ground, in the destination, at a particular location. This obviously changes how they think, what they theorize about (or not) and what they publish. These two very different ways of conducting research and formulating strategies and ideas inform the material available to the public – and, ultimately, what is known about PPT. The PPT partnership deserves applause for publishing material widely, material that is available to practitioners and not just to the academic circle and, as such, for spreading PPT ideas. It is maybe a shame that this material and indeed the PPT name has been misused by well-known organizations wanting to hop onto the bandwagon.

Tourism as a Tool for Poverty Reduction?

The pros and cons of tourism as a development tool have been discussed widely over the past 30 years (e.g. Bryden, 1973; de Kadt, 1979; Britton, 1982; Wilkinson, 1987; Rao, 2002) and the arguments put forward in the research probe by Scheyvens mirror these well-known discussions – but do they indeed add anything new?

Both tourism academics and practitioners are all well aware that tourism can contribute to employment and income generation, inter-sectoral linkages, entrepreneurship and even nature protection, while, at the same time, tourism can also increase inequalities, lead to the displacement of communities and contribute to conflict over scarce resources – to name just a few of the widely discussed negative impacts of tourism development.

Many academics are pessimistic about the role of tourism in development and voice this frequently with their own 'tinted-view' of how global structures should be revolutionized (e.g. Duffy, 2002; Scheyvens, 2007a; Schilcher, 2007). The fundamental question of how far tourism *does* and *can* contribute to poverty reduction, however, has often fallen by the wayside to make space for fundamentally idealistic discussions about the impacts of the global capitalist systems and wider political structures. These are obviously important discussions but it is slightly unclear so far how they aid poverty reduction on the ground.

This re-focusing on the very specific and pragmatic questions has been at the heart of the early PPT work carried out by the ODI and IIED in the late 1990s. While being consistently aware of the negative (and positive) impacts of tourism, the PPT partnership set out to find ways in which the tourism industry – which was arriving at great speed in many developing countries – could possibly be changed to increase the contribution it could make to poverty reduction. This is a considerable departure aimed at finding practical solutions rather than continuing with esoteric discussions about the 'bad and failing' world we live in.

The PPT partnership seems to be very aware that (a) it is very difficult (if not impossible) to reach the poorest with tourism initiatives alone and (b) that tourism development will ultimately also benefit those already better off as PPT strategies work best in a developed tourism destination. But is that a reason not to pursue PPT strategies as the research probe might suggest? The emphasis for PPT has always been on increasing net benefits to the poor – this is the *raison d'être* for PPT. When impact assessment studies were carried out (Ashley, 2006; Ashley & Mitchell, 2007; Mitchell & Faal, 2007; Harrison & Schipani, 2007) they showed a variety of benefits to the poor, benefits that, prior to PPT interventions, were not available. While these studies by no means claim that poverty is being reduced on a grand scale they show improvements – which should be an essential goal of any PPT initiative. But probably most crucially, these findings have shown that it would be possible to scale-up PPT interventions.

Scheyvens seems to argue that PPT is supportive of neoliberal ideologies and thereby neglects the all important equity issues. This accusation, however, might be misplaced, given that PPT is *not* about theorizing or subscribing to a particular ideology (directly or indirectly), but essentially about attempting to find workable market intervention strategies that enable the poor to participate in the industry – an essential means for (re)distribution. The discussion of equity is frequently linked to fundamental changes to the structures of global tourism and the wider political system, which would require an ideological change that will give greater powers to developing countries. While this would obviously be very commendable, it is questionable how far PPT initiatives could possibly achieve this without support from the global system, the industry and policy makers. One of the obvious challenges related to tourism development is that market forces cannot equitably distribute benefits and costs, and, therefore, the public sector will have to step in via policies that attempt to redistribute some of the excesses of a market and private sector-led tourism industry. Much of the work done by PPT practitioners focuses on working within the existing structures, with the aim of changing operating practices of both the private and public sectors. This does not in any way imply that PPT is blind to the global structures characterizing this 'unfair' relationship – but it means that PPT is attempting to slowly change practice which will result in benefits to the poor.

The Role of the Tourism Private Sector

Scheyvens claims that 'the major players in this industry, as in any industry, are still concerned with profit maximization so we need to consider whether PPT is just "window dressing", or tokenistic,. . .intended mainly to reduce costs and/or enhance the positive publicity for the agencies concerned'. Reading this statement, one wonders how realistic the author is. Surely, the private sector needs to be aware of its marketability and ensure profits are being maximized, given the tight competition in this industry – anything else would be unsustainable, both for the enterprise and the destination. To mind come endless small-scale, 'green' community-based tourism (CBT) projects that have failed to deliver because (a) they were initiated by non-commercially minded but oh-so-well-meaning individuals who had limited knowledge of the workings of the tourism industry; (b) they were often heavily reliant on donor funding and ultimately collapsed after funding seized; and (c) they focused on producing 'alternatives' rather than working with the mainstream tourism industry. The failures of these CBT adventures are documented elsewhere (Goodwin, 2006).

Should one really criticize the industry for being commercially minded? Or to be very blunt: does it really make a difference whether it is seen as 'window dressing' if the impacts are positive? One would expect any business person to make decisions based on expected returns for their business – be it increased marketability, image-raising, risk aversion, licence to operate, cost-cutting or philanthropy – all of which might be classified as 'window dressing' by some.

Tourism businesses are not development agencies – and why should they be? The tourism industry is fiercely competitive as characterized by extremely tight profit margins among the key operators and a highly competitive operating environment among the plethora of small owner-operated enterprises. Current 'demands' frequently made on tourism businesses include the reduction of carbon emission, health and safety issues and the respect for human rights – let alone helping countries to get out of poverty which many might not regard as the responsibility of the industry but as a governance issue. *The business of business is business* as Milton Friedman might have argued – and this is maybe exactly the strength that the private sector can bring into the equation, whether they are large players or niche operators.

Amongst the development community there seems to be a consensus today that the private sector can play a considerable role in the fight against poverty by helping to speed up economic development through its core business activities in the workplace, the marketplace, along the supply chain, through their social investment and philanthropic activities, and through their engagement in public policy dialogue and advocacy.

As in other industries, tourism businesses have so far mainly focused their CSR efforts on environmental issues and philanthropic contributions. Environmental considerations have been incorporated by the tourism industry for over a decade, be it to cut costs or to satisfy a greening customer group. A large number of tourism businesses are also involved in philanthropic activities or what might be termed 'business engagement in the community', such as donating money to support good causes or alternatively encouraging their customers to do so. This is widespread among the industry – with relatively little song and dance made about it.

Much less attention has, however, so far been paid to managing their economic multipliers and impacts along local and global value chains to contribute to local economic development and poverty reduction in destinations in developing countries. Here the aim is for businesses to work through every aspect of their activities, to ask whether development impacts can be improved. This urgently requires assistance from PPT practitioners and academics as the industry cannot do it alone. The value-chain approach seems to offer considerable potential for PPT, in particular, as it focuses on wider local economic development that goes beyond simply working with a narrowly defined tourism industry. Several studies have found that possibly the best opportunity for the poorer segments of society is being integrated into the tourism supply chain via sourcing in terms of food supplies, arts and crafts or services rather than just direct employment in tourism businesses (Ashley, 2006; Mitchell & Faal, 2007).

And this is probably where a key challenge, but also opportunity, for PPT lies in the future and I congratulate research organizations such as ODI and IIED and technical assistance organizations such as SNV and GTZ, to name but a few, for tackling these innovative and challenging issues. Thus I come back to the title of the research probe with a question: is there actually much rhetoric, and, if so, whose?

5.3

Pro-Poor Tourism: Is There Value Beyond 'Whose' Rhetoric?

David Harrison

Regina Scheyvens' lead paper contains much of value, but there are problems with her critique of Pro-Poor Tourism (PPT). The preamble, though, is acceptable: tourism obviously makes a major contribution to the economies of many less developed countries (LDCs), but this is the standard defence for most kinds of tourism. Equally evidently, the most strident advocates of tourism's poverty-alleviating qualities often come from (national and international) non-government organizations (NGOs), donors and aid agencies, though sometimes with academic support (Jamieson, 2003). Noticeably, though, any necessary association of PPT with community-based tourism (CBT) has been specifically denied by PPT advocates, some of whom suggest that CBT is singularly unsuccessful in alleviating poverty (Goodwin, 2006; Mitchell & Muckosy, 2008).

The 'promise of PPT', then, is something of a misnomer, and many of the ensuing criticisms are most appropriately directed at development organizations and self-interested industry players who might be described as PPT's fellow-travellers. In other respects, too, to take the 'harsh realities' in turn, the critique is flawed.

First, is *international tourism founded on inequalities?* Unfortunately, this writer is misquoted as supporting this assertion; instead, it was argued that wealth disparities in destination societies 'are often highlighted by international tourism' (Harrison, 2001a: 252), which is very different. In fact, most international tourism is within developed areas, where resident and tourist are more or less of equal status. In LDCs, disparities in wealth are often more pronounced, but they invariably pre-existed tourism and (as in the Caribbean) enabled those with financial and cultural capital to advance the new industry (Harrison, 2001b: 28–33). To deduce from this that there is somehow a *logical* connection between international tourism and wealth discrepancies in destination areas, and that international tourism is founded on inequalities, is inappropriate.

The more general (implied) argument is that any capitalist development which involves wealth discrepancies (and is there any other?) is unacceptable, and thus to take a position which, as a former colleague noted long ago, effectively says 'that capitalism ought to be nice' (Smith, 1983: 74). That it frequently is not is no argument against tourism per se or any other driver of economic and social change.

Secondly, Scheyvens suggests that *the figures don't add up*, but to link leakages from tourism in LDCs with dependency on metropolitan countries is disingenuous. How could it be otherwise? It is eminently sensible to promote import substitution, but the 'harsh reality' is that small, resource-poor LDCs are unlikely to ever compete on equal terms with more developed manufacturing economies! Furthermore, leakage figures are notoriously misleading and unreliable. Much depends on the initial outlay (50% of US $250 spent at the Sheraton or Inter-Continental is surely economically preferable to 90% of US $20 at a cheap guest house), on how calculations are done and on what is included and excluded. As Mitchell and Ashley (2007: 1) (both involved in PPT projects) argue, ignoring out-of-pocket spending at destinations (which critics of high leakages often do) is simply unacceptable, as is considering expenditure made outside a destination, e.g. on marketing, packaging and flights, to be a 'leakage'. This, they say, is 'like suggesting that staff and accommodation costs for serving a cappuccino in a London café are "leakage" from coffee plantations in Ethiopia!'.

We should indeed look very carefully at how the statistics are portrayed by PPT advocates (see first paragraph of page 127 of this chapter) and no *less* carefully at those used by their critics.

Thirdly, the accusation that PPT is a form of *'window dressing'* is common and not without substance (Mowforth & Munt, 2003: 184–185). 'Greenwashing' has undoubtedly occurred; the involvement of tourism (and other) corporations in corporate social responsibility (CSR) has been exaggerated, and often prompted by self-interest. In general, too, much PPT has focused on 'minor reforms for a marketed tourism sector...' though this, surely, has some merit, and to assert it is (only?) 'to deflect criticisms and prevent unwelcome regulation' (see Higgins-Desbiolles (2006) claim on page 128 of this chapter) is an unwarranted demonization of all such programmes and their instigators.

Fourthly, Scheyvens asks: *Are small benefits sufficient and sustainable?* The implied response is that they are not, and then follow several quite separate simplistic criticisms, which have to be countered equally briefly:

- As PPT is defined, wealthier factions in a destination might actually benefit more than the poor in a PPT programme. This, though, is fully accepted by PPT's founders!
- The poor might be only passive participants. Such passivity is not advocated by PPT advocates, but arguably there is no *necessity* for them to participate at all levels in CBT (Tosun, 2000; Simpson, 2008).

- PPT does not take account of environmental, social or cultural impacts of growth-oriented strategies of tourism development. This is probably so, but the central concern of PPT practitioners is to incorporate the poor as far as possible in existing tourism markets.

In short, PPT advocates are accused of taking positions to which they freely admit or of not focusing on priorities they have never defined!

Fifthly, *businesses are in existence to make profits, not to service the poor*. As the first priority of any business *has* to be survival this is stating the obvious and there is no *a priori* reason to attribute a greater (or *lesser*) commitment of tourism entrepreneurs to alleviate poverty than operators of other businesses. However, as Scheyvens notes, environmentally- or socially-friendly business practices make good business sense. Indeed, this is so *even if they are performed by people whose motives might be questionable* (an issue that merits wide debate). In the case of tourism, which is reliant on hospitality skills, there are strong pragmatic reasons for having destination residents on your side, and while CSR in tourism (as in other industries) may prompt only a small proportion of trans-national corporations (TNC) activities, communities in LDCs are also sometimes assisted by owners of resorts and other tourist establishments in ways that are frequently unpublicized. Undoubtedly more *could* be done, but it is unclear why tourist establishments, to an extent greater than any other kind of business, should be upbraided for their failure to do more.

Finally, to suggest that 'neoliberal logic. . .was behind the move to make poverty-alleviation the leading development agenda in the 1990s' is an unworthy attempt of Scheyvens to besmirch PPT by associating it with what (for *some)* is an unacceptable political and economic ideology. There is no evidence PPT advocates were or are committed to neoliberalism (unless working with capitalists is an offence in itself, in which case most of us are guilty), or of widespread hostility to governments taking an enabling role in tourism development. Indeed, the literature on tourism development and planning commonly accords the state a key role (Hall, 2005), and in many LDCs (including several in South-East Asia and China), the state *is* a major player (Wood, 1997; Hall, 2001). And the choice of SNV as an exemplar of neoliberal policy is especially strange, given its prominent role in assisting governments to intervene in tourism in such LDCs as Laos (Harrison & Schipani, 2007).

Other references in this section of the probe are similarly open to question: that poverty-alleviation is not simply a matter of meeting very basic needs is fully accepted by PPT proponents (Harrison, 2008: 857), and labour exploitation in some tourism enterprises is not an indictment of all tourism (any more than noting unsafe working conditions in some coal mines necessitates closing all mines).

Scheyvens' critique of PPT, then, is problematic because it illogically and incorrectly identifies tourism with inequality, implicitly seeks a greater commitment to poverty-alleviation by tourism entrepreneurs than representatives of other industries and continuously switches the critique from PPT's key writers to NGOs, aid agencies and other fellow travellers, and from tourism in general to smaller and more self-contained PPT projects.

This does not make PPT the greatest innovation since sliced bread. This writer agrees that PPT practitioners do not focus only on the poor, and has argued, at length (Harrison, 2008), they have no distinct theory or methods, and have no necessary association with any one type of tourism. Lacking a distinctive approach to tourism as a development tool, they operate largely outside the academic community, usually in conjunction with aid donors and NGOs, but with the (necessary) support of a few highly capitalistic international tourism organizations in developing relatively small tourism projects. In essence, PPT is 'a stated concern with the poor – a moral injunction – [and] a movement, an incipient pressure group, which consistently runs the risk of being hijacked by those who seek to claim the high moral ground. . .and the more moral product' (Harrison, 2008: 865).

Ultimately, though, PPT is a soft target, and most criticisms expressed in Scheyvens' research probe would be accepted by its key proponents and more appropriately directed at their followers. More positively, however, it is to the considerable credit of PPT practitioners that they have successfully redirected attention to the needs of the poor, a central concern in early approaches to tourism and development (de Kadt, 1979: xii) but one more recently neglected.

And then what? First, it is about time it was recognized that 'tourism' is an umbrella term for all kinds of related activities, carried out by individuals and collectivities. As a non-actor, 'it' does nothing and to blame 'tourism' for 'causing' this or that impact is unproductive reification.

Secondly, although critics might deplore capitalism in all its forms and guises, it is a fact of life – a 'harsh reality' – that production for profit is the dominant *motif* of our age. So far, critics of modern tourism (sic) have not produced a feasible alternative; whether or not we advocate PPT, we all operate within the constraints of the system. Indeed, even *Tourism Recreation Research* must at least break even!

Thirdly, in development studies generally, the role of the state has re-emerged as a central concern (Leftwich, 2000). As Lockwood (2005: 3–4) indicates: 'Neither liberalization nor *dirigisme* is guaranteed to lead to industrialization and sustainable economic growth – what counts is the quality of the intervention, and, therefore, the nature of the state'. More specifically, in a *developmental* state, tourism – *virtually any kind of tourism* – can be used to benefit the poor, but where governments are not actively committed to increasing the welfare of citizens, and are characterized instead by graft, greed and corruption, no tourism, whether or not under the PPT label, will be a successful tool for development.

5.4

Pro-Poor Tourism, Climate Change and Sustainable Development

Paul Peeters

The critical probe on Pro-Poor Tourism (PPT) by Scheyvens (see 5.1) is most welcome, as it asks generally the right questions which should have been answered long ago to substantiate the common optimistic views of governments and tourism officials. A generally overlooked problem of scale is that of the impact of PPT on climate change. This oversight is also the case in the probe. Therefore, in my reaction to the probe, I will not only try to show why PPT is at odds with sustainable development, but also why the opposition against mitigating climate change of tourism by some PPT proponents may lead to an increase of poverty.

The term 'sustainable development' has been coined by the Bruntland Report (World Commission on Environment and Development, 1987: 43) as 'development that meets present needs without compromising the ability to meet needs in the future'. The roots of the idea of sustainability go back to the Report of the Club of Rome (Meadows *et al.*, 1972). It simply means that the global human population has to respect the limits of finite resources on earth, because ignorance will inevitably lead to collapse of global human society. Collapses of all kinds of societies have happened frequently and were more often than not caused by resource mismanagement (Diamond, 2005; Harris, 2007). Therefore environmental sustainability should be understood in terms of 'staying below certain limits'.

Climate change is a typical example of stressing a resource (the climate) beyond its limits by emitting greenhouse gases, causing an increased concentration of CO_2 in the atmosphere by more than 50% over its long-term historical value. There is no precise consensus about a 'safe level' of climate change and greenhouse gas emissions, but it seems the limits are decreasing with time. Where Schellnhuber *et al.* (2006) considered a CO_2 concentration of 450–550 parts per million (ppm) to be low enough to keep the temperature rise below the 'dangerous level' of 2°C, a slightly later publication by Meinshausen *et al.* (2006) concludes that 450 ppm should be the goal. The most recent publications recommend reducing the concentration of CO_2 to 350 ppm by the end of the 21st century, from its current value of 385 ppm. To achieve this, greenhouse gas emissions should decline by 3% per year after 2015 (see Parry *et al.*, 2008; Hansen *et al.*, 2008: Fig. 6a). Whatever the outcome of this discussion, it is clear that tourism will need to reduce its CO_2 emissions just as other sectors.

The share of tourism's CO_2 emissions has been estimated at about 5% in 2005 (UNWTO-UNEP-WMO, 2008). Total tourism CO_2 emissions were 1170 million tonnes (Mton) and it is projected by UNWTO that these will increase by a factor 2.6 to 3060 Mton in 2035. If tourism is to be on track of a global emissions reduction path, required to avoid dangerous climate change, this would have led to a scenario with a reduction in 2035 of almost 28% with respect to the emissions in 2005. In other words, the 'business-as-usual' CO_2 emissions are 3.6 times higher than the sustainable emissions. Clearly, current tourism is not developing in a sustainable way, and the tourism sector will have to do a major job by reducing its emissions against current trends.

The average emission of a tourist trip in 2005 was 274 kg, but the variety is large. Whereas an average domestic trip within the developing world accounts for just 73 kg, one in the developed world adds up to 258 kg and the average international trip contributes 676 kg. The average long-haul trip between a developed and developing country emits 1990 kg for air transport alone (UNWTO-UNEP-WMO, 2008: 135) and some 200–400 kg for accommodation and leisure activities, hence up to 2390 kg per trip. In other words, PPT is at the wrong end of the scale. And mainly transport by air is to blame for this.

What options are there to reduce air transport emissions? The first is to increase the fuel efficiency of aircraft: less petrol burnt per seat-kilometre simply means less CO_2 emissions. The technology of modern jetliners has a long history of saving fuel for good operational and economic reasons and new long-haul jet aircraft designed in 2000 are 70% more fuel efficient as the first jets in the 1950s (see Penner *et al.*, 1999). This is an average reduction of 2.8% per year, but this reduction is not evenly distributed over this period, with more than 6% per year in the first decade down to less than 2% in the last. Therefore, most scenarios use future reduction percentages of less than 2% per year (e.g. 1.2–2.2%, Lee *et al.*, 2001; most scenarios presented by Penner *et al.*, 1999 use 1.4%). Based on a more precise regression of the same fuel-efficiency data that incorporates the reduction of improvement per year apparent from historical data, Peeters and Middel (2007) find that in 2040 the reduction for long-haul aircraft will end at 35% with respect to 2000.

A second option is to increase the number of seats in an aircraft. On average, long-range aircraft are not filled with seats up to the maximum certified level. Some 25% reduction of fuel per seat-mile might be reached, pending the acceptance of the market of low seat-pitch long-haul travel (Peeters, 2007). Improved air traffic management, shorter routes, etc., may add another few percent on long-haul flight (Williams *et al.*, 2007). With growth in traffic of 4–5% on average, it is clear that technology, seat-pitch reduction or better air traffic management cannot reduce emissions.

Whereas the sustainability problem is large at the individual trip scale, it becomes huge at a global scale. Nawijn *et al.* (2008: 4) put it in this way: 'The average tourist in a developing country spends approximately US$ 550 per visit (WTO, 2002: 27). Poverty can be defined as an annual income of US$ 730 (WTO, 2002: 19). To raise the income of 2.7 billion of the poorest with US$ 730 per year per head would require 18 billion Western tourists and as many long-haul round trips, assuming 20% of the Western tourists' spending benefits the poor. These require about 220,000 billion passenger kilometres by air transport. Current aviation produces 3,500 billion travellers kilometres per year (Peeters, 2005). This growth would triple the total human emissions of greenhouse gases, responsible for climate change, of all economic sectors together.'

Clearly, also in environmental terms the 'numbers do not add up' and large-scale PPT is strongly incompatible with the desire to avoid dangerous climate change. This is specifically a problem affecting tourism based on air transport. For example, by 2040–2060, the growing air transport CO_2 emissions of the EU will be larger than the total (reducing) emissions cap of EU emissions (Bows *et al.*, 2007). The range is caused by different assumptions for air transport growth rates and technological improvement. This either means that in the medium term future political pressures on the air transport sector will accumulate and continued growth (of emissions and volume) will not be possible anymore, or it means that humanity is unable to avoid dangerous climate change. The first leads to dwindling opportunities for PPT, as it depends so much on long-haul air travel. The second means that the whole range of devastating impacts of climate change may evolve. For example, between 1.1 and 3.2 billion additional people will suffer from serious water stress (shortages), the majority of whom are living in Africa, Asia and Latin America (Parry *et al.*, 2008). Most of these people will likely be driven into deep poverty. This means that we may expect some two billion more poor people due to unchecked climate change by 2050. With a tourism share of 5% to 10% of CO_2 in the future, the tourism sector is responsible for 100 to 200 million extra poor people. Proponents of tourism and air transport often argue that because of PPT, air transport should be stimulated, not hampered (see examples in Gössling & Peeters, 2007), which, in other words, ultimately means that because of PPT we cannot solve the climate change problem.

To conclude, PPT, when based on current practice of North–South tourism, is incompatible with sustainable development, because the growing greenhouse gas emissions of the concomitant long-haul air transport cannot be fully countered by technology. Therefore, this kind of PPT cannot play an important role in alleviating poverty but would even seriously impede solutions for climate change. While this conclusion is well documented in the scientific literature, the question whether tourism really helps to alleviate poverty is rather inconclusive as, for example, shown by Wattanakuljarus and Coxhead (2008). If 'dangerous climate change' is not checked, this will cause a much larger increase in poverty than the reduction PPT will ever be able to realize. A solution would be a strong refocus of PPT on domestic tourism to redistribute wealth in poor countries and regional tourism between neighbouring countries. In those cases where this is not economically viable, we must accept that PPT will not be a solution for poverty in those countries.

Acknowledgements

I want to thank my colleagues Jeroen Nawijn and Jos van der Sterren for their most helpful comments on this essay.

Concluding Remarks

This research probe set out to investigate whether there was value beyond the rhetoric on PPT. The aim was not to criticize those who devised the concept of PPT nor their intentions, but rather to question whether the claims being made about PPT by a wide range of actors, from development practitioners to conservationists and tourism industry organizations, were realistic. Undoubtedly many people, this author included, have been attracted to the prospect that such a large global industry could deliver significant benefits to the poor.

There are now increasing calls by agencies concerned with poverty-alleviation for the private sector to fill more developmental roles, partly because 'governments and their international arms. . .have failed in their attempts to rid the planet of under-development, widespread inequalities and poverty' (Hopkins, 2007: 2). This was evident at the Fourth High Level Forum on Aid Effectiveness at Busan, Korea, in 2011 where government donors declared that business was an 'equal partner' in development. There will thus be more, not less, pressure for businesses to work in a pro-poor manner in the future, and given the importance of tourism to many developing economies, there is likely to be particularly high expectations of this sector. Thus at this juncture it is critical that we develop a deeper understanding of both the potential and risks associated with tourism businesses playing a stronger role in pro-poor development endeavours.

Dorothea Meyer rightly noted that we should not expect tourism businesses to set aside their profit motive, and nor should we expect them to start acting like development agencies. I fully concur with this point: yet advocates have at times made sweeping claims for the ability of tourism to transform the lives of the poor, often without honest reflection on the possibilities for businesses to do this. In the future, those of us who see hope in the concept of PPT may be able to learn something by engaging more with literature on Corporate Social Responsibility (CSR). CSR is not just about philanthropy or charitable donations (Ashley & Haysom, 2006). In fact, Hopkins (2007) argues that for a business to be involved in development it should work directly with both public institutions and local partners to devise sustainable projects. This accords with David Harrison's comment that for tourism to be able to contribute to poverty-alleviation, governments need to be 'actively committed to increasing the welfare of citizens'; it further suggests that in some countries with specific political contexts, it may be very difficult for tourism to improve the lives of the poor.

Earlier in this chapter, David Harrison noted that PPT does not take account of the environmental, society or cultural impacts of growth – but he does not consider this a problem, as 'the central concern of PPT practitioners is to incorporate the poor as far as possible in existing tourism markets'. I would query this acceptance. Quite simply, poverty is not just an economic phenomenon so anyone calling their efforts 'pro-poor' should be tackling the multidimensional nature of poverty (see Holden *et al.*, 2011, and Chapter 2 of Scheyvens, 2011). This is a challenge to the wide range of agencies promoting tourism as a tool for poverty-alleviation.

We should also be challenged to take a broader approach to research on tourism and poverty-alleviation. As is evident in two recent special journal issues on tourism and poverty (volume 8(3) of *Tourism Planning and Development*, and volume 14(3) of *Current Issues in Tourism*), much research still focuses on rural, often nature-based, tourism. If tourism academics genuinely want to understand tourism's potential to improve the lives of the poor, we may need to think a little more deeply about who the poor are and where they live. In particular, more studies need to be located in urban settlements where over half the world's population is concentrated. Also in line with the need to broaden our approach, Paul Peeters points out that there is significant tension between the desire of some PPT proponents to encourage growth of arrivals in poorer countries and the impacts of this travel on the global environment. His suggestion is to refocus on how domestic tourism can contribute to poverty-alleviation (see e.g. Ghimire, 2001; Scheyvens, 2007b).

A final challenge for all of us comes from Dorothea Meyer, who asserted the need for more collaboration and communication between academics and practitioners working in this field. Stakeholders on both sides of the fence could benefit from coming together more, for example, via conferences or when developing PPT programmes, to share their experiences and knowledge.

Discussion Questions

(1) Given that they are not development agencies or charities, to what extent can we expect tourism businesses to operate in ways which directly benefit the poor?

(2) In what ways has consumer pressure led to more socially and environmentally-responsible initiatives by the tourism industry?

(3) It appears that the PPT literature has focused largely on helping to alleviate rural poverty. Apart from offering tours of favelas/shanty towns, in what ways could tourism in urban areas offer sustainable livelihood options for the poor?

(4) Is a pro-poor approach to tourism likely to impact significantly on the extent and severity of poverty, or will it only ever deliver small-scale, tokenistic benefits?

(5) Can PPT work in countries or contexts where there are major institutional and structural inequalities in place, e.g. where corruption is rife, gender and racial discrimination are accepted and the industry is dominated by local elites and wealthy expatriates?

(6) Can tourism involving long-haul air transport ever be seen as sustainable?

(7) Poverty is a multidimensional concept, thus PPT should aim to deliver more than just economic benefits to the poor. Provide examples of potential social, cultural and political benefits of tourism for the poor.

(8) Is it more difficult to get large-scale businesses (e.g. hotels and resorts) or small-scale businesses (e.g. homestay accommodation) to make changes in a pro-poor direction?

Notes

(1) One notable exception is the articles published in the special issue of *Current Issues in Tourism* on pro-poor tourism (Volume 10, issues 2&3, 2007).

(2) Feedback from NZAID staff after a presentation summarizing NZAID's past approach to funding tourism, August 2006.

References

Akama, J. (1996). Western environmental values and nature-based tourism in Kenya. *Tourism Management* 17(8): 567–574.

Ashley, C (2006). Participation by the poor in Luang Prabang tourism economy: Current earnings and opportunities for expansion. *ODI Working Paper 273*. London: Overseas Development Institute.

Ashley, C. and Haysom, G. (2006). From philanthropy to a different way of doing business: Strategies and challenges in integrating pro-poor approaches into tourism business. *Development Southern Africa* 23(2): 265–280.

Ashley, C. and Mitchell, J. (2007). Assessing how tourism revenues reach the poor. *ODI Briefing Paper.* London: Overseas Development Institute.

Ashley, C. and Roe, D. (2002). Making tourism work for the poor: Strategies and challenges in Southern Africa. *Development Southern Africa* 19(1): 61–82.

Bennett, O., Roe, D. and Ashley, C. (1999). *Sustainable Tourism and Poverty Elimination: A Report for the Department of International Development*. London: Deloitte and Touch, IIED and ODI.

Bows, A., Anderson, K. and Peeters, P. (2007). Technology, scenarios and uncertainty. Paper presented in E-Clat Technical Seminar on Policy Dialogue on Tourism, Transport and Climate Change: Stakeholders Meet Researchers, March 15, 2007. Paris: Eclat.

Briedenham, J. (2004). Corporate social responsibility in tourism: A tokenistic agenda? *In Focus* 52: 11.

Britton, S. (1982). The political economy of tourism in the Third World. *Annals of Tourism Research* 9(3): 331–358.

Brohman, J. (1996). New directions in tourism for the Third World. *Annals of Tourism Research* 23(1): 48–70.

Bryden, J. (1973). *Tourism and Development: A Case Study of the Commonwealth Caribbean*. Cambridge: Cambridge University Press.

Chok, S., Macbeth, J. and Warren, C. (2007). Tourism as a tool for poverty alleviation: A critical analysis of 'Pro-poor Tourism' and implications for sustainability. *Current Issues in Tourism* 10(2&3): 144–165.

Clancy, M. (1999). Tourism and development: Evidence from Mexico. *Annals of Tourism Research* 26(1): 1–20.

De Kadt, E. (Ed) (1979). *Tourism: Passport to Development?* New York: Oxford University Press.

Diamond, J. (2005). *Collapse: How Societies Choose to Fail or Survive.* London: Penguin Books.

Dieke, P. (1993). Tourism and development policy in the Gambia. *Annals of Tourism Research* 20: 423–449.

Duffy, R. (2002). *A Trip Too Far*. London: Earthscan.

Duong, L. B. (2002). *Vietnam Children in Prostitution in Hanoi, Hai Phong, Ho Chi Minh City and Can Tho: A Rapid Assessment*. Bangkok: International Labour Organization.

E-Turbonews (2007). 'Aerial highway' critical for poor countries', November 18 2007. Online at: http://forimmediaterelease.net/pm/853.html. Accessed 18 March 2008.

Ghimire, K.B. (2001) *The Native Tourist.* London: Earthscan.

Ghimire, K.B. and Li, Z. (2001). The economic role of national tourism in China. In K.B. Ghimire (ed) *The Native Tourist* (pp. 86–108). London: Earthscan.

Goodwin, H. (2006). Community-based tourism: Failing to deliver? *id21 insights* 62. London: ODI.

Gössling, S. and Peeters, P.M. (2007). "It does not harm the environment! An analysis of industry discourses on tourism, air travel and the environment. *Journal of Sustainable Tourism* 15(4): 402–417.

Gössling, S., Peeters, P. and Scott, D. (2008). Consequences of climate policy for international tourist arrivals in developing countries. *Third World Quarterly* 29(5): 873–901.

Hall, C.M. (2001). Tourism and political relationships in Southeast Asia. In P. Teo, T.C. Chang and K.C. Ho (eds) *Interconnected Worlds: Tourism in Southeast Asia* (pp. 14–26). Oxford: Pergamon Press.

Hall, C.M. (2005). The role of government in the management of tourism: The public sector and tourism policies. In L. Pender and R. Sharpley (eds) *The Management of Tourism* (pp. 217–230). London: Sage.

Hall, C.M. (2007). Pro-poor tourism: Do 'tourism exchanges benefit primarily the countries of the south'? *Current Issues in Tourism* 10(2&3): 111–118.

Hall, D. and Brown, F. (eds) (2006). *Tourism and Welfare: Ethics, Responsibility and Sustained Well-being*. Wallingford: CABI.

Hansen, J., Sato, M., Kharecha, P., Beerling, D., Masson-Delmotte, V., Pagani, M., Raymo, M., Royer, D.L. and Zachos, J.C. (2008). Target atmostpheric Co_2: Where should humanity aim? *Eprint Arxiv(0804.1126)*.

Harris, G. (2007). *Seeking Sustainability in an Age of Complexity*. Cambridge: Cambridge University Press.

Harrison, D. (2001a). Afterword. In D. Harrison (ed) *Tourism and the Less Developed World: Issues and Case Studies* (pp. 251–263). Wallingford: CAB International.

Harrison, D. (2001b). Tourism and less developed countries: Key issues. In D. Harrison (ed) *Tourism and the Less Developed World: Issues and Case Studies* (pp. 23–46). Wallingford: CAB International.

Harrison, D. (2008). Pro-poor tourism: A critique. *Third World Quarterly* 29(5): 851–868.

Harrison, D. and Schipani, S. (2007). Lao tourism and poverty alleviation: Community-based tourism and the private sector. *Current Issues in Tourism* 10(2&3): 194–230.

Higgins-Desbiolles, F. (2006). More than an "industry": The forgotten power of tourism as a social force. *Tourism Management* 27: 1192–1208.

Holden, A., Jonne, J. and Novelli, M. (2011). Tourism and poverty reduction: An interpretation by the poor of Elmina, Ghana. *Tourism Planning and Development* 8(3): 317–334.

Hopkins, M. (2007). *Corporate Social Responsibility and International Development: Is Business the Solution?* London: Earthscan.

International Institute for Environment and Development (IIED) (2001). Pro-poor tourism: Harnessing the world's largest industry for the world's poor. Online at: http://www.iied.org/docs/wssd/bp_tourism_eng.pdf. Accessed 20 March 2008.

Jamieson, W. (2003). *Poverty Alleviation through Sustainable Tourism Development*. Bangkok: United Nations Economic and Social Commission for Asia and the Pacific.

Lee, J.J., Lukachko, S.P., Waitz, I.A. and Schafer, A. (2001). Historical and future trends in aircraft performance, cost and emissions. *Annual Review Energy Environment* 26: 167–200.

Leftwich, A. (2000). *States of Development: On the Primacy of Politics in Development*. Cambridge: Polity.

Leijzer, M. (2006). SNV WTO partnership. Presentation at the WTO General Assembly 2005. The Netherlands: SNV.

Lockwood, M. (2005). *The State They're In: An Agenda for International Action on Poverty in Africa*. Burton-on-Dunsmore: ITDG Publishing.

Mansperger, M. (1995). Tourism and cultural change in small-scale societies. *Human Organization* 54(1): 87–94.

Meadows, D.H., Meadows, D.L., Randers, J. and Behrens, W.W. (1972). *The Limits to Growth: A Report for the Club of Rome's Project on the Predicament of Mankind*. New York: Universe Books Publishers.

Meinshausen, M., Hare, B., Wigley, T.M.M., Van Vuuren, D., Den Elzen, M.G.J. and Swart, R. (2006). Multi-gas emissions pathways to meet climate targets. *Climatic Change* 75(1): 151–194.

Mitchell, J. and Ashley, C. (2007). 'Leakage claims: Muddled thinking and bad for policy? In *Opinion*. London: Overseas Development Institute.

Mitchell, J. and Faal, J. (2007). Holiday package tourism and the poor in the Gambia. *Development Southern Africa* 24(3): 445–464.

Mitchell, J. and Muckosy, P. (2008). A misguided quest: Community-based tourism in Latin America. In *Opinion*. London: Overseas Development Institute.

Mitchell, J. and Phuc, L.C. (2007). *Final Report on Participatory Tourism Value Chain Analysis in Da Nang, Central Vietnam.* London: Overseas Development Institute.

Mowforth, I. and Munt, M. (2003). *Tourism and Sustainability: Development and Tourism in the Third World* (2nd edn). London: Routledge.

Nawijn, J., Peeters, P. and Van Der Sterren, J. (2008). The ST-EP programme and least developed countries: Is tourism the best alternative? In P.M. Burns and M. Novelli (eds) *Tourism Development: Growth, Myths and Inequalities* (pp. 1–10). Wallingford: CAB International.

Pacific and Asia Travel Association (PATA) (2008). Developing world leads global tourism growth. PATA News Extra, 10 March 2008. Online at: http://www.pata.org/patasite/index.php?id=1303. Accessed 18 March 2008.

Parry, M., Palutikof, J., Hanson, C. and Lowe, J. (2008). Squaring up to reality. *Nature Reports Climate Change* 68.

Peeters, P.M. (2005). Climate change, leisure-related tourism and global transport. In M.C.H.A.J. Higham (ed) *Tourism, Recreation and Climate Change* (pp. 247–259). Clevedon: Channel View Publications.

Peeters, P.M. (2007). *Report on the Environmental Performance Class of Airlines Flying at Swedish Airports.* Breda: NHTV Cstt.

Peeters, P.M. and Middel, J. (2007). Historical and future development of air transport fuel efficiency. In R. Sausen, A. Blum, D.S. Lee and C. Brüning (eds) *Proceedings of an International Conference on Transport, Atmosphere and Climate (Tac)* (Oxford, United Kingdom, 26–29 June 2006) Oberpfaffenhoven: Dlr Institut Für Physic Der Atmosphäre: 42–47.

Penner, J.E., Lister, D.H., Griggs, D.J., Dokken, D.J. and McFarland, M. (eds) (1999). *Aviation and the Global Atmosphere: A Special Report of IPCC Working Groups I and III.* Cambridge: Cambridge University Press.

Plüss, C. and Backes, M. (2002). *Red Card for Tourism? 10 Principles and Challenges for a Sustainable Tourism Development in the 21st Century.* Freiburg: DANTE (NGO network for sustainable tourism development).

Rao, M. (2002). Challenges and issues for tourism in the South Pacific Island States: The case of Fiji Islands. *Tourism Economics* 8(4): 401–429.

Richter, L.K. (2001). Tourism challenges in developing nations: Continuity and change at the millennium. In D. Harrison (ed) *Tourism and the Less Developed World: Issues and Case Studies* (pp. 47–59). New York: CABI Publishing.

Schellnhuber, J., Cramer, W., Nakicenovic, N., Wigley, T. and Yohe, G. (eds) (2006). *Avoiding Dangerous Climate Change.* Cambridge: Cambridge University Press.

Scheyvens, R. (2007a). Exploring the poverty-tourism nexus. *Current Issues in Tourism* 10(2&3): 231–254.

Scheyvens, R. (2007b). Poor cousins no more: Valuing the development potential of domestic and diaspora tourism. *Progress in Development Studies* 7(4): 307–325.

Scheyvens, R. (2011). *Tourism and Poverty*. London: Routledge.

Schilcher, D. (2007). Growth versus equity: The continuum of pro-poor tourism and neoliberal governance. *Current Issues in Tourism* 10(2&3): 166–193.

Simpson, M.C. (2008). Community-benefit tourism initiatives: A conceptual oxymoron? *Tourism Management* 29: 1–18.

Smith, S. (1983). Class analysis versus world system: Critique of Samir Amin's typology of development. In P. Limqueco and B. McFarlane (eds) *Neo-Marxist Theories of Development* (pp. 73–86). London: Croom Helm.
Sofield, T. (2003). *Empowerment for Sustainable Tourism Development*. Oxford: Pergamon.
Sofield, T., Bauer, J., De Lacy, T., Lipman, G. and Daugherty, S. (2004). *Sustainable Tourism – Eliminating Poverty: An Overview.* Australia: Cooperative Research Centre for Sustainable Tourism.
Todaro, M. and Smith, S. (2003). *Economic Development.* Harlow: Pearson Education.
Tosun, C. (2000) Limits to community participation in the tourism development process in developing countries. *Tourism Management* 21(6): 613–633.
Tourism Concern (2012) 'Sun, sand, sea and sweatshops'. Online at: http://www.tourismconcern.org.uk/index.php?page=sun-sand-sea-sweatshops. Accessed 13 March 2012.
United Nation World Tourism Organization (UNWTO) (2005). *Report of the World Tourism Organization to the United Nations Secretary-General in preparation for the High Level Meeting on the Mid-Term Comprehensive Global Review of the Programme of Action for the Least Developed Countries for the Decade 2001–2010.* Madrid: World Tourism Organization. Online at: http://www.unohrlls.org/UserFiles/File/MTR/Agency_inputs/worldtourisminput.pdf. Accessed 20 March 2008.
United Nations Conference on Trade and Development (UNCTAD) (1998). Developing countries could target tourism to boost economic growth Online at: http://www.unctad.org/Templates/Webflyer.asp?docID=3243&intItemID=2068&lang=1. Accessed 20 March 2008.
UNWTO-UNEP-WMO (2008). *Climate Change and Tourism: Responding to Global Challenges.* Madrid: UNWTO.
Wattanakuljarus, A. and Coxhead, I. (2008). Is tourism-based development good for the poor? A general equilibrium analysis for Thailand. *Journal of Policy Modeling* 30(6): 929–955.
Wilkinson, P.F. (1987). Tourism in small island nations: A fragile dependency. *Leisure Studies* 6(2): 127–146.
Williams, V., Noland, R.B., Majundar, A., Toumi, R. and Ochien, W. (2007). Mitigation of climate impacts with innovative air transport management tools. In P.M. Peeters (ed) *Tourism and Climate Change Mitigation: Methods, Greenhouse Gas Reductions and Policies* (pp. 91–104). Breda: NHTV.
Wood, R.E. (1997). Tourism and the state: Ethnic options and constructions of otherness. In M. Picard and R.E. Wood (eds) *Tourism, Ethnicity, and the State in Asian and Pacific Societies* (pp. 1–34). Honolulu: University of Hawai'i Press.
World Commission on Environment and Development (1987). *Our Common Future.* Oxford: Oxford University Press.
WTO (2002). *Tourism and Poverty Alleviation.* Paris: WTO.
Zhao, W. and Ritchie, B. (2007). Tourism and poverty alleviation: An integrative research framework. *Current Issues in Tourism* 10(2&3): 119–143.

Further Reading

Hunt, C. (2011). Passport to development? Local perceptions of the outcomes of post-socialist tourism policy and growth in Nicaragua. *Tourism Planning and Development* 8(3): 265–279.
Mitchell, J. and Ashley, C. (2010). *Tourism and Poverty Reduction: Pathways to Prosperity.* London: Earthscan.
Mowforth, M. and Munt, I. (1998). *Tourism and Sustainability: Development and New Tourism in the Third World.* London: Routledge.
Scheyvens, R. (2011). *Tourism and Poverty*. London: Routledge.

Chapter 6
Volunteer Tourism: Is it Benign?

Daniel Guttentag, Jim Butcher and Eliza Raymond

Context

This chapter, comprising three research probes, examines whether volunteer tourism (VT) is genuinely a beneficial, altruistic form of tourism. The majority of the research on VT has promoted the sector's benefits without much rigorous critical analysis, so the articles in this research probe are intended to inject an increased level of critical debate into discussions about the sector. Guttentag's initial paper questions the purported benefits of VT (the work that the volunteers achieve, the personal changes that the volunteers experience and the cross-cultural exchange that transpires) to demonstrate that such benefits are merely potential, not inevitable, consequences of VT. Butcher's paper then takes a broader look at the 'small is beautiful' philosophy that is behind much VT, arguing that this mindset prevents VT projects from making substantial contributions in the communities where the projects are based. Finally, Raymond's paper offers a variety of recommendations and a table of guidelines to assist VT organizations in creating projects that avoid some of VT's recognized shortcomings.

6.1

Volunteer Tourism: As Good as it Seems?

Daniel Guttentag

Volunteer tourism (VT) has been widely praised as an optimal form of tourism that is beneficial for everyone involved. In VT, tourists supposedly are no longer uncaring hedonists, but rather compassionate ambassadors of goodwill, and host communities supposedly are no longer objects of exploitation and commodification, but rather respected equals and grateful recipients of needed assistance. In other words, VT has been positioned as the antithesis of mass tourism and all of the problems frequently associated with it. As Brown (2005) stated, 'The volunteer vacation purports an infusion of an ideological divergence from the market-driven priorities of mass tourism' (p. 493). Although alternative tourism has been assailed by numerous critiques (e.g. Cohen, 1989; Butler, 1990; Wheeller, 2003), the subsector of VT has remained mostly unblemished, maintaining its image as tourism at its very best – tourism that encompasses such buzzword ideals as sustainability, empowerment, local development, community participation, environmental conservation and cross-cultural exchange.

Numerous studies have identified and described various benefits that can be derived from VT (e.g. Crabtree, 1998; Wearing, 2001; Broad, 2003; Brown & Morrison, 2003; Ellis, 2003; Singh & Singh, 2004; Brown, 2005; Jones, 2005; McGehee & Santos, 2005; Clifton & Benson, 2006; Zahra & McIntosh, 2007; Lepp, 2008; McIntosh & Zahra, 2008; Ruhanen *et al.*, 2008; Wearing *et al.*, 2008), but these benefits often have been accepted unquestioningly, with VT receiving meagre critical assessment. The apparent benefits of VT certainly should not be disregarded, but it is vital to recognize that such benefits are potential – not inevitable – consequences of VT. In fact, there is reason to believe that such benefits may be far less common than much of the VT research suggests. Furthermore, VT even has the potential to produce negative impacts on the individuals and communities involved (Guttentag, 2009). The predominant focus on VT's benefits in existing research has, therefore, resulted in a troublingly incomplete image of VT that may be used to encourage it in host communities. A more complete and accurate image of VT can be provided through a critical analysis of the benefits that VT purportedly offers: the work that the volunteers achieve, the personal changes that the volunteers experience and the cross-cultural exchange that occurs among the volunteers and the hosts.

The Work That the Volunteers Achieve

The work that the volunteers accomplish represents a seemingly intrinsic benefit of VT. Because VT often involves volunteers from developed countries working in underdeveloped countries (Higgins-Desbiolles & Russell-Mundine, 2008: 187; Sin, 2009: 495–496), VT projects seem to offer a wonderful form of charity for underprivileged communities. Nevertheless, for a form of tourism alleged to be particularly sustainable, the long-term impacts and potential unintended consequences of VT projects have received scant attention.

For example, VT projects may foment dependency, as host communities learn to rely on external sources of assistance, meaning immediate gains can end up subverting a community's capacity to develop sustainably. Dependency also renders host communities extremely vulnerable because VT projects may be discontinued at any time. McGehee and Andereck (2008) found dependency was a major concern for local organizations that the authors researched in West Virginia and Tijuana, and the organizations experienced varying levels of success convincing volunteer tourists not to give free handouts.

The work that the volunteers perform also may reduce local job opportunities. By definition, volunteers provide labour freely, so naturally they may undermine locals competing to offer those same labour services. This essential phenomenon has been observed on a larger scale in Africa where food aid (Dugger, 2007) and donated clothing (Matheson, 2000; Frazer, 2008) have sometimes destroyed local markets for those same products, thereby impairing development. Although huge aid shipments clearly differ from the work achieved in VT projects, the similarities are close enough that it would be unwise for the VT sector to ignore such lessons and risk repeating the same mistakes. It is undoubtedly possible, for instance, that local English teachers, construction labourers or other workers could encounter a decreased demand for their services in the face of a steady supply of volunteers eager to perform the same jobs for free.

Furthermore, volunteers may be incapable of performing their jobs adequately. Many projects have no prerequisite skills (Brown & Morrison, 2003: 77) and it is incorrect to assume that volunteers possess some innate ability to perform jobs like teaching English or constructing houses. This issue is further exacerbated because volunteers may remain for only a brief duration, may be unable to communicate in the local language and may be unfamiliar with the local culture. As the coordinator of a VT project in Argentina explained, 'When we bring an intern without strong Spanish skills, it is unavoidably going to be a burden rather than an asset to the organization' (Raymond, 2008: 55).

Such potential issues with VT projects are not inevitable, yet the issues should not be dismissed as merely improbable outcomes associated with poorly planned projects that exhibit obvious deficiencies. For example, even when tourists perform volunteer medical work, which many would see as having unassailable merit, it should not be viewed as inherently beneficial. This work may provide short-term benefits, but the efforts may also engender dependency on outside personnel and resources, undermine confidence in local healthcare providers and compete directly with such local providers (Montgomery, 1993; Bishop & Litch, 2000; DeCamp, 2007; Bradke, 2009). Moreover, the quality of care that these volunteers provide has been criticized for a variety of reasons: the volunteers often possess little knowledge of the local culture and language; volunteer groups sometimes permit individuals without appropriate medical training to dispense basic medical care; the volunteers have no accountability; the volunteers may put their egos above the best interests of the patients, feeling that the normal standards of care do not apply; the volunteers cannot provide the long-term care that is sometimes necessary (e.g. after certain surgeries); and volunteer groups often do not associate with local healthcare providers, which increases the chances that inappropriate care will be given either by the volunteers or subsequently by the local providers (Bishop & Litch, 2000; Roberts, 2006; Wall *et al.*, 2006).

Despite such concerns, projects like those providing free medical care likely will receive strong local support. In fact, the limited research examining the attitudes of host communities has generally found that they view VT fairly positively (e.g. Clifton & Benson, 2006; McIntosh & Zahra, 2008; McGehee & Andereck, 2009). Nevertheless, it is erroneous to assume that VT projects inherently enjoy widespread local support.

Myriad studies have investigated volunteer tourists' motivations (e.g. Wearing, 2001; Broad, 2003; Galley & Clifton, 2004; Stoddart & Rogerson, 2004; Rehberg, 2005; Campbell & Smith, 2006; Clifton & Benson, 2006; Pike & Beames, 2007; McIntosh & Zahra, 2008; Söderman & Snead, 2008), and the studies have repeatedly found that the volunteers are motivated by personal reasons in addition to altruism. These studies generally have accepted this finding without much concern, as volunteers' motivations are irrelevant when evaluating the impacts of the projects. However, such reasoning ignores the idea that volunteers' motivations influence volunteers' preferences, and these preferences influence the selection and design of projects as project operators strive to attract volunteers. As Lorimer (2008) found during interviews with managers of VT conservation projects,

'Managers know from past experience which projects work and sell well, they continuously gauge and channel volunteer enthusiasms and then seek to establish or solicit similar ventures' (p. 9). In this scenario, a host community's needs may be superseded by the desires of the volunteers. Such a situation is worrisome because the volunteers may hold opinions on relevant issues like development and conservation that are inconsistent with the needs and wishes of the host communities. For example, when participating in a Guatemalan VT project, Vrasti (2009) found, 'Never is the rhetoric of "small is beautiful" questioned. Never does it cross the minds of volunteer tourists that their ideals may be at odds with those of locals' (p. 21).

Even more troublesome, the project operators' goals – as unrelated to the volunteers' motivations – similarly may contrast with host communities' goals. This concern seems particularly germane for conservation projects, and especially those run by non-governmental organizations (NGOs), which is ironic because NGO VT operators are often perceived as superior to commercial operators (e.g. Lyons & Wearing, 2008). However, sometimes NGOs unwaveringly promote conservation against the wishes of local communities (e.g. Kinan & Dalzell, 2005; Butcher, 2007: 70–71). As Butcher (2007) described when discussing NGOs and ecotourism development, in some cases 'community participation amounts to participation in a pre-existing agenda, rather than in determining the agenda' (p. 74). This limited community participation was experienced first-hand by Matthews (2008) as she participated in a VT sea turtle conservation project in Costa Rica and found that many locals expressed little support or even resentment towards the project. Such attitudes certainly do not signify that conservation efforts should be abandoned, but the situation clearly refutes the idea that VT projects invariably receive high levels of community participation and support.

The Personal Changes that the Volunteers Experience

Regardless of what volunteer tourists actually accomplish, many researchers have praised VT for providing the volunteers with an opportunity to experience positive personal transformations. As Wearing (2001) stated, 'The most important development that may occur in the volunteer tourist experience is that of a personal nature, that of a greater awareness of self' (p. 2). The diverse personal changes that volunteers may experience include enhanced personal awareness, increased confidence, greater self-contentment (Wearing, 2001), personal growth, a rejection of materialism (Brown, 2005), increased interpersonal skills, increased problem-solving skills, increased communication skills (Jones, 2005), a broadened perspective on life, a greater sense of social justice and responsibility (Zahra & McIntosh, 2007), identity development (Matthews, 2008), a discovery of self (Lepp, 2008), and a development of self (Wearing *et al.*, 2008). However, it is only sensible to focus on these forms of personal development after the benefits of the volunteer work itself have been established. If the work is somehow detrimental to a host community, then the volunteers' personal transformations become benefits earned at the expense of the host community. In other words, VT ends up producing the exact situation its proponents oppose – tourists exploiting locals for the tourists' own personal gains.

Moreover, the significant personal transformations that volunteers may experience should not be perceived as inevitable. In fact, the very idea that personal traits are flexible enough to be transformed by brief tourist experiences, yet persistent enough to be maintained thereafter, is somewhat contradictory (Brookes, 2003). Furthermore, personal traits are not necessarily absolute, but rather situational (Brookes, 2003). In other words, it is incorrect to assume the personal changes that volunteers experience during a project will inevitably remain once the tourists return to their previous lives. For example, Sin (2009) researched volunteer tourists working in South Africa and found, 'While there was a sense amongst respondents. . . that they felt a greater consciousness towards particular societal issues, respondents were not necessarily able or willing to commit to further volunteering activities in other contexts' (p. 494).

The Cross-Cultural Exchange

Volunteers' personal transformations may result in part from the cross-cultural exchanges enabled by VT, which are perceived as beneficial to both the volunteers and the hosts. For example, McIntosh and Zahra (2008) stated, 'With volunteer tourism, more intense rather than superficial social interaction can occur; a new narrative between host and guest is created; a narrative that is engaging, genuine, creative and mutually beneficial' (p. 179). VT is perceived as an effective catalyst for such favourable intercultural interaction partly because VT can create an environment in which power is shared equally between tourists and hosts. As Wearing (2001) explained, 'The power balance between tourist and host can be destabilized. . .and tourists spaces constructed for genuine exchange which will benefit all the selves involved' (p. 172). However, in actuality, an environment in which one privileged group is donating their time and another underprivileged group is receiving assistance is not particularly conducive to producing an equal-power relationship. In fact, this aspect of VT has even led some to posit the activity as a form of neo-colonialism. For instance, one sending organization's director criticized, '[Some] providers reinforce a colonial attitude that development is something that educated people from rich countries do to poor people who know no better. They perpetuate the notion that Africa, Asia and Latin America are playgrounds for young people to experience "real life"' (Brodie, 2006).

Additionally, the impacts of the cross-cultural exchange may not always be desirable. For example, volunteers may observe poverty and suffering up close, and it is suggested that this experience can offer the volunteers a better perspective on their own lives and possibly inspire action against global inequalities (e.g. McGehee & Santos, 2005; Zahra & McIntosh, 2007; Lepp, 2008). However, many volunteers actually appear to rationalize or even romanticize their surrounding poverty by focusing on the happiness that the hosts exhibit (e.g. Simpson, 2004; Pike & Beames, 2007; Raymond & Hall, 2008). As one volunteer working in Ghana commented, 'These people lack of lot of things financially, but the riches they've got inside themselves is priceless' (Pike & Beames, 2007: 152). Unfortunately, this 'poor-but-happy' mentality can excuse poverty instead of inspiring opposition to it (Simpson, 2004).

Cross-cultural exchange also has been lauded as a way to foment greater cultural respect and reduce stereotypes (e.g. Jones, 2005; Lepp, 2008). This outcome appears logical and it is supported by studies on 'intergroup contact theory', which generally have found that intergroup contact reduces prejudices (Pettigrew & Tropp, 2006). Likewise, in tourism it has been found that the closer interactions provided by ecotourism can improve tourists' attitudes toward their hosts (Pizam *et al.*, 2002), even though mass tourism may have the opposite effect (e.g. Milman *et al.*, 1990; Anastasopoulos, 1992). However, several VT studies have found that the experience actually may lead volunteers to reinforce their pre-existing cultural stereotypes (e.g. Raymond & Hall, 2008; Sin, 2009). Such reinforcement may occur if the volunteers witness behaviours confirming pre-existing stereotypes, and even disconfirming observations may be discounted. For instance, Raymond and Hall (2008) found, 'Several [volunteers] implied that the positive relationships they had developed with individuals from different countries were simply "exceptions to the rule"' (p. 536).

Moreover, VT has been posed as conducive to cross-cultural exchange because of the close contact between volunteers and hosts, but this close contact may also produce undesirable cultural changes. For example, changes may occur inadvertently through the 'demonstration effect' as hosts are influenced by affluent foreign tourists exhibiting their own customs and items of wealth (Wall & Mathieson, 2006). On the other hand, short-term missionary trips, which are growing in popularity and represent a significant subset of VT, may be specifically intended to invoke changes in the host culture. Degrees of evangelism certainly vary between different groups, but there is little question that many groups see proselytizing as a key feature of their trips (Fanning, 2009). Unsurprisingly, many host communities appear to resent being submitted to such proselytizing (e.g. Rohde, 2005; McGehee & Andereck, 2008).

Research Biases

The purported benefits of VT have received considerable praise and comparatively little scepticism in part because of apparent biases in the VT research. Numerous studies have investigated the motivations of the volunteers or examined the benefits of the projects, but far fewer studies have examined VT from a host perspective or submitted VT to a high level of critical scrutiny. Therefore, it comes as no surprise that the research primarily has found reasons to support VT.

Furthermore, much of the VT research has relied on evaluations made by volunteers, yet these evaluations are extremely vulnerable to biases. As Vittersø *et al.* (2000) explained, tourists' vacation assessments can be influenced by cognitive dissonance because, 'Having used a considerable amount of time and money to visit an attraction, it might for example be difficult for some persons to admit that the visit was a failure' (p. 433). For instance, on a Southeast Asian adventure tour, Bowen (2001) found, 'Tourists sought to justify and re-justify the decision to undertake the tour in question' (p. 55). Applied to VT, and combined with potential social desirability biases, is it really surprising that volunteers being interviewed or surveyed often state that they have experienced positive personal transformations or developed meaningful connections with their hosts?

Although far fewer studies have evaluated VT from the perspective of host communities, such studies also may be influenced by biased evaluations. For instance, Daly (2009) participated in an El Salvadorian project and found that host community leaders involved in the project exhibited positive attitudes towards it, but her own observations and those of a local project facilitator indicated that some significant concerns were being overlooked. Daly offered various possible explanations for such oversights, including that locals may be wary of criticizing a project to foreign researchers who resemble the volunteers, or that hosts who benefit in any way from a project may fear that criticisms could cause the project to be terminated.

Conclusion

The benefits of VT clearly are not inevitable, yet neither is it inevitable that VT will fail to provide benefits. For example, some problems with VT originate with sending organizations, so improvements made by such organizations could produce significant improvements throughout the sector (e.g. Raymond, 2008; Raymond & Hall, 2008). Moreover, there is no question that VT already has offered many benefits to volunteers and host communities around the globe, and sometimes these benefits trump all other concerns. For instance, it would be extremely unfortunate if medical assistance that could be offered by volunteer tourists was withheld because of outsiders' fears of potential problems like dependency formation. In other words, the issues this paper has raised are not a complete rejection of VT's possible benefits, but rather a caution that these benefits cannot be taken for granted. VT requires no less critical evaluation than any other form of tourism and advocating it as inherently beneficial denies the opportunity for potential issues to be identified and corrected. On the other hand, by recognizing the potential benefits of VT as possibilities and not inevitabilities, these benefits can hopefully be made more common.

6.2

Volunteer Tourism May Not be as Good as it Seems

Jim Butcher

Much volunteer tourism is characterized by a highly circumspect view of economic development in parts of the world in which it is most needed. As Guttentag mentions, 'small is beautiful' is all too often taken as gospel, and a hostility to big business is meat and drink to some of the foremost academic advocates of volunteer tourism.

Volunteer tourism projects often seek to combine conservation and development at the level of the village or villages, in the name of sustainable development. Tourism is seen as able to achieve this 'symbiosis' as people will come to enjoy nature, and local people will be able to benefit from living 'sustainably' without too much change to their environment and way of life.

Yet this vision of development all too often precludes, or is hostile to, forms of economic development that might alter people's relationship to their environment in any substantial way. I have yet to read of volunteers helping in the building of big dam projects to facilitate electrification of towns and villages, or in the construction of cement works so that shanty towns can be replaced with safer buildings.

So the development politics behind much volunteering are important – what the purpose of the volunteering is and the wider conception of development adopted are key issues. I would argue, in similar vein to the lead article, that volunteering and the debates around it are often characterized by a romantic view of poverty and, in the academic discussion, a strong post-development outlook.

Also, it is striking that there is a discussion about 'volunteer tourism' at all. Although one or two academics have suggested differently, there is little chance that Voluntary Service Overseas (VSO) volunteers from the UK or US Peace Corps volunteers in the 1960s would have seen themselves or been seen as tourists in any sense. One commentary even draws a parallel between volunteer tourists and volunteers in the Spanish Civil War! Yet such comparisons only serve to show what is distinctive about today's volunteer tourism and today's politics.

In the past, political views of the developing world from the political Left and Right had the shared ambition to transform poverty into prosperity. Economic development, technology and societal transformation, be it through the dynamism of the market or the collective strength of socialism, were seen as the way forward. Post-colonial societies generally shared this ambition. Volunteering in the past was likely to be connected to a wider perspective on development in this way.

Volunteer tourism today is a personal and lifestyle strategy to make a difference to the world. It eschews politics in the sense previously understood. Indeed, it is part of a retreat from politics into the realm of lifestyle. The personal transformation of the tourist is often deemed as important as the benefits to the host population. This is an individualistic, narcissistic and incredibly limited approach to politics.

Yet, in a sense, volunteer tourism is just at the principled extreme in a wider debate about 'ethical tourism' – how we can 'make a difference' through what we buy, in this case on holiday, has been part of the political scene for two decades. Volunteer tourism just takes this to its logical conclusion and structures a holiday directly around helping conservation or the local community.

Modern theories of 'life politics', 'lifestyle politics' and 'ethical consumption' talk up the possibilities for aspects of lifestyle and consumption to provide a new point of connection between the individual and the political sphere. The lack of any substantial benefit for the hosts from the large majority of volunteer tourists suggests that this is illusory. The lack of any expectation of much benefit, of much change, suggests such approaches are incredibly limited as moral or political strategies to change things for the better.

Having said this, one danger in this discussion is that the desire to make a difference itself is dismissed as naive, self-centred or arrogant. It is important to distinguish between the impulse to act upon the world, 'agency' if you like, and the political and ethical climate that shapes the understanding of and responses to social problems. To criticize the latter is not a slight on the former impulse.

This bears mentioning as there is a great deal of cynicism mixed in with the commentaries on different types of tourism. Mass tourism has long been subject to a degree of cynicism, and what I would call an 'anti-people' perspective. This sees human beings, and their mobility, pessimistically, as an environmental and cultural burden, rather than seeing holiday-making and mobility in general as a part of human progress and worthy of celebration. This critique is premised on the view that too many people do too much travelling, and that this, and the businesses that provide it, are ultimately damaging to people, cultures and environments. It is a view that is highly circumspect as to the ability of societies to develop economically – cultures and environments are characterized as fragile in the face of too many people or too many of the wrong sort. Such a view tends to lead to a pessimistic attitude towards all types of tourism.

Hence proposed 'alternative' forms of leisure travel originally put forward as 'ethical' options, are quickly problematized as they, too, contribute to and spread the 'burden' of international tourism. As Guttentag observes, whilst cynics have knocked 'alternative tourism' from its pedestal, volunteer tourism has emerged as an ethical option thus far beyond reproach.

Yet how long before it goes the way of alternative tourism, ecotourism and others, and becomes subject to an 'anti-people' critique? I think it is important that criticisms of volunteer tourism be clear about what is being criticized.

Volunteering is not the problem and neither is tourism. For me, it is the politics and political claims behind and around international volunteer tourism that are problematic and at times quite reactionary.

6.3

Volunteer Tourism: Looking Forward

Eliza Raymond

'Volunteer Tourism: As Good as it Seems?' provides an important critique of the often assumed benefits of volunteer tourism (VT). The author approaches these benefits as potential but not inevitable and thus provides a much-needed deconstruction of the frequently cited advantages of VT. By bringing together research from a variety of projects and countries, the author is able to address the perception that VT is inherently 'good' and highlight the importance of VT receiving the same level of critical analysis as other forms of sustainable tourism.

I congratulate Guttentag for also addressing some of the unintended consequences of VT. If VT is to be perceived as a model of 'best practice' in tourism (Wearing, 2004), then I believe it is essential that such consequences are further unravelled. The author argues that 'if the work (of volunteers) is somehow detrimental to a host community, then the volunteers' personal transformations become benefits earned at the expense of the host community'. In my opinion this statement is central to the future of VT because if VT exploits locals for tourists' own personal gain, then it can no longer be seen as a benign and mutually beneficial form of tourism.

So how can we create a form of VT where the needs of the host communities, volunteers and VT organization are all met? In this article, I will give examples of measures which I believe need to be taken by VT organizations seeking to create a form of VT which is economically, socially and environmentally sustainable (for other examples, see also Comlámh, 2007; Raymond & Hall, 2008; Raymond, 2008; Wearing, 2001).

The Matching Process

It's about getting the volunteers to fit in and match their skills with the needs of the community rather than just as some companies do, getting the volunteers and then dumping them on a project.

Voluntour SA, Director

I would argue that the ideal approach to volunteer tourism is 'bottom-up', where VT organizations identify the needs of the host country and then find suitable volunteers to match these needs. However, Table 6.1 suggests that where VT organizations do take a 'top-down' approach, this can still benefit all stakeholders if volunteer tourists are carefully matched with appropriate work.

Another important step which can be taken by VT organizations is to find host organizations that will genuinely benefit from the presence of volunteer tourists. This can be a challenge because many organizations value the credibility or financial contributions that international volunteers bring and therefore may accept volunteers without considering what work the volunteers will be able to realistically contribute towards.

It is, therefore, important for VT organizations to develop strong, honest and equal relationships with host organizations (Comlámh, 2007; Simpson,

Table 6.1 The matching process

	Project		*Placement*	
	Multi project	*Single Project*	*Work-focused*	*Project-focused*
Description	Volunteers spend 1 or 2 days on a variety of different projects	Volunteers complete one specific project over a set period of time	Volunteers act as an extra member of staff within the host organization	Volunteers develop and implement a specific project with the host organization
Work Example	Manual work	Seasonal data collection	Assistants to staff	Project development and implementation
Programme Example	ISV – Volunteers involved in a variety of activities such as: beach clean-ups, potting seedlings, planting trees with local school children and constructing footpaths	Earthwatch – Volunteers assist scientists with data collection for their research	AVIVA – Volunteers work in a centre for seabirds, assisting staff with a variety of tasks e.g. cleaning, feeding and releasing birds	FSD – Volunteers are placed in a local NGO and develop and implement a specific project with the NGO
Suitable for short-term volunteers (0–4 weeks)	Yes	Yes (depending on specific project)	Generally no (due to training and 'settling in' time)	Generally no (due to training, 'settling in' time and time required to develop and implement an appropriate project)
Suitable for volunteers without specific skills	Yes	Yes (unless specific skills required which cannot be taught rapidly)	Knowledge of local language important. Other skill requirements should depend on the host organization and the job	Knowledge of local language important. Other skill requirements should depend on the host organization and the job
Suitable for groups (5+)	Yes	Yes (depending on specific project)	No (unless the host organization is large or run primarily by volunteers)	No (too many volunteers could disrupt the organization, appear imperialistic and/or be a burden to the organization)

Table 6.1 Continued

	Project		*Placement*	
	Multi project	*Single Project*	*Work-focused*	*Project-focused*
Suitable for one-off volunteer projects	Preferable to have a regular flow of volunteers so that new volunteers can build on work of past volunteers	Yes (although with some projects it may be necessary to consider who will maintain it once it is completed)	No (regular flow of volunteers needed to maintain continu-ous numbers of staff)	Yes
Key benefits for the host organizations	-Energy, enthusiasm and motivation - Combined manpower (more work can be achieved) -Potential source of funding	- Energy, enthusiasm and motivation - Combined manpower (more work can be achieved) -Potential source of funding	-Energy, enthusiasm and motivation -Paid staff can work on other priorities - Potential source of funding	-Energy, enthusiasm and motivation -New ideas (and skills) -Positive change through project implementation -Potential source of funding
Key benefits for the volunteer	-Interesting due to variety - Opportunity to learn a variety of skills superficially - Opportunity to learn about a variety of aspects relating to the host country - A good introduction to volunteering	- Sense of achievement, especially where there is a visible, completed product - Opportunity to develop a strong connection with the project - Opportunity to develop a specific skill	-Opportunity to integrate into the host organization and culture -Opportunity to learn in detail about the way in which the host organization works and take part in a variety of work -Opportunity to develop durable relationships with members of the host organization	-Opportunity to integrate into the host organization and culture -Sense of achievement, especially where there is a visible, completed product - Opportunity to develop durable relationships with members of the host organization -Opportunity to develop a specific skill

Source: Raymond (2007)

2007). 'What we do with every project is go and evaluate it first. We won't just say "OK yeah sure, put it on the website, off you go". We actually go and visit them and do a proper evaluation' (personal communication, AVIVA Programme Director). Subsequently, when host organizations and VT organizations do decide to work together, it is crucial that they maintain frequent communication. Regular evaluations of the project can also be beneficial in order to ensure that the programme is meeting the expectations of all involved and to identify any changes that need to be made.

You Won't Save the World!

VT organizations can also play a central role in developing appropriate expectations amongst their volunteers. Simpson (2007) argues that the greatest source of dissatisfaction for volunteers usually occurs when they are not involved in the work which they planned and paid to do. VT organizations, therefore, not only need to ensure that volunteers are aware of what they will gain from the programme, but also of what they can expect to contribute. Otherwise, there is a risk that volunteer tourists will have an idealistic impression regarding what they will achieve during their programme and they may feel frustrated when they realize the limits of their ability to contribute. The following quote highlights the importance of preparing volunteers for the fact that they will not 'save the world':

> The other issue is ideological. Here wanting to do something isn't the same as being able to achieve it. . .It was challenging for her [the volunteer] to adapt to the realities of being here. . .She had the belief that she could change the social situation but I explained to her: 'You will not change it. The State cannot change it. The country cannot change it. How are you going to change it in two months¿'. (personal communication, Host Organization)

In addition to having realistic expectations, VT organizations can also help to shape volunteers' attitudes. Previous research has suggested that volunteers can sometimes inappropriately take on the role of 'expert' and this can be perceived as maintaining and reinforcing power inequalities between developed and developing countries (e.g. Griffin, 2004; Simpson, 2005a, 2005b; Roberts, 2004; Wearing, 2001, 2004). If volunteer tourism is to move away from such criticisms, existing literature argues that volunteer tourists should be encouraged to take on the roles of learner and guest (Butcher, 2005; Comlámh, 2007; IVPA, 2007; Simpson, 2007; Wearing, 2001).

Moving Forward

I support Guttentag's call for more studies to examine VT from a host perspective. This will be essential in developing a stronger critique of VT and a better understanding of how VT organizations can take a more pro-active role to ensure that VT develops appropriately. The author identifies some key challenges with such research including the reluctance of host communities to criticize VT to foreign researchers who resemble volunteers. Due to the fact that some VT organizations pay projects to host volunteers I believe this will add to such challenges as projects will not wish to criticize an organization which supports them financially. I would suggest that these challenges could be addressed by having a member of the local community conducting the research. Alternatively, a visiting researcher would need to take part in participant observation or extended research which would allow time for trust and relationships to develop.

I also suggest that further research into the role of VT organizations will be crucial to ensure that these organizations take responsibility for the impacts of their programmes. Some organizations (e.g. Comlámh) have begun to develop 'codes of conducts' for VT organizations and through further developing such ideas, we can place pressure on VT organizations to take an active role in creating mutually beneficial programmes.

Concluding Remarks

The three chapters that form this research probe critically analyze the volunteer tourism (VT) sector from unique perspectives, collectively contributing to a more complete picture of VT in which some of its shortcomings are given their due attention. All three papers concur with the general notion that, despite its allure, VT should not be perceived as 'beyond reproach'. Guttentag's paper illustrates that VT's supposed benefits cannot be taken for granted, and Butcher's paper highlights the limitations associated with the prevailing philosophy behind much VT. Raymond then looks toward the future by proposing recommendations that may help to avoid some of the sector's acknowledged shortcomings. This final step can only occur once the existence of problems has already been accepted, so together these probes underscore the importance of engaging in such critical analysis and discussion.

It is hoped that these probes will help to inspire and guide further critical examination of the VT sector, in which any assumptions about its benefits are spurned from the outset. The sector continues to steadily grow and diversify, so there will be myriad opportunities for valuable future research. In particular, the papers have underlined the importance of increased research on the host perspective, as any potential benefits derived from the sector should start with the host communities.

Increased critical examination of VT will inevitably beget disagreements about VT's shortcomings, including, even, what exactly these shortcomings are. For instance, Butcher criticizes the sector's impacts for being too limited, while simultaneously expecting that others will criticize the sector's impacts for being too substantial. Such looming debates will clearly raise some very challenging questions, but they are precisely the types of questions that researchers must grapple with when considering a form of tourism with such lofty ideals. Future research will undoubtedly allow for more precision and clarity within the discussions surrounding VT, but disagreements will surely remain. Nevertheless, of critical importance is not that a perfect consensus be reached but rather that VT continue to be treated as a subject for genuine critique, which will allow the necessary debates surrounding it to continue.

Discussion Questions

(1) Guttentag describes three categories of potential benefits that volunteer tourism (VT) can provide: the work that is achieved, the personal changes that the volunteers experience and the cross-cultural exchange that occurs. Are any of these potential benefits more important than the others? If so, how should such differences be taken into account when VT projects are being planned?

(2) Butcher points out that VT projects frequently blend development with conservation, but he argues that this approach can thwart meaningful economic development. Can VT genuinely serve as a tool for advancing substantial economic development? What, if anything, could be done to make VT more effective in this regard? If VT is incapable of fostering substantial economic development, is that a serious concern?

(3) Raymond's table on 'The matching process' details the key benefits associated with four different types of VT projects. What would the potential key negative impacts associated with these four different types of projects be?

(4) Should a VT project be undertaken, provided it enjoys widespread local support, even if research strongly suggests that the project will cause long-term negative impacts (e.g. dependency) in the host community? Alternatively, should a VT project be undertaken, despite being largely opposed by the local community, provided the project is working towards an ostensibly noble goal (e.g. protection of an endangered species)?

(5) VT projects can be found all over the world in all different types of communities. What characteristics of a community or destination would make it a good or bad location for a VT project?

(6) VT projects involve myriad types of work. What types of work, if any, are especially likely to result in either positive or negative impacts?
(7) It is noted that a multitude of studies have investigated volunteer tourists' motivations, whereas far fewer studies have investigated VT from a host perspective. What are some possible reasons for this apparent imbalance?
(8) As with other areas of tourism, the future of the VT sector will see continued evolution. What are the best-case and worst-case scenarios for this evolution? What are the most important steps that should be taken to ensure that VT's evolution is favourable?

References

Anastasopoulos, P.G. (1992). Tourism and attitude change: Greek tourists visiting Turkey. *Annals of Tourism Research* 19(4): 629–642.

Bishop, R.A. and Litch, J.A. (2000). Medical tourism can do harm. *British Medical Journal* 320: 1017.

Bowen, D. (2001). Antecedents of consumer satisfaction and dis-satisfaction (CS/D) on long-haul inclusive tours – A reality check on theoretical considerations. *Tourism Management* 22(1): 49–61.

Bradke, A.J. (2009). The ethics of medical brigades in Honduras: Who are we helping? Unpublished MA thesis, University of Pittsburgh, USA. Online at: http://etd.library.pitt.edu/ETD/available/etd-07262009-200714/unrestricted/Bradke.pdf. Accessed 9 December 2009.

Broad, S. (2003). Living the Thai life – A case study of volunteer tourism at the Gibbon Rehabilitation Project, Thailand. *Tourism Recreation Research* 28(3): 63–72.

Brodie, J. (2006). Are gappers really the new colonialists? *The Guardian*, 26 August. Online at: http://www.guardian.co.uk/travel/2006/aug/26/gapyeartravel.guardiansaturdaytravelsection. Accessed 15 December 2009.

Brookes, A. (2003). A critique of Neo-Hahnian outdoor education theory. Part one: challenges to the concept of "character building". *Journal of Adventure Education and Outdoor Learning* 3(1): 49–62.

Brown, S. (2005). Travelling with a purpose: Understanding the motives and benefits of volunteer vacationers. *Current Issues in Tourism* 8(6): 479–496.

Brown, S. and Morrison, A. (2003). Expanding volunteer vacation participation: An exploratory study on the mini-mission concept. *Tourism Recreation Research* 28(3): 73–82.

Butcher, J. (2005). The impact of international service on host communities in Mexico. *Voluntary Action* 7(2): 101–113.

Butcher, J. (2007). *Ecotourism, NGOs and Development: A Critical Analysis*. New York: Routledge.

Butler, R. (1990). Alternative tourism: Pious hope or Trojan Horse? *Journal of Travel Research* 28(3): 40–45.

Campbell, L. and Smith, C. (2006). What makes them pay? Values of volunteer tourists working for sea turtle conservation. *Environmental Management* 38(1): 84–98.

Clifton, J. and Benson, A. (2006). Planning for sustainable ecotourism: The case for ecotourism research in developing country destinations. *Journal of Sustainable Tourism* 14(3): 238–254.

Cohen, E. (1989). "Alternative tourism" – A critique. In T.V. Singh, H.L. Theuns and F.M. Go (eds) *Towards Appropriate Tourism: The Case of Developing Countries* (pp. 127–142). New York: Peter Lang.

Comlámh (2007). Code of good practice/volunteer charter. Available at: http://www.volunteeringoptions.org/index.php/plain/volunteer_charter. Accessed 4 January 2007.

Crabtree, R.D. (1998). Mutual empowerment in cross-cultural participatory development and service learning: Lessons in communication and social justice from projects in El Salvador and Nicaragua. *Journal of Applied Communication Research* 26(2): 182–209.

Daly, K. (2009). Community perspectives on North-South volunteer programs in El Salvador. Prepared for the 2009 Congress of the Latin American Studies Association, Rio de Janeiro, Brazil, 11–14 June 2009. Online at: http://lasa.international.pitt.edu/members/congress-papers/lasa2009/files/DalyKatherine.pdf. Accessed 9 December 2009.

DeCamp, M. (2007). Scrutinizing global short-term medical outreach. *Hastings Center Report* 37(6): 21–23.

Dugger, C.W. (2007). Charity finds that U.S. food aid for Africa hurts instead of helps. *New York Times*, 14 August. Online at: http://www.nytimes.com/2007/08/14/world/americas/14iht-food.4.7116855.html?_r=1. Accessed 13 December 2009.

Ellis, C. (2003). Participatory environmental research in tourism: A global view. *Tourism Recreation Research* 28(3): 45–55.

Fanning, D. (2009). Short term missions: A trend that is growing exponentially. *Trends and Issues in Missions*. Online at: http://digitalcommons.liberty.edu/cgi/viewcontent.cgi?article=1003&context=cgm_missions. Accessed 9 December 2009.

Frazer, G. (2008). Used-clothing donations and apparel production in Africa. *The Economic Journal* 118(532): 1764–1784.

Galley, G. and Clifton, J. (2004). The motivational and demographic characteristics of research ecotourists: Operation Wallacea volunteers in Southeast Sulawesi, Indonesia. *Journal of Ecotourism* 3(1): 69–82.

Griffin, T. (2004). A discourse analysis of UK sourced gap year overseas projects. Unpublished MA thesis, University of the West of England, UK.

Guttentag, D.A. (2009). The possible negative impacts of volunteer tourism. *International Journal of Tourism Research* 11(6): 537–551.

Higgins-Desbiolles, F. and Russell-Mundine, G. (2008). Absences in the volunteer tourism phenomenon: The right to travel, solidarity tours and transformation beyond the one-way. In K. Lyon and S. Wearing (eds) *Journeys of Discovery in Volunteer Tourism* (pp. 182–194). Cambridge, MA: CABI Publishing.

International Volunteer Programs Association (IVPA) (2007). IVPA principles and practices. Online at: http://www.volunteerinternational.org/index-principles2.htm. Accessed 4 December 2007.

Jones, A. (2005). Assessing international youth service programmes in two low income countries. *Voluntary Action: The Journal of the Institute for Volunteering Research* 7(2): 87–100.

Kinan, I. and Dalzell, P. (2005). Sea turtles as a flagship species: Different perspectives create conflicts in the Pacific Islands. *MAST* 4(1): 195–212.

Lepp, A. (2008). Discovering self and discovering others through the Taita Discovery Centre Volunteer Tourism Programme, Kenya. In K. Lyon and S. Wearing (eds) *Journeys of Discovery in Volunteer Tourism* (pp. 86–100). Cambridge, MA: CABI Publishing.

Lorimer, J. (2008). The scope of international conservation volunteering from the UK. Environment, Politics and Development Working Paper Series, Department of Geography, King's College London, Year 2008, Paper 3. Online at: http://access.kcl.clientarea.net/content/1/c6/03/95/42/LorimerWP3.pdf. Accessed 9 December 2009.

Lyons, K. and Wearing, S. (2008). Volunteer tourism as alternative tourism: Journeys beyond otherness. In K. Lyons and S. Wearing (eds) *Journeys of Discovery in Volunteer Tourism* (pp. 3–11). Cambridge, MA: CABI Publishing.

Matheson, I. (2000). Charity killing Zambia's textile industry. *BBC News*, 25 December. Online at: http://news.bbc.co.uk/2/hi/africa/1076411.stm. Accessed 13 December 2009.

Matthews, A. (2008). Negotiated selves: Exploring the impact of local-global interactions on young volunteer travellers. In K. Lyon and S. Wearing (eds) *Journeys of Discovery in Volunteer Tourism* (pp. 101–117). Cambridge, MA: CABI Publishing.

McGehee, N. and Andereck, K. (2008). 'Pettin' the critters': Exploring the complex relationship between volunteers and the voluntoured in McDowell County, West Virginia, USA, and Tijuana, Mexico. In K. Lyon and S. Wearing (eds) *Journeys of Discovery in Volunteer Tourism* (pp. 12–24). Cambridge, MA: CABI Publishing.

McGehee, N. and Andereck, K. (2009). Volunteer tourism and the "voluntoured": The case of Tijuana, Mexico. *Journal of Sustainable Tourism* 17(1): 39–51.

McGehee, N. and Santos, C. (2005). Social change, discourse and volunteer tourism. *Annals of Tourism Research* 32(3): 760–779.

McIntosh, A. and Zahra, A. (2008). Journeys for experience: The experiences of volunteer tourists in an indigenous community in a developed nation – A case study of New Zealand. In K. Lyon and S. Wearing (eds) *Journeys of Discovery in Volunteer Tourism* (pp. 166–181). Cambridge, MA: CABI Publishing.

Milman, A., Reichel, A. and Pizam, A. (1990). The impact of tourism on ethnic attitudes: The Israeli- Egyptian case. *Journal of Travel Research* 29(2): 45–49.

Montgomery, L.M. (1993). Short-term medical missions: Enhancing or eroding health? *Missiology: An International Review* 21(3): 333–341.

Pettigrew, T.F. and Tropp, L.R. (2006). A meta-analytic test of intergroup contact theory. *Journal of Personality and Social Psychology* 90(5): 751–783.

Pike, E.C.J. and Beames, S.K. (2007). A critical interactionist analysis of 'youth development' expeditions. *Leisure Studies* 26(2): 147–159.

Pizam, A., Fleischer, A. and Mansfeld, Y. (2002). Tourism and social change: The case of Israeli ecotourists visiting Jordan. *Journal of Travel Research* 41(2): 177–184.

Raymond, E. (2007). Making a difference? Good practice in volunteer tourism. Masters thesis, University of Otago, NZ.

Raymond, E. (2008). 'Make a difference!': The role of sending organizations in volunteer tourism. In K. Lyon and S. Wearing (eds) *Journeys of Discovery in Volunteer Tourism* (pp. 49–60). Cambridge, MA: CABI Publishing.

Raymond, E. and Hall, C. (2008). The development of cross-cultural (mis)understanding through volunteer tourism. *Journal of Sustainable Tourism* 16(5): 530–543.

Rehberg, W. (2005). Altruistic individualists: Motivations for international volunteering among young adults in Switzerland. *Voluntas: International Journal of Voluntary and Nonprofit Organizations* 16(2): 109–122.

Roberts, M. (2006). Duffle bag medicine. *The Journal of the American Medical Association* 295(13): 1491–1492.

Roberts, T. (2004). Are western volunteers reproducing and reconstructing the legacy of colonialism in Ghana? An analysis of the experiences of returned volunteers. Unpublished MA thesis, Institute for Development Policy and Management, Manchester, UK.

Rohde, D. (2005). Mix of quake aid and preaching stirs concern. *New York Times*, 22 January. Online at: http://www.nytimes.com/2005/01/22/international/worldspecial4/22preach.html?_r=1. Accessed 14 December 2009.

Ruhanen, L., Cooper, C. and Fayos-Solá, E. (2008). Volunteering tourism knowledge: A case from the United Nations World Tourism Organization. In K. Lyon and S. Wearing (eds) *Journeys of Discovery in Volunteer Tourism* (pp. 25–35). Cambridge, MA: CABI Publishing.

Simpson, K. (2004). 'Doing development': The gap year, volunteer-tourists and a popular practice of development. *Journal of International Development* 16(5): 681–692.

Simpson, K. (2005a). Broad horizons? Geographies and pedagogies of the gap year. Unpublished PhD thesis, University of Newcastle, UK.

Simpson, K. (2005b). Dropping out or signing up? The professionalisation of youth travel. *Antipode* 37(3): 447–469.

Simpson, K. (2007). Ethical volunteering. Online at: http://www.ethicalvolunteering.org/index.html. Accessed 3 January 2007.

Sin, H.L. (2009). Volunteer tourism – "Involve me and I will learn"? *Annals of Tourism Research* 36(3): 480–501.

Singh, S. and Singh, T.V. (2004). Volunteer tourism: New pilgrimages to the Himalayas. In T.V. Singh (ed) *New Horizons in Tourism: Strange Experiences and Stranger Practices* (pp. 181–194). Cambridge, MA: CABI Publishing.

Söderman, N. and Snead, S. (2008). Opening the gap: The motivation of gap year travellers to volunteer in Latin America. In K. Lyon and S. Wearing (eds) *Journeys of Discovery in Volunteer Tourism* (pp. 118–129). Cambridge, MA: CABI Publishing.

Stoddart, H. and Rogerson, C. (2004). Volunteer tourism: The case of Habitat for Humanity South Africa. *GeoJournal* 60(3): 311–318.

Vittersø, J., Vorkinn, M., Vistad, O.I. and Vaagland, J. (2000). Tourist experiences and attractions. *Annals of Tourism Research* 27(2): 432–450.

Vrasti., W. (2009). Love and anger in a small place: Ethnographic improvisation on the politics of volunteer tourism. Paper presented at the annual meeting of the ISA's 50th Annual Convention "Exploring the Past, Anticipating the Future", New York, 15 February 2009. Online at: http://www.allacademic.com/meta/p313978_index.html. Accessed 9 December 2009.

Wall, G. and Mathieson, A. (2006). *Tourism: Change, Impacts and Opportunities*. Toronto: Pearson Prentice Hall.

Wall, L.L., Arrowsmith, S.D., Lassey, A.T. and Danso, K. (2006). Humanitarian ventures or 'fistula tourism?': The ethical perils of pelvic surgery in the developing world. *International Urogynecology Journal* 17(6): 559–562.

Wearing, S. (2001). *Volunteer Tourism: Experiences that Make a Difference*. New York: CABI Publishing.

Wearing, S. (2004). Examining best practice in volunteer tourism. In A.R. Stebbins and M. Graham (eds) *Volunteering as Leisure/Leisure as Volunteering: An International Assessment* (pp. 209–224). Oxfordshire, UK: CABI Publishing.

Wearing, S., Deville, A. and Lyons, K. (2008). The volunteer's journey through leisure into the self. In K. Lyon and S. Wearing (eds) *Journeys of Discovery in Volunteer Tourism* (pp. 63–71). Cambridge, MA: CABI Publishing.

Wheeller, B. (2003). Alternative tourism – A deceptive ploy. In C. Cooper (ed) *Classic Reviews in Tourism* (pp. 227–234). Clevedon: Channel View Publications.

Zahra, A. and McIntosh, A.J. (2007). Volunteer tourism: Evidence of cathartic tourist experiences. *Tourism Recreation Research* 32(1): 115–119.

Further Reading

Benson, A.M. (ed). (2011). *Volunteer Tourism: Theoretical Frameworks and Practical Applications*. New York: Routledge.

Butcher, J. (2003). *The Moralisation of Tourism: Sun, Sand... and Saving the World¿* New York: Routledge.

Butcher, J. (2007). *Ecotourism, NGOs and Development: A Critical Analysis*. New York: Routledge.

Lyons, K. and Wearing, S. (eds). (2008). *Journeys of Discovery in Volunteer Tourism*. Cambridge, MA: CABI Publishing.

Sharpley, R. (2009). *Tourism Development and the Environment: Beyond Sustainability¿* London: Earthscan.

Stebbins, A.R. and Graham, M. (eds). (2004). *Volunteering as Leisure/Leisure as Volunteering: An International Assessment.* Oxfordshire, UK: CABI Publishing.

Wearing, S. (2001). *Volunteer Tourism: Experiences that Make a Difference*. New York: CABI Publishing.

Chapter 7

Tourism and Welfare: Seeking Symbiosis

Derek Hall and Frances Brown, Jim Butcher, David Fennell and Carson L. Jenkins

Context

Within the context of increasing and widening interest, much rhetoric yet little practical application, ideals of welfare, ethics and responsibility offer potentially important dimensions to an understanding, discussion and application of 'sustainability'. But tourism and tourism academics have arrived at the point of acknowledging this somewhat later than a number of other sectors and disciplines.

The core of this chapter, the first paper, attempts to act as a springboard for the critical discussion and evaluation of this direction in tourism development and research. It addresses some of the ways in which the tourism and travel sector has reacted to perceived requirements for a better ethical underpinning and sense of responsibility towards the sector's stakeholders.

That such perceptions may not be fully shared is reflected in the subsequent three short responses. All agree that important questions are raised over the sector's ability to sustain stakeholders' well-being. They diverge, however, on the type and level of analysis for, and relevance, significance and realism of, the propositions advanced. Indeed, each of the three commentaries approaches the issue debated from a different perspective and in so doing they collectively illumine the heterodox character of tourism studies that can be both a strength and a potential weakness for this fragmented multi-sectoral arena (see also the Post-colonialism and Education chapter in this respect).

Thus, while the commentaries place considerable emphasis upon the need for academics to have a critical understanding of deep ethics (Fennell), and of the actual nature of being 'ethical' (Butcher), in order to sensibly and effectively pursue critical analysis of welfare and responsibility issues, it is also claimed that much of the tourism business views itself as being under no obligation to embrace welfare considerations since its main purpose is to generate profits (Jenkins).

While reading this chapter therefore, one could well be posing the question: can any of us afford to sustain the sometimes yawning chasm between tourism academics, who enjoy talking to each other, and tourism business practitioners, who usually claim to be too busy to take notice of academic tourism research?

7.1

The Tourism Industry's Welfare Responsibilities: An Adequate Response?

Derek Hall and Frances Brown

It has long been recognized that the global tourism industry is fiercely competitive and dominated by transnational corporations, mainly based in developed countries. These organizations leverage power over the suppliers of the tourism product, potentially creating unequal exchange and power relationships. They are forced to compete through international mergers and acquisitions, and are able to survive on small margins because of substantial economies of scale. This results in continuous new product development and aggressive marketing through lower prices. The resulting instability of the sector makes it difficult for companies to plan for a more sustainable future, and 'against such a background taking steps to behave more responsibly has traditionally received a predictably low priority' (Miller, 2001: 590). At the same time, pressure on companies to take responsibility for the environmental and social damage that tourism can cause, and for the well-being of the tourists who pay its wages, is growing, not just among community and lobby groups, but also within the wider public. Despite the difficulties, there is a general perception that companies have absorbed the message that they need to think more about the ethics of their actions, and that they are taking steps to behave more responsibly. This research probe offers some observations on the tourism industry's response to the increasing awareness of the need for ethical practice and on the nature of responsibility exhibited.

Corporate Social Responsibility

The currently popular concept of corporate social responsibility (CSR) represents the objective of forging stronger connections between business and society and allowing companies to take a direct role in improving the business environment (Laing, 2004). It is claimed to be a:

> commitment by business to behave ethically and contribute to economic development while improving the quality of life of the workforce and their families as well as of the local community and society at large. (World Business Council for Sustainable Development, quoted in Kalisch, 2000: 2)

This would appear ideally suited to benefit the welfare of tourism industry stakeholders. Certainly, acknowledgement of CSR and the perceived need to respond to ethical considerations are increasingly expressed in tourism business practice. Yet there are important questions relating to the choice of social responsibilities that tourism businesses can sensibly undertake and the ways in which they can be measured and justified (Henderson, 2007).

Some specialist operators are unhappy with such terms as 'responsible' and 'ethical', believing that the former can falsely raise expectations, while the latter may be vague and difficult to define (Weeden, 2005: 238). Kalisch (2002) argues that CSR represents a recognition that tourism sustainability cannot be achieved unless corporate bodies take greater responsibility towards society in general. The World Travel and Tourism Council (WTTC, 2002: 1) articulated the *raison d'être* of CSR in tourism partly in terms of poverty-alleviation and social equity, and this appears to be consonant with one of the other more recent fashionable marketing concepts, pro-poor tourism (e.g. see Butcher, 2003; UNWTO, 2004; Hall & Brown, 2006).

Corporate social responsibility is in danger of becoming just another fashion and a cynical means of conveying the impression that the corporate world willingly embraces ethical concern and acceptance of its moral responsibilities (Briedenham, 2004). For such approaches to represent a fundamental re-evaluation of tourism's role in relation to global equity there needs to be an – as yet apparently unrealistic – significant commitment to directly address the structural causes of global inequity (Chok *et al.,* 2007).

Can stakeholders and companies negotiate a responsible path, then, or is there a need for independent or regulatory organizations to guide and exert pressure for tourism development to be pursued for the collective good (however defined)¿ The CSR literature gushingly suggests that companies should learn to be inspired by their customers and other stakeholders, and to combine this inspiration with the confidence to take socially responsible products beyond niches and into the mainstream. Once these products are in the mainstream, we are told, the evidence shows that they are unlikely to be rejected, and other companies may well imitate them (Laing, 2004: 58; Burgess, 2003). These aspirations are based on a number of contested observations and assumptions (Hall & Brown, 2006: 160):

(1) that ethical production and consumption are growing and diversifying, linked to a desire for quality products and services not constrained by

socio-economic boundaries. Yet, the rhetoric of enthusiasm for working for and purchasing services from companies with responsible practices is in conflict with the reality of often poor working conditions part-driven by a widespread desire for low prices and convenience;

(2) that relatively limited stakeholder pressure can inspire wider-ranging beneficial effects. In practice, most information available to consumers and employees is imperfect; decisions are not well informed by evidence of ethical and sustainable practice, although increasing amounts of useful information are being made available in the public domain through the internet;

(3) that independently-verified information on services and products generates a positive response from consumers and has benefits for companies. However, the different cultures in which tourism business takes place have varying conceptions of ethics and responsibility (e.g. Yaman & Gurel, 2006). This renders the concept of 'independence' arbitrary in a global tourism context.

Pro-poor and fair trade tourism should represent important elements in the CSR debate, shifting the emphasis from short-term 'doing good' to finding 'win-win' situations (Ashley & Haysom, 2005). Locke (2003) recognizes a four-fold typology of CSR approaches that demonstrates how these are apparently becoming more proactive (Table 7.1). Advocates of Pro-poor tourism approaches, for example, would see companies shifting from the 'philanthropic' to the 'encompassing' mode. But while the table implies a clear and positive progression, in reality most of the tourism industry has not gone far enough 'across the columns' to make much difference.

Table 7.1 Approaches to Corporate Social Responsibility

Minimalist	*Philanthropic*	*Encompassing*	*Social activist*
❑ Basic stakeholder support ❑ Addressing aspects that are generally human resource-oriented ❑ Tokenistic	❑ Project specific ❑ Related to specific issues relevant to the particular organization ❑ Donations and gifts ❑ Seeks to change	❑ Looks beyond the immediate business stakeholder group to the broader community ❑ Embedded in company values and management style ❑ Seeks to lead change	❑ Approach is the foundation of the business ❑ Business is a catalyst for change ❑ Seeks to effect change in others

Sources: Locke, 2003; Ashley and Haysom, 2005; Hall & Brown, 2006: 161.

Assessing Progress

The tourism industry in many respects mirrors the way in which contemporary business organizations represent a wide range of forms, activities, linkages and senses of corporate ethos, from the highly rigid and bureaucratic to the highly flexible and *ad hoc*. Fisher (2003) draws a distinction between surface and deep approaches to ethics in understanding the difference between the rhetoric concerning ethics and actual business practice. She argues that a surface approach to ethics, which is associated with self-interest, will not promote ethical behaviour, while a deep approach, motivated by the desire to do the right thing, does have the potential to do so. The difference between the rhetoric and business practice suggests that most businesses either intentionally or unintentionally adopt a surface approach to ethics.

It is, therefore, important that a company should be assessed against a number of criteria in order that its ethical position and sense of (social and environmental) responsibility can be determined. Such ethical benchmarking criteria are brought together in Table 7.2. It should be borne in mind, however, that few elements of social citizenship are based solely on 'rights', and that all imply responsibilities.

Tour Operators' Role

Tour operators provide an important link between tourism supply and demand, facilitating the circulation of products and information between the two (Curtin & Busby, 1999), and representing a critical pressure point in the mass tourism system. They trigger the actions and responsibilities of other tourism stakeholders (Budeanu, 2005), in terms of supply chains, consumer behaviour and destination response. Many would argue that tour operators should take their responsibilities more seriously.

Indeed, calls by the UN Commission on Sustainable Development for 'voluntary initiatives' in support of sustainable tourism development, which would 'preferably exceed' any relevant standards (CSD, 1999), presented an important external stimulus to action. At the same time, negative publicity – not least accrued in tourists' and employees' blogs – has seen 20 operators (including TUI, Thomas Cook and Accor) establish, with UNWTO support, a *Tour Operators Initiative for Sustainable Tourism Development* website (http://www.toinitiative.org) to detail company case studies of claimed 'sustainable' and 'responsible' activity. These feature under headings that include integrating sustainability into business, supply chain management and co-operation with destinations.

Miller's (2001) examination of the role of corporate social responsibility in the global tourism industry identified and evaluated a number of factors influencing the responsibility of tour operators' positions. He found that

Table 7.2 Ethical benchmarking criteria for tourism companies

Criterion	*Issues*
Access and equity considerations	• Can the company represent, facilitate and accommodate a wide range of disabilities? • What is the company's policy on maternity rights and childcare facilities? • Does the company pay men and women equally? • Is there positive discrimination in favour of any particular (minority or disadvantaged) group? • Is there an equitable promotions policy? • Does it have a scheme to monitor the number of people it recruits from traditionally disadvantaged sections of the population? • Does it positively encourage survival and physical access for such groups? • Are its premises accessible to the (variously) disabled?
Client rights	• Does the company fully respect the rights of its clients – tourists – as recognized by international conventions? • Does it endeavour to provide full and impartial information? • Does the company make client surveys and questionnaire responses available? • Does the company respond promptly and effectively to client complaints?
Employees' rights	• Does the company respect its employees' right to belong to a trade union? • Is there a constructive dialogue with the workforce? • Does management receive disproportionate benefits? • Are employees asked to work unacceptably long hours? • Are employees asked to work in unhealthy conditions or put at risk of injury? • Does the company have a proportionate part-time and seasonal workforce complement? • Is there encouragement and support for employee mobility?
Human rights	• Does the company trade with countries or organizations with a poor human rights record? • Does the company research how tourism planning and development processes are executed? • Does the company positively support the participation of local people in deciding the nature and scale of tourism developments?

Table 7.2 Continued

Criterion	*Issues*
Exploitation of developing countries	• Does the company exploit developing countries, for example, by driving down wage and price levels? • Does the company ring-fence employment roles in developing countries for Western nationals?
Environmental	• What is the company's attitude to local sourcing, energy saving, renewable energy and recycling, conservation, organic agriculture, pollution and climate change? • Does the company have ethical codes of environmental behaviour for its employees, sub-contractors and clients?
Animal welfare	• Does the company respect animal welfare and avoid inflicting suffering on animals? • Does the company positively contribute towards species conservation?

Source: Hall & Brown, 2006: 162

smaller companies better understood the destinations to which they took their clients and so had a heightened awareness of destination issues and problems, albeit coupled with an inability to provide solutions to them. Larger tour operators were seen as being financially able to take remedial steps, but were so removed from the destination that they often lacked awareness of the issues and problems found there. There were, of course, exceptions to this simplistic generalization.

Yet, it is clear that tour operators need to monitor in greater depth company performance and provide accessible outcomes of such monitoring to validate their claims for destination responsibility. Few tour operators appear willing to take action without external pressure to do so. Gordon and Townsend (2001) found inaction justified by operators on the grounds of such constraints as tourist attitudes and health and safety liability.

Although market surveys repeatedly indicate that consumers regard the environment as an important consideration, the gap between what respondents idealize and their actual behaviour is often substantial. Thus, views are held that destination marketing which promotes 'sustainable tourism' probably results in more business (Berry & Ladkin, 1997), while received wisdom suggests that market advantage forces companies to improve their ethical performance. Conversely, as more tourism companies see the need to adopt the marketing clothes of social responsibility, so the concept loses its ability to provide market advantage, merely preventing companies from becoming uncompetitive; it may thus be seen as a necessary extra cost. In this way, the over and indiscriminate use of 'social responsibility' in tourism business literature renders the concept relatively meaningless and leaves stakeholders justifiably cynical, in much the same way as has befallen the fate of 'sustainability' marketing.

Codes of Conduct

One earlier response to the criticisms of tourist and tourism company behaviour has been the creation of codes of conduct or behaviour. Three general guiding principles emerged in these: understanding the culture visited, respecting and being sensitive to the host population and treading softly on the host environment. Again, companies were able to emphasize how such codes were being applied to the behaviour of their clients while minimizing adoption within the realm of their own business conduct. Evidence of codes' effectiveness is scant in the absence of significant research (Cole, 2007).

Even at face value, a wide range of critical issues surround such codes in tourism: their numbers and indiscriminate application have devalued their intrinsic worth; better coordination of and consistency between codes is needed to avoid confusing and conflicting messages being communicated; there is little data available on their effectiveness; the ethical purpose of a code may be subordinated to the role of a marketing rather than a monitoring tool; few codes offer measurable criteria or conform to a widely accepted set of standards; and, most obviously from the foregoing, the existence of a code is no guarantee of ethical behaviour (Hall & Brown, 2006: 170). The WTTC has established a database of codes of conduct for the travel and tourism industry, yet what is really needed is an appropriate reference system that can help improve industry credibility (Mowforth & Munt, 2003).

Within the tourism industry there is evidence to suggest that codes of practice and conduct have been employed to deflect responsibility from company management either to employees or, more especially, to tourists. That is not to argue that tourists should be absolved of responsibility for their behaviour in destination environments. But it suggests that codes of behaviour drawn up for tourists (although often not by tourists nor with such 'stakeholder' representation) may reflect an element of lateral displacement of responsibility both by tourism companies and by host destination authorities.

At a wider level, a global code of ethics might be viewed as a frame of reference for the responsible and sustainable development of world tourism (Weeden, 2005: 235). Yet, the global code as evolved by the WTO (2001) only tinkers with the hegemonic structure of tourism, the roots of which lie in the profit-driven global economic system that largely disregards social costs (e.g. D'Sa, 1999).

This raises the question of how far the tourism industry(ies) wields power in a wide range of markedly different contexts, and whether it can realistically be a force for global welfare enhancement through deliberate policies and explicit acts. Superficially, there appears to be something of a paradox here that only serves to constrain positive action. On the one hand, we are often told that tourism does not have a loud voice or strong representation in government in the tourism generating, usually most developed, countries (e.g. McKercher, 1993). This is perhaps reflected in the way in which 'tourism', if it is explicitly named at all, may be located within different state departments and ministries in different countries, and may even be moved from one to another over time.

On the other hand, in less developed countries and regions where tourism may play a much more significant, even dominant, role in the local economy, the power and influence of tourism, derived externally and/or from within, may be substantial. Using Barbados as an example, Pugh (2003) invokes the concept of symbolic power to illustrate this reality drawing on a relatively local example. Symbolic power is:

> a legitimating form of power which involves the consent or active complicity of both dominant and dominated actors. (Hillier & Rooksby, 2002: 8)

In Barbados, there appears to have been a 'democratic' consensual approach to the work of the National Commission for Sustainable Development (NCSD). Yet a sustainable development policy prepared by the NCSD in 1998, notionally the product of participatory procedures, did not include measures to control the direct environmental impacts of tourism because of the underlying role of its symbolic power. It was 'agreed' that tourism should not be questioned, despite the fact that all interest groups on the NCSD were supposed to discuss the environmental effects of each other's activities (Pugh, 2003). Many members of the tourism sector did not feel the need to attend meetings of the NCSD, knowing that their actions would not be seriously questioned or checked by the Commission. As one member of the NCSD put it: 'this is Barbados...tourism gets special treatment, and I believe that it should...tourism is our bread and butter' (Pugh, 2003: 128). For Pugh (2003: 129), 'This creates the impression of rational consensus, where in fact the embodied and unconscious effects of symbolic power are really at work below the surface'.

Conclusions

In this short piece we have looked at some of the ways in which the tourism and travel industry has reacted to perceived requirements for a better ethical underpinning and sense of responsibility towards the industry's stakeholders. The sector, or at least various elements of it, has adopted notions of corporate social responsibility and ethical codes of behaviour and conduct. Yet there remain many shortcomings and inconsistencies that can have a negative effect on the welfare of tourism workers, destination residents and environments, and tourists themselves. Small individual companies often have a clear understanding of the welfare problems entailed by tourism but lack the capability to do anything about them, while the more powerful majors are too distant from sites and issues to be aware of the requirement for remedial action. This raises question marks over the industry's ability to sustain stakeholder well-being. Two points can be made. First, without some form of legally enforceable regulation of industry activities, companies' responses to ethical challenges are likely only to be effective at the margins. Second, short of major global structural change, tourism will find it difficult to be an important element in poverty-reducing development policies.

7.2

'Ethical' Travel and Well-being: Reposing the Issue

Jim Butcher

Derek Hall and Frances Brown note the rise of Corporate Social Responsibility (CSR) in business, and rightly note that this mostly takes the form of a discussion about ethical tourism in the academic literature. Books from Smith and Duffy (2003) and Fennell (2006b) are probably the main contributions here, the latter especially arguing that an examination of philosophy is much needed to better inform the discussion.

The emphasis on ethics extends well beyond the academic discussion. The recent title, *The Final Call: In Search of the True Cost of Our Holidays* (Hickman, 2007), written by the ethical lifestyle columnist for *The Guardian* newspaper in the UK, is a notable case in point. Echoing the themes in the paper under discussion here, he argues that mass tourism has ignored social responsibilities, and he promotes conservation, cultural sensitivity and restraint on economic growth as the route to a more ethical industry. Calls for ethical benchmarking and regulation are the consequence of this viewpoint.

I want to advance a different perspective. Rather than the need to set down and enforce a set of ethical criteria, what is more constructive for those concerned with tourism's contribution to welfare is to look at the character of the ethical debate, notably what is taken to be 'ethical', and what assumptions are being made and accepted in this discussion.

Welfare Enhancement

The paper focuses on the 'welfare responsibilities' of the industry. No one could disagree with increased welfare, so on the face of it a greater regard for this would appear to be indisputably good.

The authors ask sceptically whether tourism can 'realistically be a force for global welfare enhancement through deliberate policies or explicit acts' (Hall & Brown, 2008). Well, yes it can. Spain's remarkable economic growth from the 1960s owes something to the development of a successful mass tourism industry which remains the country's biggest earner of foreign exchange. The tourism industry brought jobs and investment to some of the poorest areas in the south of Spain, areas in which people now enjoy far better levels of education, healthcare and income.

Paradoxically, the discussion of ethical tourism is unlikely to champion Torremolinos or Fuengirola as positive examples. Rather, they are more likely to be invoked as examples of unethical mass tourism. The tone of the paper, as with virtually all commentaries on ethical tourism, is that large-scale tourism developments and big business are more likely to be damaging to welfare, and that smaller-scale green niches are the way forward.

The problem in this discussion is that welfare has been divorced from, and is often counter posed to, economic growth. Look at post-war Europe, or South Korea since 1960, and it is clear that increased welfare, measurable through healthcare, literacy, infant mortality and general standard of living, correlates fairly well with economic growth. China's recent development of tourism would not pass the paper's ethical audit, but is contributing to economic growth that generates welfare – China's performance against the Millennium Development Goals is far better than, say, countries in sub-Saharan Africa in which ecotourism is often championed as an 'ethical' alternative. Given that, it seems perverse that so much of the discussion on socially responsible, ethical tourism focuses on small green tourism projects that make a virtue of changing very little at all.

It is pretty clear that what is taken to be ethical, or addressing 'welfare', is a product of what British philosopher Professor Simon Blackburn terms the 'ethical climate' (2002, introduction) as opposed to it being a given ethical imperative standing above contestation. The substantial separation of welfare from economic growth is part of this climate. That is not to argue that economic growth is sacrosanct, but that it is badly neglected in much discussion of ethical tourism.

Measuring Ethics

The paper produces a table of ethical benchmarking criteria, and refers to the role of voluntary or statutory regulation of the industry to promote welfare. Again, I think that such a code, presented here as a universal imperative for the industry, is to miss the point.

Take, for example, the 'environmental' criterion, which refers to the need to check the company's attitude to 'recycling, conservation, organic agriculture, pollution and climate change' (Table 7.2).

Many, myself included, would argue that a commitment to modern scientific techniques in agriculture (something that many ecotourism integrated conservation and development projects stand squarely against) provides far greater possibilities for the populations of developing countries than any amount of ethical benchmarking and auditing of tourism. Yet the paper does reflect much opinion on ethical tourism in the literature and amongst niche companies and NGOs. For example, many of the environmental NGOs involved in ecotourism stand against genetically modified technology, and some projects explicitly link tourism to promoting subsistence, 'organic' agriculture in the name of supporting 'local knowledge' (Butcher, 2007: Chapter 5).

Likewise, to question a company's commitment to renewable energy may not be particularly progressive in countries that lack an electricity grid.

Ethical tourism is an etiquette, a discourse, a sensibility that is pervasive across many involved in writing about and organizing what most people would regard as 'ecotourism'. It is a stifling etiquette. Certain niches are casually associated with sustainable development and with being ethical (or at least having the potential to be these things) and through this association acquire the moral high ground in the debate. To question these claims then becomes being 'unethical'. Drop the etiquette and we would be left with a more frank debate on what are political and economic options.

Companies and trade bodies have become attuned to this ethical climate and it has become the focal point in discussions of CSR generally. For example, witness the way the International Year of Ecotourism of 2002, co-sponsored by the World Tourism Organization, declared that ecotourism 'embrace(s) the principles of sustainable development' (UNEP/WTO, 2002: 1), that as such it can 'provide a leadership role' to the rest of the industry and that it plays an 'exemplary role' in 'generating economic, social and environmental benefits' (UNEP/WTO, 2002: 7). Green tourism awards and trade bodies such as The Travel Foundation, in promoting CSR, adopt the same terms of reference in their pronouncements on what is and is not ethical or sustainable. It is hard to see what lessons the Spanish Costas, the Portuguese Algarve or the island of Malta have to learn from 'ethical' ecotourism – certainly very few concerning well-being or development.

Ecotourism

The promotion of green ethics in discussions of tourism and well-being is especially prominent with regard to the developing world. The paper under discussion makes reference to the 'win–win' scenario, often invoked in relation to tourism in these countries, through which the environment is conserved and the community can enjoy increased welfare. This is the basis for the ethical credentials of small-scale ecotourism and related niches.

Leaving aside the very poor record of such projects, and, in the case of NGO initiatives, their inability to survive when funding expires, what are the assumptions underlying such ethical claims?

Firstly, the projects claim to be participatory – to involve local communities in design and operation. This is in keeping with the wider trend towards 'Participatory Rural Appraisal' and similar attempts to factor in the community in rural development. If the community is involved, then welfare is likely to be addressed (and indeed participation, if it is regarded as an entitlement to democratic rights, may be regarded as benefiting welfare in and of itself (Sen, 2001)).

But, invariably, the priorities of NGOs, companies and aid agencies are decided prior to participatory processes in the developing world. Certainly, some involvement in deciding how a project should operate might yield some small benefits. Substantially, though, in practice participation in small-scale tourism simply involves local communities in renegotiating the terms of their poverty. So while participation invokes agency – the neopopulist aim of 'doing development for oneself' – it masks a profoundly narrow set of options for rural communities. Small-scale tourism promises small-scale benefits – but nothing beyond, as further development damages the 'sustainable' and 'harmonious' relationship between the local community and the environment.

Secondly, ethical tourism is taken in the paper to be tourism that promotes conservation (see Table 7.2). In fact, the claim made for tourism is that this can occur at the same time and in the same place as welfare benefits. The tourists will pay to experience pristine nature, the local people will be incentivized to protect it and sustain traditional ways of living in harmony with nature, and this ensures the tourists will keep coming – a sort of virtuous circle or, as some have it, a 'symbiosis' between people and nature (Goodwin, 2000).

Yet the circle is a static one – it may yield limited development, but it simultaneously caps that development within strict, localized natural limits. This is completely alien to development as has taken place in the developed countries.

Thirdly, welfare is promoted, it is argued, by deference to local knowledge and tradition. The local culture is treated with great sensitivity, but at the same time the aspiration for change is not on the agenda. The invocation of local knowledge as reflecting local agency reifies local knowledge and creates 'culture' as a cage from which there is no escape. To start with local knowledge in development is often worthy, but to treat it as the *basis for* development, as most ecotourism projects do, is quite another thing (Butcher, 2007: Chapter 5).

So ecotourism, the exemplar of ethical tourism, the type of tourism most often held up as a normative goal for all due to its ability to improve welfare and save biodiversity, is perhaps not so virtuous. You may disagree, but the point is that the terms of the ethical debate do not encourage a frank exchange. What too often passes for debate is Wheeller's (1994) view, ironic and pessimistic in equal measure (we need to save the planet but we are just acquisitive...and the green alternatives are no different), or academic nitpicking (e.g. whether fishing is ethical or not).

Climate Change and Ethical Consumption

If one holds the view that climate change is the biggest issue facing humanity, then ethical action would be to not fly on holiday, to reduce the carbon footprint as much as possible. This is a quandary for promoters of ethical tourism. For example, Tourism Concern, like other advocates of a new, ethical tourism, have for a long time argued that travelling in the right way to the right places can make a difference. As consumers, the choice may now be to travel to assist people in a destination, but contribute to carbon emissions, or to stay at home to the short-term detriment of the community but with the long-term future of the world in mind.

The paper does not really address this – climate change is another item on a long list. Yet it has trumped all other 'impacts' in terms of its importance in contemporary political debate.

However, perhaps this quandary reveals that as consumers (and a significant part of corporate social responsibility is directed at promoting ethical consumption or appealing to ethical consumers) our options are limited. If we fly it is a problem. If we don't, people miss out on prospective revenue. Similarly, if we buy a coral necklace from a beach vendor in Thailand the seller is better off, but we encourage damage to the coral. If we do not, fearing for the survival of the coral, he is no better off.

These are not 'win–win scenarios'. In fact, they only reveal that ethical consumption is a limited moral and ethical universe. In this universe we are creators of a 'human footprint', users of finite resources – we are rarely presented as agents capable of creative solutions and the development of new resources. For example, preoccupied with our human (carbon) footprint, it is easy to lose sight of the possibilities for carbon-free flying within a few generations.

The extension of travel, holidays – and more importantly the potential for more to partake of them through development – is part of progress that we would be misguided in reigning in through an ethical climate that everywhere sees natural limits (the poorer the area, the greater the biodiversity, the more severe the limit) and baulks at the ability of human societies to overcome them.

Finally, this is not an unqualified thumbs up for the free market or big companies. However, the separation of welfare from economic growth and the association of ethical tourism with small-scale initiatives that promote a localized harmony between people and nature together mean that the terms of the debate are skewed away from a rational assessment of the contribution of the tourism industry to human welfare.

7.3

Tourism Ethics Needs More Than a Surface Approach

David Fennell

I applaud Frances Brown and Derek Hall for taking the time to address an area of study that, at least in my view, is essential to the future advancement of tourism on many levels. The following few remarks are offered for the purpose of helping to stretch our thinking on many of the ethical themes that have emerged here and in the literature over the past 15 years. Although tourism ethics is still very much at an incipient stage, there are important contributions being made that will resonate with both the applied side of the 'industry' as well as the theoretical realm, and where a critical nexus can have far-reaching implications. Ethics, and the related concept of responsibility, have gained momentum in response to the failure of more traditional approaches (e.g. impacts) to fix the many problems that plague tourism. As such, the incorporation of ethics into tourism research comes packaged with a broad array of theoretical tools that provide opportunities to look at tourism problems from a number of different perspectives.

Concerning the divide between industry and non-industry perspectives on tourism, I'm not sure I agree with the authors when they argue that we need [only¿], 'observations on the tourism industry's response to the increasing awareness of the need for ethical practice...' in moving forward with the issue of ethics. I'm not discounting the importance of industry feedback on such an issue, but I don't think the industry is prepared or indeed equipped to make these changes or assessments alone (i.e. we choose to be ethical if ethics pays). Increasingly there is the recognition that, just like business in general, there is the need for a dynamic tension between business as a practice (and business ethics) and business education. Letting industry, therefore, dictate what is and is what is not ethical in tourism is just as counterintuitive as making decisions about the tourism industry solely from an academic standpoint. In this, we are well behind many other fields/disciplines and the influence they have on ethical practices in everyday life.

The authors further observe that 'Corporate social responsibility is in danger of becoming just another fashion'. Indeed, perhaps in the same manner of a range of different responses in tourism that were initially designed to change the nature of the industry, but yet have somehow failed in their attempts. Alternative tourism comes to mind, as well as ecotourism and now more recently pro-poor (as noted by the authors) and responsible tourism. The authors call for a 'significant commitment to directly address the structural causes of global inequity'. But how many times have we seen this exact call? Numerous. The reason that these concepts fail to affect positive change for tourism is that we have been unsuccessful at understanding that at the heart of these types of tourism, in time and space, are human beings. For whatever reason, we hope, imagine or wish that because of an orientation towards another newer form of tourism, humans (e.g. tourists) are about to change their innate characteristics – people will somehow be (or become) different by virtue of their association with this new 'caste'. Simply stated, we in tourism studies have chosen to ignore research on human nature, especially in regards to the inherent tendency to balance costs and benefits as individuals (individual tourists, service providers and local people) and in interactions with others (Fennell, 2006a, 2006b). No new form of tourism is suddenly about to change human nature. So, we can introduce any number of new flashy terms in tourism, but if we fail to understand *why* we act the way we do, then we are doomed to relive the failures of the past. As such, we tend to want to treat the symptoms of the problem rather than the problem itself. CSR, codes of ethics, the role of tour operators, and so on, are merely band-aid solutions of more firmly entrenched mysteries to tourism theorists. We place our faith in proximate causation instead of ultimate causation.

The authors argue, citing Weeden (2005), that responsible tourism falsely raises expectations, while ethical tourism is difficult to define. I would argue that in fact it is the opposite as regards the point on definitional difficulties. 'Responsible' especially in tourism suffers by virtue of a dysfunctional or non-existent theoretical or conceptual basis, whereas we know what ethics is and how it has been defined and discussed for some 2500 years. If we in tourism have trouble making any distinction along these lines, it is because of a failure to look outside of our rather insular walls in gaining any perspective of the bigger picture – which may be symptomatic of the need to gain respect through 'disciplinary' distinction at the expense of interdisciplinary knowledge that would help clear away the fog.

I think we need to place more thought into the issue surrounding the fact that, 'a surface approach to ethics, which is associated with self-interest, will not promote ethical behaviour, while a deep approach, motivated by the desire to do the right thing, does have the potential to do so.' This is a central matter to tourism ethics and one that deserves more than passing comment. Unfortunately, the article fails to consider anything close to a deep approach. Ethicists would argue that some attempt needs to be made in tying the discourse to an understanding of the meaning of 'good' (teleology) or 'right' (deontology) in tourism, as philosophically loaded terms; to understand how virtue ethics, hedonism, or utilitarianism, or the Golden Rule, Kantian ethics or social contract ethics could help us navigate through the many difficult turns we encounter in such a dynamic and interactive sector of the economy. In the absence of a deep approach, the authors chose to favour the shallow through a consideration of codes of ethics, tour operators and corporate social responsibility in a rudimentary fashion. In fairness, the focus of the article is on the tourism industry, yet there so much more at hand that could have been used to push the discussion forward into new territory. And as for the aspect of self-interest noted above, and in reference to my previous comments here, an understanding of self-interest is an important part of who we are as human beings, for better or for worse, whether we choose to do the right thing (i.e. to suppress the need to be self-interested in favour of some level of altruism) or to do what's right for ourselves (because we are on vacation and deserve it). This has been nicely articulated by Wheeller (1994: 648) who observes that, 'We are out for ourselves. It is a question of what is best for me and if someone…else pays the cost, then too bad as long as I get the benefits…isn't this Darwin's survival of the fittest?' I don't see how our research has even come close to addressing this first-order need.

Finally, I appreciate the fact that the authors have devised a method by which to assess the social and environmental responsibility of a given tourism company via ethical benchmarking. Criteria such as access and equity, client rights, employee rights, human rights, exploitation of developing countries, environmental, and animal rights are included along with accompanying issues. The bigger question, however, lies in how such a scheme is to be implemented and by whom. Certification schemes in tourism are fraught with philosophical and operational problems (Jamal *et al.*, 2006) and the implementation of an ethical system as such is bound to be more contentious. In this regard I would urge the authors to read

Mamic (2004) who would no doubt provide considerable scope to the ideas being presented here. So, when the authors argue that 'there are important questions relating to the choice of social responsibilities that tourism businesses can sensibly undertake and the ways in which they can be measured and justified', it may take more time to arrive at an acceptable stock of these benchmarking criteria beyond examples related to whether or not a company exploits developing countries. Defining exploitation will no doubt prove challenging.

I find myself tracing back to the title of this work in deciding whether there has in fact been an adequate response to what is posited as an inadequate response by the industry. We have such a long way to go, all of us, in our efforts to address many of the difficult questions raised by Hall and Brown.

7.4

Tourism and Welfare: A Good Idea and a Pious Hope

Carson L. Jenkins

Hall and Brown are to be congratulated in bringing to our attention some of the most important – and controversial – questions relating to the social and ethical responsibilities of tourism companies. Somehow, there is a growing perception that 'responsible tourism' and 'ethical' practices are parameters for business practice. The fact that there is no consensus on how these factors should be defined and measured does not seem to have limited their significance. Hall and Brown (2008) note, quoting Henderson (2007), there are important questions relating to the choice of social responsibilities which tourism businesses can sensibly undertake, and ways in which they can be measured and justified. If *welfare* includes impacts on the wider society and not just in the sphere of tourism business activity, then what are these impacts and what benchmarks will we use to evaluate them? In response to the article there are seven points to be made.

First, tourism is a business and the tourism company, similar to any other, has to earn profits to survive. It, therefore, follows that the *primary* responsibility of company management is to survive at a profit level appropriate to reward the owners and/or shareholders and with some margin to facilitate re-investment. There can be no doubt that the primary stakeholders are the company's investors. Without an appropriate and acceptable level of profit, the business will not survive, and a failed business, however ethically and responsibly-managed, cannot generate any economic, community or welfare benefits. A Friedmanite economist might argue that the company has only responsibility to its shareholders, and any other considerations are extraneous to the purpose of the company. Not many development advocates would subscribe to this view but it does emphasize that profit comes first in the agenda and the distribution of that profit is very much a secondary consideration.

Second, and encapsulated in Kalisch's statement, is that 'tourism sustainability cannot be achieved unless corporate bodies take greater responsibility towards society in general'. This statement has no substance in fact or experience. Even a cursory examination of international tourism arrival figures (the usual proxy for demand) since 1950 indicates a continuous average growth rate of more than 4% per annum on a global basis. The growth of long-haul tourism and tourism to developing countries has also increased and all indications are that it will continue to do so. From this statistical base it is clear that tourism has been economically sustainable. What should be appreciated is that business sustainability is an economic and financial criterion. Recently, notions of sustainability have been broadened to consider *impacts*, positive and negative, economic and non-economic on society. But to assert that tourism can only be sustainable with societal approval is a claim yet to be substantiated whereas it contradicts existing statistical evidence. It is perhaps unfortunate that the sustainability debate has been centred exclusively on tourism's impacts whereas what drives tourism is market demand. Many tourists choose a destination sight unseen – that is, they may never have visited before. If they do not receive an acceptable welcome from local people because of their dissatisfaction with tourism and tourists, the visitors might not recommend it to friends. But this does not mean that tourism at the destination will stop as other, first-time visitors will arrive. Where is the evidence (case studies of destinations¿) that demonstrates that sustainability is affected by local reactions to the type, scale and locations of tourism development. Residents may object to many aspects and characteristics of tourism development but in many cases it is the economic development advantages which prevail.

Third, and especially relevant to developing countries, is the constant search for investment funds. With the globalization of business, including tourism business, many governments recognize the dangers inherent in certain types of projects but lack either the expertise or countervailing power to modify or limit the investor's requirements. Often investment is a 'take it or leave it' decision for governments. The wide range of investment incentives available in both developed and developing countries attest to the competition for available funds. In such competitive circumstances it is often difficult to impose welfare considerations on investors for risk of the investment going elsewhere.

Fourth, relates to the nature of tourism stakeholders. Some writers on business practice refer to the concept of *business in the community*. At its simplest, the concept suggests that the company is not only part of the *business* community but also part of the wider community, including the local and national society. Therefore, corporate objectives should be aligned with social objectives including ethical business practices. However ethical or responsible business practices are, they will have, at best, a marginal contribution to addressing economic and social inequities in a country. Many of the dramatic claims for tourism such as the UNWTO's ST-EP (Sustainable Tourism-Eliminate Poverty) slogan is marketing hyperbole with little basis in reality. As noted above, tourism has been growing on a global basis for the last 60 years and poverty is still with us! As Hall and Brown (2008) and Hall note, without a consensus to change the world economic system global inequities will remain. The failure of the Doha round of trade negotiations is indicative of the continuing trade tensions between the developed and developing worlds. It follows that tourism may have some localized welfare impacts but to broaden that claim to a country and societal basis may not be attainable – a pious hope.

Fifth is the irrelevance of the CSR criteria listed in Table 7.2. Companies, including tourism companies, operate in the environment within which they are located. The basic tenet of globalization is comparative advantage. Does Nike pay its workers in China US level wages and benefits? Do any companies operating in foreign locations pay home-base levels of remuneration and benefits to local employees? In both cases, the answer to the question is no! The Western-centric criteria listed in Table 7.2 are a utopian exercise – a wish list. The criteria have no evaluatory explanation or measurement benchmarks; these might be developed and used, but could result in the business moving to less demanding environments, a position which most governments would not accept. In Europe it is noteworthy that the British Government has opted out of the part of the European Community's Social Charter limiting the number of hours a person can work each week on the grounds that it is 'bad for business'. The newly elected President of France has similarly reacted by introducing incentives for workers to break the limit on working hours per week. Government views on what is good for the economy do not always accord with welfare activists' views of what is good for the worker.

Sixth, a recognized feature of the tourism sector is that it is not an industry! It is essentially a combination of multi-sectoral activities as our continuing failure to define tourism exemplifies. The lack of definition, the fragmented nature of tourism and the large number of small companies in many countries tends to make representation difficult. In many countries, tourism does not have a strong lobby base to advocate the positions of the sector. In these circumstances, the tourism sector is often a policy-taker and not a policy-maker. The national economic, social, cultural and political aims provide the parameters within which tourism companies operate. Against this background, it is difficult to envisage how tourism companies can initiate welfare programmes except for their own employees. Is it part of their role and responsibilities to extend these functions to the wider society?

Seventh, and as already suggested in this response, tourism companies are no different from any other businesses. To expect them to adhere to Codes of Conduct and a litany of CSR requests is unrealistic except where such actions may generate financial advantage or good public relations. Concepts such as ecotourism, equitable distribution of benefits, fair trade and sustainability objectives are easy to accept but are very difficult if not impossible to achieve. It may be that CSR will join other tokens in the academic dustbin. This comment does not advocate tourism or any other companies adopting unethical or irresponsible business practices. It may be that the company's overseas location provides a comparative cost advantage compared with the home base which will generate a level of profit to raise employees' pay and other benefits above prevailing local norms. There may be good financial reason for doing so, e.g. reduce labour turnover, increase productivity and improve employees' loyalty to the company. These *business decisions* should not be mistaken for *welfare* actions which, arguably, are not a business obligation.

Concluding Remarks

The core discussion on the role, nature and importance of welfare, ethics and responsibility within tourism development has argued that this area of analysis offers potentially important dimensions for an understanding, discussion and application of 'sustainability'. The three subsequent commentaries have approached the issues addressed and have supplemented them, from differing viewpoints and nuanced positions.

Critical for Butcher in producing a coherent evaluation of the propositions advanced is an agreed, grounded understanding of what being 'ethical' actually is and an understanding of its implications. This is clearly an important starting point. He observes the 'stifling etiquette' of aspects of ethical tourism, such as benchmarking, and exemplifies his perspective through a critical discussion of 'ecotourism'. In particular Butcher notes that welfare and economic growth are often conceived of separately, although in fact they may be inextricably related, not least in benefits derived by the populations of developing countries from economic growth mechanisms. This relationship is seen as badly neglected in much of the discussion of ethical tourism.

Fennell argues the need for academics to have a critical understanding of 'deep ethics', and points out that, while 'ethics' has a strong conceptual basis, 'responsibility' does not. In the relationship between ethics and tourism business, he points to the need for a dynamic tension between business practice and business education, a sentiment that is echoed in Cohen's commentary in the Post-colonialism and Education chapter in this book. Fennell recognizes that human nature is innately egocentric.

This neatly elides with Jenkins' main thrust that much of tourism business views itself as being under no obligation to embrace welfare considerations. Businesses are seen as having at best a marginal role in addressing social and economic inequalities. Their main purpose is, of course, to generate profits. And within the context of global business competition, comparative advantage is a central requirement of profit generation.

Influencing the conception of, and need for, such comparative advantage, and ways to attain it, not least when considering the implications of climate change, is, therefore, a major challenge. As Butcher suggests, climate change is arguably the most critical issue facing the planet, and tourism/travel is central to any debate in terms of both mitigating contributions to greenhouse gases and adapting to the potential critical impacts of climate change on its practices. The roles of welfare, ethics and responsibility must therefore surely be at the forefront of climate change action research for sustainable development strategies. Central to establishing a meaningful understanding of tourism's role is to get away from the cosy pretence that sustainability begins and ends at the destination.

As multi-sectoral and fragmented as tourism may be, a responsible – ethical – recognition that sustainable development strategies must begin at home and with ourselves, however we may conceive of ethics and responsibility, is a fundamental prerequisite for the continued 'welfare' of our planet.

Discussion Questions

(1) What are the key components of welfare enhancement in tourism?
(2) What is different about an ethical approach to tourism?
(3) What are the most significant implications of the differences between 'deep' and 'shallow' ethical approaches to tourism?

(4) Examine real life examples where the application of corporate social responsibility has had an impact on tourism.
(5) Clarify the differences between, and significance of, 'pro-poor' and 'fair trade' tourism.
(6) What are the main advantages and shortcomings of codes of conduct in tourism?
(7) Explain and evaluate the contention that ethical tourism is 'a stifling etiquette'.
(8) Discuss the contention that, if tourism is a business whose prime aim is to make a profit, then the economic dimension is key to welfare enhancement in tourism.

References

Ashley, C. and Haysom, G. (2005). *From Philanthropy to a Different Way of Doing Business: Strategies and Challenges in Integrating Pro-Poor Approaches into Tourism Business.* Pretoria: ATLAS Africa.

Berry, S. and Ladkin, A. (1997). Sustainable tourism: A regional perspective. *Tourism Management* 18(7): 433–440.

Blackburn, S. (2002). *Being Good* (2nd edn). Oxford: Oxford University Press.

Briedenham, J. (2004). Corporate social responsibility in tourism – A tokenistic agenda? *In Focus* 52: 11.

Budeanu, A. (2005). Impacts and responsibilities for sustainable tourism: A tour operator's perspective. *Journal of Cleaner Production* 13(2): 89–97.

Burgess, J. (2003). Sustainable consumption: Is it really achievable? *Consumer Policy Review* 13(3): 78–84.

Butcher, J. (2003). *The Moralisation of Tourism: Sun, Sand ... and Saving the World?* London: Routledge.

Butcher, J. (2007). *Ecotourism, NGOs and Development: A Critical Analysis*. London: Routledge.

Chok, S., Macbeth, J. and Warren, C. (2007). Tourism as a tool for poverty alleviation: A critical analysis of 'pro-poor tourism' and implications for sustainability. *Current Issues in Tourism* 10 (2–3): 144–165.

Cole, S. (2007). Implementing and evaluating a code of conduct for visitors. *Tourism Management* 28(2): 443–451.

Commission on Sustainable Development (CSD) (1999). *Report of the Commission on Sustainable Development*. New York: Seventh Session, 19–30th April.

Curtin, S. and Busby, G. (1999). Sustainable destination development: The tour operator perspective. *International Journal of Tourism Research* 1: 135–147.

D'Sa, E. (1999). Wanted: Tourists with a social conscience. *International Journal of Contemporary Hospitality Management* 11(2–3): 64–68.

Fennell, D.A. (2006a). Evolution in tourism: The theory of reciprocal altruism and tourist-host interactions. *Current Issues in Tourism* 9(2): 105–125.

Fennell, D.A. (2006b). *Tourism Ethics*. Clevedon: Channel View Publications.

Fisher, J. (2003). Surface and deep approaches to business ethics. *Leadership and Organization Development Journal* 24(2): 96–101.

Goodwin, H. (2000). Tourism and natural heritage – A symbiotic relationship? In M. Robinson, J. Swarbroke, N. Evans, P. Long and R. Sharpley (eds) *Reflections on*

International Tourism: Environmental Management and Pathways to Sustainable Development (pp. 97–112). Sunderland: Business Education Publishers Limited.

Gordon, G. and Townsend, C. (2001). *Tourism: Putting Ethics into Practice*. Teddington, UK: Tearfund.

Hall, D. and Brown, F. (2006). *Tourism and Welfare: Ethics, Responsibility and Sustained Well-being*. Wallingford, UK: CABI.

Hall, D. and Brown, F. (2008). The tourism industry's welfare responsibilities: An adequate response? *Tourism Recreation Research* 33(2): 213–218.

Henderson, J.C. (2007). Corporate social responsibility and tourism: Hotel companies in Phuket, Thailand, after the Indian Ocean tsunami. *International Journal of Hospitality Management* 26(1): 228–239.

Hickman, L. (2007). *The Final Call: In Search of the True Cost of Our Holidays*. London: Guardian Books.

Hillier, J. and Rooksby, E. (eds) (2002). *Habitus: A Sense of Place*. Aldershot, UK: Ashgate.

Jamal, T., Borges, M. and Stronza, A. (2006). The institutionalization of ecotourism: Certification, cultural equity and praxis. *Journal of Ecotourism* 5(3): 145–175.

Kalisch, A. (2000). Corporate social responsibility in the tourism Industry. *Fair Trade in Tourism Bulletin* 2: 1–4.

Kalisch, A. (2002). *Corporate Futures: Social Responsibility in the Tourism Industry*. London: Tourism Concern.

Laing, S. (2004). *Consuming for Good? The Role of Consumers in Driving Responsible Business*. Edinburgh: AGENDA and the Scottish Consumer Council.

Locke, R. (2003). *Note on Corporate Citizenship in a Global Economy*. Boston, MA: Sloan School of Management and Department of Political Science, Massachussetts Institute of Technology.

Mamic, I. (2004). *Implementing Codes of Conduct: How Businesses Manage Social Performance in Global Supply Chains*. Sheffield: Greenleaf.

McKercher, B. (1993). The unrecognized threat to tourism: Can tourism survive 'sustainability'? *Tourism Management* 14(2): 131–136.

Miller, G. (2001). Corporate responsibility in the UK tourism industry. *Tourism Management* 22(6): 589–598.

Mowforth, M. and Munt, I. (2003). *Tourism and Sustainability: New Tourism in the Third World* (2nd edn). London: Routledge.

Pugh, J. (2003). A consideration of some of the sociological mechanisms shaping the adoption of participatory planning in Barbados. In J. Pugh and R.B. Potter (eds) *Participatory Planning in the Caribbean: Lessons from Practice* (pp. 118–137). Aldershot, UK: Ashgate.

Sen, A. (2001). *Development as Freedom*. Oxford: Oxford University Press.

Smith, M. and Duffy, R. (2003). *The Ethics of Tourism Development*. London: Routledge.

UNEP and WTO (2002). *The Quebec Declaration on Ecotourism*. Quebec: UNEP and WTO.

UNWTO (2004). *ST-EP (Sustainable Tourism – Eliminating Poverty)*. Madrid: UNWTO.

Weeden, C. (2005). Ethical tourism. In M. Novelli (ed) *Niche Tourism: Contemporary Issues, Trends and Case Studies* (pp. 233–245). Amsterdam: Elsevier.

Wheeller, B. (1994). Egotourism, sustainable tourism and the environment – A symbiotic, symbolic or shambolic relationship? In A.V. Seaton (ed) *Tourism: The State of the Art* (pp. 647–654). Chichester: John Wiley.

World Travel and Tourism Council (WTTC) (2002). *Corporate Social Leadership in Travel and Tourism*. London: WTTC.

WTO (2001). *Global Code of Ethics for Tourism*. Madrid: WTO.

Yaman, H.R. and Gurel, E. (2006). Ethical ideologies of tourism marketers. *Annals of Tourism Research* 33(2): 470–489.

Further Reading

Cole, S. and Morgan, N. (eds) (2010). *Tourism and Inequality: Problems and Prospects*. Wallingford: CABI.

Copeland, B.R. (1991). Tourism, welfare and de-industrialization in a small open economy. *Economica* 58: 515–529.

Giannoni, S. (2009) Tourism, growth and residents' welfare with pollution. *Tourism and Hospitality Research* 9(1): 50–60.

Kumar, P. (2004). Economic impact of tourism on Fiji's economy: Empirical evidence from the computable general equilibrium model. *Tourism Economics* 10(4): 419–433.

Lane, B. (2005). Sustainable rural tourism strategies: A tool for development and conservation. *Interamerican Journal of Environment and Tourism* (RIAT) 1(1): 20–27.

McCool, S.F. and Moisey, R.N. (2008). *Tourism Recreation and Sustainability: Linking Culture and the Environment*. Wallingford: CABI.

Nowak J., Sahli M. and Sgro P. (2003). Tourism, trade and domestic welfare. *Pacific Economic Review* 8: 245–258.

Reeder, R. and Brown, D. (2005). Recreation, tourism and rural well-being. United States Department of Agriculture, *Economic Research Report*.

Reisman, D. (2010). *Health Tourism: Social Welfare Through International Trade*. London: Edward elger.

Sheng L. and Tsui Y. (2009). A general equilibrium approach to tourism and welfare: The case of Macao. *Habitat International* 33: 419–424.

Sheng L. and Tsui Y. (2009). Taxing tourism: Enhancing or reducing welfare? *Journal of Sustainable Tourism* 17: 627–635.

Urtasun A. and Gutierrez I. (2006). Tourism agglomeration and its impact on social welfare: An empirical approach to the Spanish case. *Tourism Management* 27: 901–912.

Chapter 8

Tourism Education: Quo Vadis?

Chris Cooper, Brian Wheeller
and Lisa Ruhanen

Context

This chapter focuses on whether tourism education is equipping the future tourism workforce with the appropriate knowledge and skills to operate in the contemporary tourism sector. In particular, the chapter examines the concern that tourism education is living in the past and has not adapted to the needs of the 21st century. There are two key areas of concern. Firstly, since '9/11' and the climate change debate, are we confident that the content and balance of the curriculum is appropriate? Secondly, the 'disconnect' between educators and the sector itself has brought about a curriculum that does not meet the needs of industry and has acted as a barrier to knowledge transfer between educators and the sector. Lisa Ruhanen goes on to identify the numerous fissures that exist between educators, students and the tourism industry. She explores the notion that to remain static because 'that's the way it's always been done' will not move tourism education into the 21st century and will certainly not meet the needs and expectations of changing student cohorts. Neither will it allow us to equip our students with the skills and abilities needed when they move into the workforce. Brian Wheeller, too, identifies the deep fissures fracturing the higher education sector – which is splintering and fragmenting under duress with ideals in tourism teaching eroded by the corrosive clamour for funding, resource constraints, the restraints of assessment, promotion, demands of the 'immediate' and the intense, consuming stress exerted by the research prerogative. The chapter concludes by drawing together the threads of the debate on tourism education.

8.1

Contemporary Tourism Education: Notes from the Frontline

Chris Cooper

Introduction

This opinion piece questions the current thinking that underlies tourism education and training in the 21st century. There is no doubt that tourism education and training went through a golden age in the last quarter of the 20th century but it has to be questioned whether the sector has addressed the needs of the new century and embraced the knowledge economy. The paper examines this question from three directions.

Firstly, there is the question of whether the tourism education curriculum is *fit for purpose*. Here, there is a need to examine not only the very nature of the subject area itself, but also the many schisms that are evident – for example the balance of tourism-specific material in the curriculum versus generic approaches (tourism marketing versus mainstream marketing, for example). These question marks over the effectiveness and relevance of tourism education and training are impacting upon student numbers. In Australia, for example, the number of applicants for tourism programmes is falling in favour of more generic degrees. There is also the issue that the 21st century is experiencing a period of unexpected and rapid change, and the tourism sector is not exempt from this. This began with the terrorist attacks in New York in 2001 and continues in 2008 with the financial crisis in the world's banking system. It has to be questioned, however, whether tourism education and training is equipping future employees of the sector to deal with this environment or if, instead, it is focusing more on traditional content and material.

Secondly, we have to question why tourism educators have not propelled the tourism sector towards the knowledge economy. We identify two basic reasons for this. One is the nature of the tourism labour market and the inherent *vocational reinforcers* which mitigate against investment in the human talent of tourism. Many argue that the primary role of tourism education is to deliver qualified human resources to the tourism sector. Yet, tourism jobs are viewed as low status, there is evidence that they are low paid and non-standard (part-time and seasonal) and that employees see tourism jobs as a temporary stopgap and not a career. As a result, the sector in many countries faces both a labour and a skills shortage and has failed to embrace the fundamental principle of the knowledge economy – that of investing in human talent. The other reason is that educators and researchers represent different *communities of practice* to the tourism sector itself. This has meant that the two groups have not communicated in the past and knowledge transfer in tourism has been slow to develop.

The Tourism Curriculum: Fit for Purpose?

In the first decade of the 21st century the tourism subject area can be characterized by a number of features:

- There remain a small number of educators and subject champions, scattered across the world, commonly in schools and departments, which only have one or two tourism specialists. Of course, there are exceptions where there is critical mass in schools such as the one at the Hong Kong Polytechnic University.
- The number of programmes and tourism modules taught across the world continues to increase, particularly in the developing world and in parts of the world where tourism is a relatively new industry – such as the Middle East. Elsewhere – the UK and Australia are obvious examples here – tourism programmes and student enrolments are consolidating and there are examples of previously well-respected and leading schools being closed – Strathclyde University in Scotland is an example.
- The literature to support education continues to grow apace. Not only has the number of textbooks mushroomed from the situation in say the 1960s and early 1970s, but also the number and range of tourism journals is growing. In terms of texts these have now splintered from the core areas of introductory texts and those focusing on marketing and planning into niche areas. The number of tourism journals continues to increase despite the potential check on the growth caused by research assessment exercises where journals are increasingly ranked for quality and researchers are pressed to publish in the more highly ranked outlets and those that can boast high citation levels.
- The institutionalization of tourism education is also gradually occurring with societies such as the 'International Academy for the Study of Tourism' and government recognition of the value of tourism education and training has also grown. All this is leading to increased professionalism in the sector.

Whilst much has been written on the nature of tourism as a subject (see for example Jafari & Ritchie, 1981; and Tribe, 1997, 2002), the key for this paper is to understand the implications of the tourism subject for how it is approached in the classroom. Here we can identify three major influences:

(1) Tourism as a subject area remains without a theoretical underpinning effectively rendering it a thematic rather than a discipline-based subject.
(2) There are no real boundaries to the subject area. As a result, tourism education encompasses event management, hospitality, sport, recreation and parks, transport and beyond. This means that there is no real consensus as to what should be taught as part of tourism curricula. Although attempts have been made to establish core curricula and subject guidelines, a key question for educators remains – where do the boundaries lie?
(3) Tourism is not a pure subject such as physics; rather it is an applied subject area dealing with an aspect of human behaviour, which is essentially about mobility, and it also deals with aspects of an economic sector. This raises two issues for educators. First, in terms of mobility and human behaviour, just what is the relationship between tourism and other social sciences, such as geography or sociology, and how much should tourism be used merely as an exemplar of theories and concepts in those subjects? In other words, is tourism subservient to these subject areas and does this explain the fact that tourism is often seen as lightweight, easy for students to pass? Second, tourism education seems not to be respected by the tourism sector itself. The reasons are well known – the sector tends to be research-averse and has no respect for academic work. It is dominantly comprised of small companies run by those with no formal tourism education and this has resulted in two very separate 'communities of practice' – tourism educators and researchers, and the tourism sector itself. Of course, there are exceptions to this generalization such as Tui Travel PLC in the UK who have dedicated time and resources to engaging with tourism educators.

Tourism Education at the Frontline

Given the issues already identified in this paper, the tourism educator at the front line has a number of decisions to take:

- What to teach – where do the boundaries lie and how should the content be structured?
- How to deliver the content – given the nature of tourism as an applied subject how can case studies and examples be integrated? How can educators reach out to the contemporary student who will increasingly

demand content delivered via their i-Pod or mobile phone and virtual web-based chat rooms to replace seminar-based formats?

- How to assess students – are traditional methods such as examinations relevant or should the educator use case-based assessment or multiple-choice questions?
- How to engage with the sector – how can educators build relationships with the sector and utilize them effectively in their education programmes?

Unfortunately, and unlike more established subject areas, there is no agreement on these points, though, of course, the answer to these questions will depend upon the level of the programme and where in the world the programme is taught. It does clearly show, however, that tourism education remains immature with no common agreement on the approach, scope or underpinning philosophy.

International experience and research demonstrates that the tourism curriculum is not a *given*, but is effectively a *contested space* – contested between different academic subject areas and disciplines, and contested in terms of the role of the sector itself in influencing the curriculum. I would argue that the contest for tourism curriculum space and the associated battle-lines are clearly drawn. They include:

- Education approaches *versus* training approaches.
- Industry-driven content *versus* academic content.
- Generic content *versus* tourism-specific content.
- Advocacy for tourism *versus* education for the 'acceptability' of tourism.
- Vocational education *versus* tourism education for interest.
- Core curriculum content *versus* freedom of content choice.

And of course, this does not begin to approach other contests such as the correct placement of languages, technology, sustainability, climate change, ethics, law or other fields in the curriculum.

One solution to these dilemmas has been to develop a core curriculum for tourism (Holloway, 1995; QAA, 2008). Yet, there are dangers in blindly accepting this received wisdom. To adopt, uncritically, a core curriculum may be a mistake, particularly where the core has been developed for, say, the UK, but is adopted in Malaysia. In international settings I often see tourism curricula where the history of tourism is taught through the medium of the history of tourism in England – from the Grand Tour to Thomas Cook. This is simply not relevant to, say, programmes in Asia. This is a mistake – rooted, I think, in a lack of confidence and expertise. As a result, I am increasingly strongly of the view that tourism education should have some commonly agreed international principles, but that these should be delivered with local and regional sensitivity – and with a heightened awareness of local culture and practice.

There is now a strong degree of custom and practice relating to tourism education, particularly in the hospitality area; custom and practice that is often based upon operating environments of 30 or 40 years ago (for a summary of the state of the art see for example Airey & Tribe, 2005; Barrows & Bosselman, 1999; Hsu, 2005). I believe firmly that new tourism programmes have the opportunity (and responsibility) to be innovative and to rethink old ways of doing things. For example, the opportunities provided by technology for delivery and learning support – and engaging a fragmented and geographically dispersed sector – is an obvious opportunity here.

A New Curriculum Approach

Increasingly then, there is a need to equip tourism students with the ability to cope with a world of turbulence and rapid, unexpected change. Equally, the curriculum must build in flexibility to accommodate this change. For example, a major issue in terms of integrating climate change into the school curriculum is the lack of flexibility in the system to cope with new content areas. To cope with a changing world, the curriculum needs to find the space to prioritize generic skills such as:

- Decision-making skills.
- Problem-solving skills.
- The ability to adapt to shifting priorities.
- Leadership skills.

This allows the curriculum to focus on those areas that will deliver a workforce to the sector who can deal with the realities of tourism in the 21st century. Of course, the other side of the coin here is that the tourism-specific, sector-specific skills will need to be tuned down in the curriculum. I would argue that these deliver short-term skills and knowledge, which can be learned on the job relatively quickly, and that we do a disservice to both the student and the sector by overly focusing on these areas.

The Tourism Education Futures Initiative (TEFI) has gone some way to address these issues by examining the future of tourism education to 2030 (Sheldon *et al.*, 2007). TEFI has drafted four categories of skills that students of the future will need:

(1) Destination stewardship skills.
(2) Political and ethical skills.
(3) Enhanced human resources skills.
(4) Dynamic business skills.

For too long tourism educators have been complacent and failed to innovate in their education programmes. An example here is the fact that thinking and practice on how educators and their programmes should relate to the sector remains mired in the 1970s. Just what is the value to a student of spending 12 months in a hotel property rotating through the various departments and, in the main, learning little? After a few months, diminishing returns set in and the opportunity cost of not benefiting from further education in marketing, finance or HR must be clear. In Australia, the University of Queensland's School of Tourism has successfully broken with this tradition by establishing strong partnerships with local industry and creating competitive 'executive shadowing' placements for students (Solnet *et al.*, 2007). Also in Australia, Victoria University's round-table initiative has attempted a similar ground-breaking approach.

Finally, the traditional organization of tourism schools and departments should perhaps be rethought. Many are stand alone, others embedded within business schools whilst others are part of geography or sport departments. How should tourism schools be organized and what should their *raison d'être* be? A useful analogy here is with Starkey and Tiratsoo's (2007) view that business schools develop narratives and scripts, which impact upon how managers both think and act. They argue that business schools should move closer to a social science rather than a 'professional science' model, structured as multi-disciplinary departments, hubs or clusters, drawn from members of their institution or university (Jafari & Ritchie, 1981; Pink, 2005). This allows an innovative blending of subjects and disciplines, recognizing the post-disciplinary ideas of Coles *et al.* (2006).

Tourism Education and the Knowledge Economy

The knowledge economy brings a number of challenges to the tourism sector. From the point of view of concluding this paper, there are two key concerns:

(1) The *vocational reinforcers* inherent in the labour market prevent organizations investing in human talent in the tourism sector. A fundamental principle of the knowledge economy is that any organization is totally dependent upon the talent of its workforce to deliver competitiveness, innovation and creativity.
(2) The different *communities of practice* that are evident in educators/researchers and the tourism sector itself. This has prevented the transfer of research and knowledge to the tourism sector and so, many argue, the sector is not as competitive as it could be.

Vocational reinforcers

Despite checks and challenges, tourism in the world continues to grow and continues to generate demand for tourism jobs. A trained and educated workforce is a key to competitiveness in the tourism sector. Yet, despite the fact that tourism education and training is delivered in almost every country in the world, the supply of tourism labour has become a significant issue in most of the developed world and increasingly so in the developing world. The issue is manifesting itself not only in terms of labour shortages, but also skills shortages and the quality of tourism jobs. Tourism labour issues are most certainly impacting upon the competitiveness of many tourism destinations in a world when competition is as much upon service delivery as it is upon the endowed tourism resources of a destination, such as the natural and cultural heritage. If tourism education is to be relevant to the sector, then it is imperative that educators understand the key characteristics and requirements of the tourism labour market for their students.

The tourism labour market is characterized by two key features. First, the composition of the tourism sector is fragmented and dominated by small businesses.

Second, tourism as an activity is characterized by strong peaks and troughs in demand on a daily, weekly and seasonal basis (Riley *et al.*, 2002). This seasonality renders tourism vulnerable to changing economic conditions and to unpredictable 'shocks' to the tourism system such as natural and manmade disasters and terrorism incidents. Taking these two characteristics together, tourism employment tends to have the following characteristics:

- The tourism workforce is typified by the young, female and non-standard (part-time) workers.
- The tourism workplace is typified by challenging working conditions including below average wages, anti-social hours, poor health and safety record and isolation.
- The tourism labour market is characterized by being conservative, with strong occupational structures and traditions, low occupational mobility, high labour mobility and an unstable pay/skill relationship.

These are realities that not only contrast with the image of tourism as a glamorous profession but also result in the following linked issues that challenge the role of tourism educators and their students' perceptions:

- There are low barriers to entry for opening a tourism business, attracting workers who are unskilled in tourism. This leads to high labour turnover in, and a lack of commitment to, the sector – and mitigates against an educated workforce.
- Tourism students have portable skills and so the sector often loses labour to competing economic sectors such as IT where pay and conditions tend to be better.
- Students find it difficult to identify a clear career path in tourism.

- There is a tendency towards deskilling in the tourism and hospitality sectors due to innovations in technology, such as cooked/chilled meals.
- The sector has failed to develop the workforce for the knowledge economy and persists with archaic HR practices.
- There is a lack of commitment to training.
- There is a persistent gap between tourism educators/trainers and employers.
- Tourism is not a profession and does not have a standardized system for labour training.
- There has been a systemic failure to develop leaders for the tourism sector.

In other words, tourism educators are preparing students for a sector that struggles to accommodate them or understand their needs. The sector has been slow to recognize its human resources as an asset for investment; effectively, the points raised above act as *vocational reinforcers* to prevent the tourism sector embracing the principles of the knowledge economy (Cooper, 2006).

Communities of practice

At the end of the day, education is about delivering knowledge to a receptive audience. There is no doubt that the generation and use of new tourism knowledge for education, innovation and product development is critical for the competitiveness of both the tourism sector and destinations. In fact, researchers, consultants, the industry and government have grown tourism knowledge stocks over the last three decades – as evidenced by the ever-increasing numbers of degree programmes, discipline-specific journals, books and conferences. However, compared to other fields, tourism knowledge flows – the transfer of knowledge to the tourism sector – have been slow to develop. In fact, many commentators argue that one of the key challenges for tourism researchers is to transfer knowledge from the academic to the tourism sector. As a result of the fact that flows have been slow to develop, tourism education and research has not, until recently, been subject to a knowledge management approach and the sector is, therefore, not as competitive as it could be.

This then is a key challenge for tourism educators. The poor linkage between educators/academic research and the tourism business sector is one of the reasons that tourism has failed to recognize the importance of knowledge as a resource and embrace the knowledge economy. In many sectors of the economy, such as primary industries, knowledge transfers efficiently because the gearing between researcher and business is tight and formalized. However, in the case of tourism this is not the case and educators and researchers seldom influence the real world of tourism.

As the number of researchers and publications has grown, this has had the effect of creating a *community of practice* for tourism, with common publications and language. We can, therefore, think of at least two different *communities of practice*: (i) educators and researchers; and (ii) the tourism business sector. This may be partly to blame for the lack of gearing between the educator/researcher community and the tourism sector.

Whilst this goes some way to explaining the poor record of knowledge transfer in tourism, it can be argued that many of the prior conditions necessary for the successful transfer and adoption of knowledge are not present in tourism. This problem is related to the very nature of tourism as a sector:

- The vocational reinforcers identified above.
- Small enterprises, which are often single person or family owned. As a consequence, knowledge must be highly relevant to their operation if they are to adopt and use it.
- Fragmentation across a variety of activities.
- A lack of trust between the knowledge creators and those who might use the knowledge, due to different *communities of practice*.

Conclusion

At the beginning of this paper I questioned whether tourism education is equipping the future tourism workforce with the appropriate knowledge and skills to operate in the contemporary tourism sector. In particular, there is a concern that tourism education is living in the past and has not adapted to the needs of the 21st century. There are two key areas of concern. Firstly, since '9/11' the tourism sector has been pitched into a period of turbulence, characterized by rapid and unexpected change and there is a clear need for changes to both the content and balance of the curriculum. Secondly, the 'disconnect' between educators and the sector itself has brought about a curriculum that does not meet the needs of industry and has acted as a barrier to knowledge transfer between educators and the sector. In part, this is explained by the vocational reinforcers that have meant that tourism employers do not invest in their people and, as a result, have failed to attract the best and the brightest for the future workforce. Only by developing effective means of collaboration and communication with the tourism sector will tourism educators deliver an appropriate knowledge and skills set and will tourism itself finally enter the knowledge economy.

8.2

The Cost of Everything and the Value of Nothing

Brian Wheeller

A question for tourism academics. Do we publish our 'research' findings where we genuinely believe our work will have the greatest good, or do we opt for publication where it will most benefit our own CV and careers? Think about it. And consider the ramifications of your answer. Where do our priorities lie? With ourselves or with others? All right, it may not be that straightforward. 'Lies that life is black or white…etc'. Maybe it is more of a balancing act. But in our contemporary 'What's in it for me, now?' world I'm afraid the scales are increasingly weighing heavily in favour of self-interest (Wheeller, 2004, 2005). Attend an academic tourism conference and the ensuing networking conversations rarely, if ever, involve discussion as to who is (or is not) a good lecturer. Or who is good with tutoring the students. Rather, social discourse invariably revolves around who has published what and where. And judgements are made accordingly. Publications are paramount for peer status: crucial, too, for promotion and career trajectory.

I start with this as, unfortunately, vested interest plays a pivotal role – lying (in both senses of the word) at the heart of any (tourism) education debate. Chris Cooper's probe has not, in my opinion, given sufficient attention to this disturbing reality. It is peripheral, rather than central, in his probe which, seemingly coming at the subject from a different angle, presents a less emotive, more detached, more objective and (perversely, as a result) possibly a more unrealistic perspective.

While clearly not every relevant factor can be covered in a short probe, it would seem significant that several vital determining elements may have been overlooked or underplayed. And I suggest that those included are neither sufficiently framed in 'the personal' nor contextualized within the wider arena of (higher) education. And are those issues raised always given the individual attention that their import warrants? For example, 'publication' is mentioned, along with a number of other astute observations made by Cooper. But I am not convinced that it is given the weight it merits.

The probe, then, doesn't appear to have given sufficient importance to some of the fundamental factors determining the direction of contemporary tourism education in higher education. The result? A somewhat opaque, distorted picture of the current state of tourism education. Or, at least, the tourism education with which I am familiar. While he, in raising a series of pertinent issues, does make a number of salient points and sound judgements, nevertheless, there are possibly gaps and ambiguities in coverage and (concomitant?) questionable flaws in analysis sufficient to generate further comment and discussion here.

Many of these inconsistencies may well be 'definitional'. Cooper's piece is at times somewhat confusing on a number of levels, not least of which is the (understandable?) lack of precision as regards the parameters for discussion. Maybe I have missed it, but he doesn't specify the actual level of tourism education attainment we are discussing – further or higher, undergraduate or postgraduate? Indeed, critically, is it tourism *education* that is under review? Or is it tourism *training*? Or both? If training is incorporated into the debate then (presumably) it is conducted at a 'lower' – but, in many respects, equally if not more, important – level? So here's the rub. Are we in fact discussing tourism education as the title implies? Or tourism education and training as suggested by the first sentence?

The title misleadingly takes us in the direction of education whereas ambiguously the very first line belies this marker by immediately and prominently introducing training onto the agenda. In the immediate, it is unclear as to whether the probe regards these two, education and training, as being synonymous…which they certainly are not. Although later he does throw some (inconclusive) light on the matter, to me as a reader the distinction remains unclear.

Similarly, are we discussing Tourism? Or are we (confusingly) also embracing Hospitality? Again, though the title hints at the former, comments within the probe itself (notably with regard to placement) would have us believe the latter – leading, from a Tourism Education perspective, to possibly erroneous general conclusions. Certainly, the worthwhile opportunities the international placements afford our undergraduate tourism students at NHTV, Breda do not, I feel, resemble those referred to in the probe. Nor do those at a number of other institutions of which I am aware: the tourism placements offered as part of University of Plymouth, UK programmes being, say, a further example of good practice.

Having said that, I do fully recognize that though wary of the author's apparent vagueness as to the subject matter actually under probe, I am acutely conscious of the difficulties of achieving this 'exactness' in practice… the old maxim 'easier said than done' springs to mind here. Given the inherent complexities involved, it is this that makes debate difficult. So, little wonder that there is ambiguity and that issues remain unresolved. The response here focuses on education at the degree and postgraduate level. While the lens is clearly on tourism education (as opposed to training), attention is drawn to broader but significantly relevant contextual issues affecting the wider arena of higher education. To ignore the context within which tourism education operates, and is delivered, presents a misleading picture.

It is probably fair to say that tourism education initially centred on business/economics/management, and at most of the academic institutions where it features it has usually been taught within these departments. In the later 1980s and 1990s there was an evolutionary process where the influences of other disciplines – geography, sociology, anthropology, politics, planning, environmental studies, etc. – increased. And the pace has accelerated in the current decade during which time tourism has ever more been drawn into the 'green/climate change' debate. However, although increasing lip service is now being paid to the environmental and cultural aspects/impacts, to many in tourism their vision is still firmly set (blinkered¿) on the business agenda – revealing, fundamentally, a functional, utilitarian, 'economic' approach to the subject, while simultaneously trying to make as much (personal/professional) gain from the green sustainable veneer as they possibly can. Politicians, and those engaged in tourism marketing and development, seem particularly prone to such deceit. But those in tourism education (or should that be tourism training¿) are also susceptible. Courses/options have often been developed and introduced because 'there is a market for them' – not because it was felt there was an inherent, intrinsic societal educative 'need' for them. It is a shame, tourism education cannot be immunized against this. (Though, of course, the argument that if there is a demand/market then this sufficient 'need' is acknowledged. It's just that I do not automatically adhere to this reasoning.)

Cooper states 'Many argue that the primary role of tourism education is to deliver qualified human resources to the tourism sector'. But isn't this more the role of tourism training¿ Surely 'education' has both a far broader remit, and deeper responsibility, than merely vocational training¿ Higher education should encourage students to 'think', rather than train them to 'do'. And this includes tourism education. Personally, I have never believed that universities etc. should be institutions (directed) to train and provide students for private sector business – nor, come to think of it, necessarily for the public sector either.

However, the underlying premise of the probe is that students undertaking tourism education and training will/should end up working in tourism (and hospitality). And that, among other factors, the 'disconnect' between educators and the sector has brought about a curriculum that 'does not meet the needs of industry'. But why is this, implicitly, deemed automatically to be an unsatisfactory state of affairs – a failure? What of education for its own sake? It begs the question as to whether an independent university system's *raison d'être* is to meet the needs of industry. To meet the needs of society, maybe. But exclusively geared to serve 'industry'? I think not. Industry and society are not synonymous. Nor are their respective interests.

We should always remember that worldwide, in tourism, there are many, many menial jobs: non managerial, low paid, no prospect, no hope jobs far 'below' the aspirations of our graduates – but crucial, nevertheless, to the functioning of the tourism sector. These are the 'other' jobs. And maybe it is worth bearing in mind that when (if) these are discussed by academics it is always in the abstract, objectively, because we, our children and/or our students don't, and will never, have to do them on a permanent basis. Is this the section of the tourism labour market we are referring to when we talk of education/training fit for purpose? Of course not. No. Rather, it is exclusively the top echelons of the sector ('our' echelon) that is solely within our compass.

Cooper laments the lack of 'knowledge transfer'. He draws attention to the two very separate communities of practice – tourism educators and researchers, on the one hand, and the tourism industry itself, removed, distant, on the other. And he regrets progress on the part of tourism education as it lags behind in the knowledge transfer stakes. Admittedly, my grasp of what precisely constitutes knowledge transfer is somewhat lacking, but to me there does actually appear to have been considerable transfer of 'knowledge' between the two sectors. Unfortunately, it happens to have been in the wrong direction – from industry into education.

> Haven't the differences between education and business become blurred? Hasn't tourism education become as much a business as the tourism industry itself? Isn't higher education now driven by the same forces that power big business? And doesn't this affect our approaches to teaching and learning? The scramble for high fee-paying overseas students; strategic policies to secure research funding; 'generic' lectures and group work to reap economies of scale; two year Master's programmes concertinaed into 12 months...in the interests of education or economics? (Wheeller, 2005: 313)

Rather than the reverse requisite flow, educative waters have themselves become muddied and sullied by 'knowledge transfer', swamped by a flood of so-called business ethics.

I strongly suspect that, at least within academia, the personal career trajectory of tourism academics now drives the tourism research agenda as much as – if not more than – the actual 'needs' of tourism (that is 'tourism' and not, exclusively and restrictively, the 'tourism industry'). But what of the links with teaching? 'The accepted approach seems to be that research-led teaching is the panacea to stimulating, up-to-date teaching. Active researchers, by feeding their current work into their lectures, are the dream ticket. This, however, not only assumes that the researchers do introduce their research material into their lectures. But, fundamentally, that they do, in fact (willingly and enthusiastically), actually embrace lectures, and students, in the first place' (Wheeller, 2005: 318). However, on the contrary, the top priority for most lecturers in many institutions is to angle for as little lecturing as possible: indeed to cut contact time with students to a minimum…avoiding students to (supposedly) free up time to concentrate on 'research', be it submitting journal papers or bids for funding. It is not necessarily that they/we don't want to teach: it's more a question that they/we consider it is in one's own interests to concentrate our endeavours elsewhere – away from the students, to where personal material reward is more tangible, namely promotion through publication. And it is a material world. Despite protestations to the contrary, '…promotion is via publication and income generation...not, it must be emphasized and despite protestations from some quarters, through ability demonstrated in the lecture theatre' (Wheeller, 2005: 316).

Marginalizing students and minimizing contact time. A perverse – and I feel extremely sad – situation for any institution claiming to be educational. And this at the very time when the race to attract (entice?) high paying international tourism students is out of control. Given the clamour for overseas students (well, more accurately, their concomitant fees) we should at least have the common decency to give them the time and attention they deserve. And which they believe they are paying for – at premium rates. But here, as elsewhere in higher education, it is the income stream rather than the student that is the attraction (Hinsliff, 2009: 8). Student as commodity: income generation the name of the game.

Are students now customers? And disaffected ones, too? At a number of UK universities there have been protests by said 'customers' unhappy at being 'short changed' by cuts in teaching contact hours, growing class sizes and the increasing use of postgraduate students as teachers (Grimston, 2009). A far cry from the 'Forget all you have been taught: start to dream' of students yesteryear.

Approaches to teaching and learning in higher education are, it is suggested here, inextricably linked with the vexed demons of 'research', career trajectories, the blinkered drive for assessment, league table success and the concomitant, controversial funding implications. Tourism education is not exempt from these determinants. Or should that be 'deterrents'?

As Alderman (2009: 22) laments: 'During a working lifetime in higher education I have witnessed what I hoped never to see: the gradual and deliberate dumbing down of academic standards in the pursuit of targets and rankings. What is more, this has taken place with the enthusiastic complicity of senior managers often, indeed, at their behest.'

Lezard (2009: 29) is similarly moribund. In his article 'We've forgotten what university is for' he, too, has regrets: 'We are in the grip of a joyless, bleak utilitarianism, one whose gaze is becoming fixed not on the horizon, or the stars, but on the bottom line: the line of economy, of cost/benefit analysis, a world of bean-counters, box-ticking and assessments.' Sentiments I sadly endorse. Higher education has become an alien world of objectives and targets.

These are, I know, views that have been aired before (see Wheeller, 2005). I am aware, too, that mine is a Western, middle-aged, perspective garnered from a life mainly cocooned within the higher education sector, a career itself the result of a privileged baby-boomer, grammar school, state funded university education – possibly the very perspective, in fact, that Cooper justifiably warns us to be wary of. But I strenuously believe this threnody is worth repeating at every opportunity, on every possible platform, as valid, core, albeit unsavoury determinants (should that be detriments¿) that are often conveniently overlooked or dismissed as tangential, in most evaluations of tourism education.

Cooper broaches the notion that 'the traditional organization of tourism schools and departments should perhaps be rethought'. But by whom¿ And then questions 'How should tourism schools be organized and what should their *raison d'être* be¿'. His presumption appears to be that somehow the requisite rethinking will actually involve tourism academics – either those affected, or in the form of, say, maybe an outside 'advisory audit' with an effective, informed tourism component. This, in turn, seems to assume that tourism as a subject has political (or, indeed, academic) 'clout' in the institutional hierarchy. As if. In by far the majority of cases where tourism is taught this patently is not the case. I doubt very much that the subject carries the kudos seemingly assumed by Cooper who appears to believe tourism has sufficient muscle to dictate where it should fit into the institutional, school, departmental hierarchical system...that the subject has some definitive say in deciding its place within the fabric of faculty/school/department structure. Similarly, when introducing the role of the educator there seems to be the belief that said educator has (autonomous) control over his/her destiny...very unlikely in the current climate.

I fear here a degree of wishful thinking on the part of Cooper in terms of what might be realistically achieved. And what of the practicalities of funding¿ These, not educative ideals, determine the direction of contemporary higher education. This is unfortunate. On two counts: namely, that this situation actually exists: and that he has not, in my opinion, given sufficient attention to this disturbing reality. He mentions it, along with several other factors, but (again) does he acknowledge the full implications? Rather than examine the 'ends' to tourism education unfortunately it is, perhaps also, the 'means' that need to be addressed.

No doubt, 'the times they are a changing'. And Cooper may be correct in his assessment that some aspects of tourism education are stuck in a time-warp. But to abandon certain values in the rush for the immediate is worrying. I would, though, of course support the author's plea to move with the times. Particularly when it comes to teaching methods and content delivery. Adopting the newer teaching methods advocated by Cooper is imperative (Garner, 2009: 9). Here, in particular, I would again take the opportunity to further the cause of imagery in the teaching of tourism. In many ways, it is an increasingly visual world. And tourism, as a subject, offers abundant opportunities for the educator to embrace the visual in both content and delivery. Technology now enables the deployment of imaginative imagery in the lecture theatre. And this is to be vigorously encouraged.

And what of content? Niche markets are receiving considerable – and, in my opinion, disproportionate – attention in tourism academia. Take, most notably, eco/ego/sustainable tourism. As regards numbers, mass tourism continues to dominate globally those participating in tourism. And yet take a look back over the last 25 years or so at our academic output…whether it be in terms of articles, papers, conferences, journals, named undergraduate and postgraduate programmes, options…whatever. And where is our body of work on Mass Tourism? Then compare and contrast this with the academic resources dedicated to eco/sustainable tourism/culture/heritage/museums/the high arts, etc. I know I may well be distorting, simplifying, and/or (by default) exaggerating, certain attitudes, aspects and arguments for effect here…high versus low culture. But even so, just a cursory glance, combined with a touch of common sense, must register that these don't match up: at the very least, something is askew, out of kilter with reality, as regards priorities in respect to the content of tourism taught courses and research output. Compare the coverage given in academia to sustainable tourism with that devoted to mass (package) holidays and then think about the relative numbers involved respectively, of 'eco travellers' and mass tourists. It is generally recognized by the tourism fraternity that, within academia, our subject has suffered from being seen as 'Mickey Mouse', lightweight. But what we fail to even contemplate is that within tourism itself, the same academic 'snob' values are there to the fore. Unfortunately, mass is automatically and erroneously assumed synonymous with 'low' culture, not worthy of our esteemed attention. Far better, it would seem, to focus on the 'traveller' and our/their 'high-brow' interests rather than the tourist: we are, after all, not the mass tourist, package holiday type, are we?

I guess the point I am trying to make is the extent of 'exclusivity' in our academic approach to tourism. At least, that's how it seems to me. Laudable efforts by John Beech (University of Coventry) to establish, within the auspices of Atlas, a specialist group on Mass Tourism, is a step in the right direction and one which deserves support (but isn't getting it. True to form, dismally, only one academic turned up for the inaugural meeting). There have, of course, (absolutely) been plenty of texts on mass tourism, but comparatively speaking there has been relatively little. Or, at least, that is my contention. A disproportionate emphasis has been on more 'acceptable', more edifying and to 'us', more appealing aspects of tourism. Well, actually 'travel'. But the situation is only 'disproportionate' when one is considering the number and 'type' of tourists worldwide. It is perfectly reasonable and representative if one is taking, as the yardstick, the background, education and interests of those academics involved in the researching, reporting and publishing of the subject matter. In that context, and given the foibles of human nature, the outcome/output is hardly surprising: predictable, in fact.

Tourism education is not immune to the deep fissures fracturing the higher education sector – which is splintering and fragmenting under duress. There has been a regrettable, invidious transformation from teaching as an educative, liberal and expansive role to that of a far more limited utilitarian and functional one. Ideals in tourism teaching are being eroded by the corrosive clamour for funding, resource constraints, the restraints of assessment, promotion, demands of the 'immediate' and the intense, consuming stress exerted by the research prerogative. No real surprise here then as, to a large extent, these traits and machinations mirror and reflect the accelerating march of the 'Me, now' attitude sweeping traditional values (both good and bad) inexorably before it. Over the last few years, I have thought that more attention to, and time spent with, students would be a welcome, rewarding reversal of current trends. But with contemporary students now morphing into 'customers' and a narrow business mentality blinkering both demand and supply, I'm no longer convinced that even this shift would lift the malaise. As the twin restrictive strangleholds of assessment and league tables tighten their grip on higher education – and the confining rigours of 'cost effectiveness' bite hard – we are rapidly reaching a situation in the sector where Oscar Wilde's quip as to knowing 'the price of everything and the value of nothing' seems all too depressingly true.

8.3

Innovate or Deteriorate – Moving Tourism Education into the 21st Century

Lisa Ruhanen

Chris Cooper has embodied many of the current trends and issues that impact on a contemporary education in tourism and has highlighted the numerous fissures that exist between educators, students and the tourism industry. The issues raised are intrinsically multi-dimensional and certainly much broader than either tourism education or the education system in general, but instead are a result of broader societal, economic, political and technological changes. While there are many areas of importance raised by Cooper that could be addressed, I will explore his comments relating to the role of the educator and, in particular, assertions that 'tourism education is living in the past and has not adapted to the needs of the 21st century'.

Chris Cooper is right in advocating the need for innovativeness in tourism education and notes that 'for too long tourism educators have been complacent and failed to innovate in their education programmes'. Arguably, the basis of tourism education has changed only a little in the 40 or so years since it first entered the higher education sector in countries such as the United Kingdom, United States and Australia. For instance, Airey and Johnson (1999) chart the core body of knowledge in tourism education through the 1970s, 1980s and 1990s. Their review highlights that there has been considerable continuity and stability in the curricula. On the one hand, this may be taken as evidence that the underpinnings of a tertiary education in tourism are 'tried and tested' and that the core competencies, skills and knowledge required for graduates entering tourism-related industries have remained constant over time. In this vein, tourism educators must be doing something right as both students continue to enter the higher education system to study tourism and the industry continues to employ students on graduation. Under this scenario the adage of 'if it ain't broke, don't fix it' prevails. Certainly this favourable meeting of demand and supply can be conceived as justifying the validity and relevancy of the tourism curriculum and programmes. As such, there is no external pressure to fundamentally redress the curriculum.

While the current situation may be tenable, it is arguably a factor contributing to the complacency referred to by Cooper – complacency on part of both the educator and the educational institution. The uptake of graduates into employment is one of the key 'quality' or 'success' measures of education programmes and as long as there is some equilibrium in the demand for student places and the supply of graduates for the sector, there is little impetus for educators and institutions to implement change. Yet there have been anecdotal reports of softening demand for tourism-related programmes and in some countries the subject area has reportedly reached a stage of maturity (Airey, 2008; Spennemann & Black, 2008). The desire to remain competitive and retain market share in the often competitive and lucrative higher education sector may provide the impetus for institutions to demand a re-examination of their tourism education programmes.

An arguably more important driver for change in tourism education should be that the contemporary tourism student is faced with a much different world than their counterparts in the 1970s. In the last decade alone, world events and their impact on the tourism sector has necessitated the incorporation of new concepts into tourism curricula. Crisis/risk management is an obvious example here. Further, the advent and theoretical acceptance of the sustainable development concept which arose in the 1980s has also seen the introduction of new content centred on the triple-bottom line, ecotourism and responsible and ethical behaviour, among others. Similarly, the prevalence of computerized and online systems in the hotel and travel sectors, in particular, has necessitated changes in programme content. However, one of the most significant changes has been the internationalization of the curriculum (Barron, 2002; Daruwalla & Malfroy, 2002; Hobson, 2008; Jordan, 2008) in response to changing student cohorts, namely, the estimated 2.5 million students globally that are studying outside their home country (Altbach *et al.*, 2009).

Broader socio-economic shifts in the contemporary student cohort should also be driving change, yet there is less evidence of this to date. For instance, changing lifestyles and the need for students to balance study, work and family commitments have impacted on both the composition of the student cohort and the methods in which students engage themselves with their learning and the wider institution (Spennemann & Black, 2008). The increasingly diverse student body creates pressure to put in place new systems for academic support and innovative approaches to pedagogy (Altbach *et al.*, 2009). In discussing the Australian situation, McInnis (2003) claims that there is a need for a shift in the way higher education institutions interpret and respond to the changing needs and expectations of students. This needs to occur in recognition of students' socio-economic circumstances, which may dictate their physical presence on campus vis-à-vis the time spent in paid work, as well as the current culture of 'choice', where today's students expect choice in their learning environment.

Indeed, mammoth changes will be required across the education sector to respond to the much feared 'Generation Y' (those born between 1978 and 1994). With their attention spans of mere minutes these students will expect their teachers to be as technologically savvy as they are and are highly unlikely to be satisfied with traditional (and still prevalent) 'talk and chalk' lectures (albeit the chalk aspect replaced by PowerPoint). This cohort is commonly characterized as having low thresholds for boredom, an unwillingness to memorize information, engages in multi-tasking and prefers entertainment in their education (Manuel, 2002). Chris Cooper questions 'how can educators reach out to the contemporary student¿' This is certainly a valid question as educators will have to adapt to this cohort's needs and find ways of bridging the gap. As the research shows 'Generation Y' are more vocal about being bored, more willing to speak out against 'pointless' memorization and are also more likely to hold instructors accountable for making learning boring or interesting to them.

Chris Cooper's calls for innovation in tourism education to adapt to the needs of the 21st century echo other appeals for change. For instance, the Tourism Education Futures Initiative (TEFI), which was cited by the author, was designed to respond to the need for tourism education systems to engage in 'radical change in order to meet the challenges of the next few decades' (TEFI, 2009: 4). Emphasizing radical, as opposed to incremental, change in tourism education the architects of the TEFI initiative claim:

> Tourism educational programmes need to fundamentally re-tool and re-design – not incrementally by adding new courses – or simply by putting courses on-line – but by changing the nature of what is taught and how it is taught. Skills and knowledge sets must be redefined, structures and assumptions need to be questioned, and old ways of doing things must be transcended. (Sheldon *et al.,* 2007: 63)

As Airey and Johnson's (1999) paper clearly demonstrates, incremental change has been the norm in tourism education to date. Certainly, in the past four decades there is no evidence of the 'radical' change advocated through TEFI. Programmes have shifted slightly to incorporate new courses on 'current' trends and issues such as sustainable development, crisis management and the like, but despite the 'tsunamis of change' (Sheldon, 2009) facing tourism, there is little evidence of fundamental shifts in the underpinning ideology of a tourism programme. For instance, at the summation of the 2007 TEFI summit, Fesenmaier (2007) concluded that 'unfortunately...many of the participants really were not ready to consider the future(s) in a meaningful way'. However, this global tourism education initiative, which has the support of many of the leading academics and educators in the field, has considerable potential to move tourism education into the 21st century.

While there may be deficits at the macro level, there is much more evidence of innovation in the actual teaching and learning process. Given that it wasn't too long ago that using PowerPoint in lectures was considered cutting-edge in the delivery of course content, it is a positive step forward that we are now seeing tourism educators initiate innovative pedagogical approaches into their courses. As a result, the tourism education literature is expanding as educators report on the introduction of innovative teaching and learning approaches such as flexible delivery (Barron & Whitford, 2004; Jordan, 2008; Morgan, 2003; Moscardo, 2008; Roberts *et al.,* 2006), new approaches to assessment (Armstrong, 2004; Slaughter & Ruhanen, 2007) and experiential learning approaches designed to facilitate deeper learning and engage students in the learning process (Beard *et al.,* 2007; Concei & Skibba, 2007; Hawkins & Weiss, 2004; Hobson & Jenkins, 2009; Lee, 2007; Mengmei & Downing, 2006; Ruhanen, 2005; Wolfe, 2006; Zapalska *et al.,* 2003). Importantly, many educators are taking advantage of the opportunities offered by internet technologies and other interactive online learning environments and are developing innovative learning tools to engage with students (Bailey & Morais, 2004; Collins, 2004; Dale, 2007; Dale & Lane, 2007; Douglas *et al.,* 2007; Edelheim, 2007; Gillespie & Baum, 2002; Hill & Schulman, 2001; Hofstetter, 2004; Lowry & Flohr, 2004; Murphy *et al.,* 2009; Newman & Brownell, 2008; O'Halloran & Deale, 2006; Penfold, 2008; Schott & Sutherland, 2008; Tesone, 2004; Varini *et al.,* 2004; Williams & McKercher, 2001).

Initiatives such as these are a promising start in moving tourism education into the 21st century and do show that there is at least piecemeal innovation taking place. Arguably, the introduction of new teaching and learning methods into individual courses is the main (indeed sometimes only) means by which educators can bring about any type of change in their programmes. However, why is it that even at this level innovation is so *ad hoc*? One answer is that in the face of growing pressure on academics regarding research assessment exercises, such as that discussed earlier in the chapter by Wheeller, innovation in teaching is generally only encouraged to the extent that it does not detract from the 'important' institutional benchmarks such as research output. It has been suggested that the prominence placed on research has been to the detriment of teaching quality (Gale *et al.,* 2005; Graham, 2008; Jenkins, 1995), in addition to academic job satisfaction (Redden, 2008) and narrow interpretations of academic performance (Darbyshire, 2008).

Undoubtedly, the tourism educator faces many challenges and arguably many of these are a result of broader institutional factors and pressures. Yet, to remain static because 'that's the way it's always been done' will not move tourism education into the 21st century and will certainly not meet the needs and expectations of changing student cohorts, nor will it allow us to equip our students with the skills and abilities needed when they move into the workforce. At the core of innovation theory is the change-agent and fortunately there is growing evidence of change agents in tourism education. Initiatives like TEFI are exciting, particularly if they can be systemic in leading change. Further, the literature is growing with accounts of innovative approaches to tourism education. Yet, despite these notable objectives, there is much work still left to do to move tourism education forward and into the 21st century.

Concluding Remarks

At the beginning of this chapter, Cooper questioned whether tourism education is equipping the future tourism workforce with the appropriate knowledge and skills to operate in the contemporary tourism sector. This challenge was then followed by two rejoinders each with a particular theme. Ruhanen's theme was the 'disconnect' between educators and the sector itself which has brought about a curriculum that does not meet the needs of industry and has acted as a barrier to knowledge transfer between educators and the sector. Only by developing effective means of collaboration and communication with the tourism sector will tourism educators deliver an appropriate knowledge and skills set and will tourism itself finally enter the knowledge economy. Wheeller's theme took this further by stating that broader institutional factors and pressures heavily influence the tourism educator. This has led to a regrettable, invidious transformation from teaching as an educative, liberal and expansive role to that of a far more limited utilitarian and functional one. He goes on to say that ideals in tourism teaching are being eroded by the corrosive clamour for funding, resource constraints, the restraints of assessment, promotion, demands of the 'immediate' and the intense, consuming stress exerted by the research prerogative. In addition, the twin restrictive strangleholds of assessment and league tables are tightening their grip on higher education leaving educators with little room to manoeuvre in terms of innovation. Yet, as the Tourism Education Futures Initiative shows, solutions to these dilemmas demand innovation and creative thinking, breaking out of the traditional mould of tourism education. Only by doing this will tourism education serve the needs and expectations of changing student cohorts, equipping them with the skills and abilities needed when they move into the workforce.

Discussion Questions

(1) Is tourism education out of touch with the tourism industry?
(2) How different are the skills and knowledge needed to work in the tourism sector from other economic sectors?

(3) As two distinct 'communities of practice', how can tourism educators and the sector communicate their needs to each other?
(4) How much do the pressures of research, league tables and accreditations detract from the tourism educator's ability to be an excellent 'teacher'?
(5) What do you think will be on the tourism curriculum in 2020?
(6) Is there a difference between tourism education and hospitality education?
(7) Can you characterize the difference between tourism education and tourism training?

References

Airey, D. (2008). In search of a mature subject? *Journal of Hospitality, Leisure, Sport and Tourism Education* 7(2): 101–103.

Airey, D. and Johnson, S. (1999). The content of tourism degree courses in the UK. *Tourism Management* 20: 229–235.

Airey, D. and Tribe, J. (eds) (2005). *An International Handbook of Tourism Education.* Oxford: Elsevier.

Alderman, G. (2009). A blind eye has been turned to teaching. *Times*, 4 August: 22.

Altbach, P.G., Reisberg, L. and Rumbley, L.E. (2009). *Trends in Global Higher Education: Tracking an Academic Revolution: A Report Prepared for the UNESCO 2009 World Conference on Higher Education*. Paris: United Nations Educational, Scientific and Cultural Organization.

Armstrong, E.K. (2004). Analysis of travel writing enhances student learning in a theoretical tourism subject. *Journal of Teaching in Travel and Tourism* 4(2): 45–60.

Bailey, K.D. and Morais, D.B. (2004). Exploring the use of blended learning in tourism education. *Journal of Teaching in Travel and Tourism* 4(4): 23–36.

Barron, P. (2002). Issues surrounding Asian students hospitality management in Australia: A literature review regarding the paradox of the Asian learner. *Journal of Teaching in Travel and Tourism* 2(3&4): 23–45.

Barron, P. and Whitford, M. (2004). An evaluation of event management education: Student reflections on flexible learning processes and procedures. *Journal of Teaching in Travel and Tourism* 4(2): 19–44.

Barrows, C.W. and Bosselman, R.H. (eds) (1999). *Hospitality Management Education*. New York and London: The Haworth Hospitality Press.

Beard, C., Wilson, J.P. and McCarter, R. (2007). Towards a theory of e-learning: Experiential e-learning. *Journal of Hospitality, Leisure, Sport and Tourism Education* 6(2): 3–15.

Coles, T., Hall, C.M. and Duval, D.T. (2006). Tourism and post-disciplinary enquiry. *Current Issues in Tourism* 9(4&5): 293–319.

Collins, G.R. (2004). Augmenting web-based instruction with satellite video: A case study. *Journal of Teaching in Travel and Tourism* 4(1): 57–74.

Concei, S.C.O. and Skibba, K.A. (2007). Experiential learning activities for leisure and enrichment travel education: A situative perspective. *Journal of Teaching in Travel and Tourism* 7(4): 17–35.

Cooper, C. (2006). Knowledge management and tourism. *Annals of Tourism Research* 33(1): 47–64.

Dale, C. (2007). Strategies for using podcasting to support student learning. *Journal of Hospitality, Leisure, Sport and Tourism Education* 6(1): 49–57.

Dale, C. and Lane, A. (2007). A wolf in sheep's clothing? An analysis of student engagement with virtual learning environments. *Journal of Hospitality, Leisure, Sport and Tourism Education* 6(2): 100–108.

Darbyshire, P. (2008). 'Never mind the quality, feel the width': The nonsense of 'quality', 'excellence', and 'audit' in education, health and research. *Collegian: Journal of the Royal College of Nursing Australia* 15(1): 35–41.

Daruwalla, P. and Malfroy, J. (2002). Diversity, doctrine and discourse: Internationalising a postgraduate hospitality management subject. *Journal of Teaching in Travel and Tourism* 2(3&4): 61–79.

Douglas, A., Miller, B., Kwansa, F. and Cummings, P. (2007). Students' perceptions of the usefulness of a virtual simulation in post-secondary hospitality education. *Journal of Teaching in Travel and Tourism* 7(3): 1–19.

Edelheim, J. (2007). Effective use of simulations in hospitality management education – A case study. *Journal of Hospitality, Leisure, Sport and Tourism Education* 6(1): 18–28.

Fesenmaier, D. (2007). Tourism education futures initiative: Summary received by Daniel Fesenmaier, Temple University. Online at: www.tourismeducation.at. Accessed 21 August 2009.

Gale, T., Gilbert, R., Seddon, T. and Wright, J. (2005). 'The RQF and educational research in Australia: Implications and responses across the research community'. Paper presented at the Annual Conference of the Australian Association for Research in Education, Sydney.

Garner, R. (2009). 'Facebook generation' of teachers must be promoted. *Independent*, 23 September: 9.

Gillespie, C.H. and Baum, T. (2002). Developing a CD-ROM as a teaching and learning tool in food and beverage management: A case study in hospitality education. *Journal of Teaching in Travel and Tourism* 2(1): 41–61.

Graham, L.J. (2008). Rank and file: Assessing research quality in Australia. *Educational Philosophy and Theory* 40(7): 811–815.

Grimston, J. (2009). Academics 'afraid' to criticise standards. *Sunday Times*, 31 May: 7.

Hawkins, D.E. and Weiss, B.L. (2004). Experiential education in graduate tourism studies: An international consulting practicum. *Journal of Teaching in Travel and Tourism* 4(3): 1–29.

Hill, J. and Schulman, S.A. (2001). The virtual enterprise: Using internet based technology to create a new educational paradigm for the tourism and hospitality industry. *Journal of Teaching in Travel and Tourism* 1(2&3): 153–167.

Hinsliff, G. (2009). How do we tell the good universities from the bad? *Observer*, 2 August: 8.

Hobson, J.S.P. (2008). Internationalisation of tourism and hospitality education. *Journal of Hospitality and Tourism Education* 20(1): 4–5.

Hobson, P. and Jenkins, J.M. (2009). Experimental education in tourism and hospitality. *Journal of Hospitality and Tourism Education* 21(1): 4–5.

Hofstetter, F.T. (2004). The future's future: Implications of emerging technology for hospitality and tourism education program planning. *Journal of Teaching in Travel and Tourism* 4(1): 99–113.

Holloway, C. (1995). *Towards a Core Curriculum for Tourism: A Discussion Paper – Guidelines 1*. London: NLG.

Hsu, C.H.C. (ed) (2005). *Global Tourism Higher Education: Past, Present and Future*. New York and London: The Haworth Hospitality Press.

Jafari, J. and Ritchie, B.J.R. (1981). Toward a framework for tourism education problems and prospects. *Annals of Tourism Research* 8(1): 13–34.

Jenkins, A. (1995). The research assessment exercise, funding and teaching quality. *Quality Assurance in Education* 3(2): 4–12.

Jordan, F. (2008). Internationalisation in hospitality, leisure, sport and tourism higher education: A call for further reflexivity in curriculum development. *Journal of Hospitality, Leisure, Sport and Tourism Education* 7(1): 99–103.

Lee, S.A. (2007). Increasing student learning: A comparison of students' perceptions of learning in the classroom environment and their industry-based experiential learning assignments. *Journal of Teaching in Travel and Tourism* 7(4): 37–54.

Lezard, N. (2009). We've forgotten what university is for. *Independent*, 7 August: 29.

Lowry, L.L. and Flohr, J.K. (2004). Technology and change: A longitudinal case study of students' perceptions of and receptiveness to technology enhanced teaching and learning. *Journal of Teaching in Travel and Tourism* 4(1): 15–39.

Manuel, K. (2002). Teaching information literacy to Generation Y. *Journal of Library Administration* 36(1): 195–217.

McInnis, C. (2003). New realities of the student experience: How should universities respond? *UniNews* 12(9): 1–4.

Mengmei, C. and Downing, L. (2006). Using simulations to enhance students' learning in management accounting. *Journal of Hospitality and Tourism Education* 18(4): 27–32.

Morgan, D.J. (2003). Implementing flexible learning practices in tourism courses. *Journal of Teaching in Travel and Tourism* 3(1): 47–63.

Moscardo, G. (2008). Meeting your own deadlines: Introducing flexibility into student assessment. *Journal of Teaching in Travel and Tourism* 8(2&3): 119–138.

Murphy, J., Hudson, K., Lee, H. and Neale, L. (2009). The Google online marketing challenge: Hands on teaching and learning. *Journal of Hospitality and Tourism Education* 21(1): 44–47.

Newman, A. and Brownell, J. (2008). Applying communication technology: Introducing email and instant messaging in the hospitality curriculum. *Journal of Hospitality, Leisure, Sport and Tourism Education* 7(2): 71–76.

O'Halloran, R.M. and Deale, C.S. (2006). The scholarship of teaching and learning: Supporting teaching excellence through technology. *Journal of Hospitality and Tourism Education* 18(3): 4.

Penfold, P. (2008). Learning through the world of second life – A hospitality and tourism experience. *Journal of Teaching in Travel and Tourism* 8(2&3): 139–160.

Pink, D.H. (2005). Revenge of the right brain. *Wired* 2: 70–72

Quality Assurance Agency for Higher Education (QAA) (2008). *Hospitality, Leisure, Sport and Tourism*. London: QAA.

Redden, G. (2008). Publish and flourish, or perish: RAE, ERA, RQF, and other acronyms for infinite human resourcefulness. *Media and Culture Journal* 11(4): 1–13.

Riley, M., Ladkin, A. and Szivas, E. (2002). *Tourism Employment*. Clevedon: Channel View.

Roberts, E., Williamson, D. and Neill, L. (2006). Utilising flexible learning packages to enhance teaching effectiveness: A New Zealand case study. *Journal of Hospitality and Tourism Education* 18(3): 76–83.

Ruhanen, L. (2005). Bridging the divide between theory and practice: Experiential learning approaches for tourism and hospitality management education. *Journal of Teaching in Travel and Tourism* 5(4): 33–51.

Schott, C. and Sutherland, K.A. (2008). Engaging tourism students through multimedia teaching and active learning. *Journal of Teaching in Travel and Tourism* 8(4): 351–371.

Sheldon, P. (2009). 'Values as pillars of tourism education: Cross-cultural perspectives'. Paper presented at the BEST Education Network: Think Tank IX 'The Importance of Values in Sustainable Tourism'.

Sheldon, P., Fesenmaier, D., Woeber, K., Cooper, C. and Antonioli, M. (2007). Tourism education futures 2010–2030: Building the capacity to lead. *Journal of Teaching in Travel and Tourism* 7(3): 61–68.

Slaughter, L.J. and Ruhanen, L. (2007). Using action learning to improve assessment: A case study from an undergraduate tourism management course. *Journal of Teaching in Travel and Tourism* 7(2): 1–19.

Solnet, D., Robinson, R. and Cooper, C. (2007). An industry partnerships approach to tourism education. *Journal of Hospitality Leisure and Tourism Education* 6(1): 66–70.

Spennemann, D.H.R. and Black, R. (2008). Chasing the 'fat' – Chasing a 'fad'? The waxing and waning of tourism and tourism-related programmes in Australian higher education. *Journal of Hospitality, Leisure, Sport and Tourism Education* 7(1): 55–69.

Starkey, K. and Tiratsoo, N. (2007). *The Business School and the Bottom Line.* Cambridge: Cambridge University Press.

Tesone, D.V. (2004). Online learning communication flows: An early adopter perspective. *Journal of Teaching in Travel and Tourism* 4(1): 1–13.

Tourism Education Futures Initiatives (TEFI) (2009). White paper – A values based framework for tourism education: Building the capacity to lead. Online at: www.tourismeducation.at. Accessed 21 August 2009.

Tribe, J. (1997). The indiscipline of tourism. *Annals of Tourism Research* 24(3): 638–657.

Tribe, J. (2002). The philosophic practitioner: The curriculum for tourism stewardship. *Annals of Tourism Research* 29(2): 338–357.

Varini, K., Maggie, C. and Holleran, A. (2004). Using subject matter experts to promote e-learning at EHL: An action learning approach. *Journal of Teaching in Travel and Tourism* 4(1): 85–98.

Wheeller, B. (2004). The truth? The hole truth. Everything but the truth. Tourism and knowledge: A septic sceptic's perspective. *Current Issues in Tourism* 7(6): 467–477.

Wheeller, B. (2005). Issues in teaching and learning. In D. Airey and J. Tribe (eds) *International Handbook of Tourism Education* (pp. 309–318). Oxford: Elsevier Science.

Williams, G. and McKercher, B. (2001). Tourism education and the internet: Benefits, challenges and opportunities. *Journal of Teaching in Travel and Tourism* 1(2&3): 1–15.

Wolfe, K. (2006). Active learning. *Journal of Teaching in Travel and Tourism* 6(1): 77–82.

Zapalska, A.M., Rudd, D. and Flanegin, F. (2003). An educational game for tourism and hospitality education. *Journal of Teaching in Travel and Tourism* 3(3): 19–36.

Further Reading

Airey, D. and Tribe, J. (eds) (2005). *An International Handbook of Tourism Education.* Oxford: Elsevier.

Cooper, C., Westlake, J. and Shepherd, R. (1993). *Tourism Educator's Workbook* Madrid: UNWTO.

Hsu, C.H.C. (ed) (2005). *Global Tourism Higher Education: Past, Present and Future*. New York and London: The Haworth Hospitality Press.

Starkey, K. and Tiratsoo, N. (2007). *The Business School and the Bottom Line.* Cambridge: Cambridge University Press.

Tourism Education Futures Initiatives (TEFI) (2009). White paper – A values based framework for tourism education: Building the capacity to lead. Online at: www.tourismeducation.at. Accessed 21 August 2009.

Chapter 9

Post-Colonialism: Academic Responsibility?

Derek Hall and Frances Brown, C. Michael Hall and Erik Cohen

Context

This chapter explores the proposition that the tourism academy needs to 'return' to an educational morality based upon ethical responsibility, and to move away from the 'entrepreneurial ethics' of the marketplace. It critically assesses the position of tourism academics in relation to Hendry's notion of a bi-moral society – of apparently waning altruistic traditional morality contrasted with competitive entrepreneurial self-interest (see also the Tourism and Welfare chapter in this book) – and poses the question, who is tourism education and training ultimately for?

The nature and roles of 'Northern' consultants working in less developed societies – and the concept of boomerang aid, where finance flows from donor to recipient and back again to the donor via consultants – are drawn into comparison with the way in which 'Northern' higher education institutions have come to value, if not actually rely upon, the income derived from overseas, often 'Third World' country, students. The (English language) tourism academy tends to be focused around a relatively few highly active, white, 'Northern', and mostly male, individuals. Arguably related to this, tourism endeavour in higher education is characterized as being split between the 'intellectual' pursuit of reflective thinking, conceptual depth and critical analysis and the business-led practicalities of profit-driven tourism entrepreneurship. That tourism departments or sections may lie within a social sciences school/faculty or within a business school (and sometimes spread over both within the same institution) articulates this internal tension.

Tourism, as an emerging cluster of professions, does not have a widely accepted professional code of ethics or powerful international organs able to enforce it (Cohen). It is therefore important for us to understand positionality and the location of individual academics, as well as academic institutions, within broader webs of power, values and interests (C.M. Hall).

So should we be challenging the structures, discourses and institutional arrangements of the contemporary academy?

9.1

'Post Colonialism', Responsibility and Tourism Academics: Where's the Connection?

Derek Hall and Frances Brown

Introduction

This short piece was being drafted as the summit on climate change in Copenhagen was ending in recrimination, and in the wake of a long (and perhaps continuing) 'global' economic recession. The two, of course, were not unrelated. They expressed just how far the explicit manifestation of self-interest – of particular elites and sectors in particular societies – continued to be able to dominate the course of world events, perhaps with even more vigour than during the height of colonial imperialism in the 18th and 19th centuries.

In discussing the likely implications of the global economic downturn for sustainable tourism, Bramwell and Lane (2009: 1) highlighted the danger that increasing numbers of tourism businesses and private and public sector organizations were claiming (and now had an 'excuse') that they lacked the resources either to invest in those prerequisites (infrastructure, facilities and training) necessary for providing more sustainable products, or to be able to pay for environmental or social enhancements. 'Instead, they may focus on the short term considerations of reducing costs and securing immediate financial returns. Such responses could have highly damaging consequences for sustainability'. They, thus, raise questions concerning the ethics of tourism development activity and the morality of business self-interest in general.

The role and position of ethics in tourism and tourism education has been much debated and evaluated in recent years, albeit at a relatively superficial level (Fennell, 2006, 2008). Hendry (2004: 252–253) has contended that we live in a 'bi-moral' society, in which people order their lives in relation to two contrasting sets of principles. The first of these is associated with traditional morality. While this set of principles allows a modicum of self-interest, it emphasizes duties and responsibilities, honesty and respect towards others, and ultimately puts others' needs above those of the self.

The second set of principles is associated with entrepreneurial self-interest: it also demands obligations, but to a much more limited extent. This set of principles tends to emphasize competition rather than cooperation, in order to advance self-interest rather than to meet the needs of others. Hendry has suggested that, until recently, behaviour according to the morality of the market was accepted as necessary for economic growth and development, but was also seen as a potential threat to traditional moral order and was permitted only in carefully circumscribed circumstances that were subject to regulatory safeguards.

Although both sets of principles have always been present in society, in recent years (especially in 'Western' society), traditional moral authorities – organized religion, state, family, community, education – have rapidly lost much of their force. At least until the 2008 'credit crunch' the morality of self-interest had acquired a much greater social legitimacy, and over a much wider field of behaviour than ever before (Hendry, 2004: 253).

Fennell (2008: 224) suggests that an understanding of self-interest is a 'first order need' for our consideration of ethics in the practice of tourism. And this is our jumping-off point for raising questions about the morality of the tourism academy. Our polemic is a self-consciously ethnocentric critique of ethnocentrism, the apparent hypocrisy of which usefully complements our conclusions. From a largely UK perspective, it focuses on apparent self-interest – at national, institutional and individual levels – among tourism academics, particularly in respect of the roles of consultant and educator.

We Pay Them to Pay Us and Boast About It

A decade ago a seminal paper was published on the processes and outcomes whereby tourism consultants facilitated transfers of know-how and capacity building within foreign aid and assistance programmes to support economic, social, civil and political restructuring (Simpson & Roberts, 2000). The sustainability of consultancy inputs from advanced capitalist countries to traditional recipients of aid was being increasingly questioned, and this paper extended the debate to embrace issues associated with post-socialist Central and Eastern Europe. The paper focused on an exploration of the interactions between development theory and the practical roles of consultants at project implementation stage, and evaluated the different approaches of consultants and how these shaped relationships, processes and outcomes of such projects.

That the paper has not been as widely read as it should have been is probably mainly the result of two factors. First, its explicit focus was on practice in Central and Eastern Europe, and although its analysis and conclusions held much wider significance and application, it may not, therefore, have come to the attention of development practitioners in the 'Third World'. Second, although plans were afoot for republishing in book form the journal issue in which the paper appeared, one of the editors of the book series, himself a well-known tourism consultant, was minded not to include this paper, allegedly because it was too critical of the consultancy process.

At its best, the consultancy process can bring, transfer and disseminate much needed expertise and experience to a situation where those involved on all sides may otherwise lack both commodities. It can be a sensitive, problem-solving, empowering, facilitating, enriching and enlightening exercise for those at the point of delivery. At its worst, consultants can usurp the work of local researchers, patronize local decision-makers, override local people and impose inappropriate formulaic 'solutions' derived from experience in other cultures and environmental circumstances. Consultants may parachute in and helicopter out with insensitivity to and ignorance of local cultures, while depriving local people of a fair share of the assistance 'aid' earmarked for them by pocketing huge consultancy fees. The current authors have intimate experience of both extremes (and of several variations between them).

Evidence suggests that the consultancy process within aid and assistance programmes, while seemingly altruistic, may represent some of the worst examples of both individual and collective self-interest expressed in colonial, dependency terms.

The international NGO CharityAid claimed in 2006 that 'for the West as a whole' at least 25% of donor foreign aid budgets, some US$19bn in 2004, was spent on consultants or on research and training. To capture headlines, the NGO suggested that more than £100m of the UK's £5bn aid budget was spent on privately educating the offspring of the consultants who, it claimed, typically earned £100,000 a year (Anon, 2006; Greenhill, 2006; see also Watt & Greenhill, 2005).

A well publicized recent (April 2009) example saw Australian Prime Minister Kevin Rudd declare, in a joint press conference with his counterpart in Papua New Guinea (PNG), Sir Michael Somare, following their discussions on historical problems with aid delivery, that 'Too much has been consumed by consultants and not enough delivered to essential assistance in teaching, in infrastructure, in health services on the ground in villages across Papua New Guinea' (Anon, 2009). One PNG provincial governor depicted Australian aid as 'boomerang aid', 'because a lot of the money goes to the consultants and many of those consultants are Australians themselves'. One commentator from the think tank Institute of National Affairs suggested that 'Maybe we've been breeding more advisers to advise on the advice of previous advisers….' (Fox, 2009).

Online debates about the nature and role of consultants in such 'boomerang aid' are numerous in 'recipient' countries (e.g. Aziz, 2009; Dahal, 2009; for an extreme example see Cockburn, 2009). Ethical critiques of the consultancy process by consultants themselves have been less notable, or at least less accessible, but Janine Wedels' (1999a, 1999b) critique of US policy in the Balkans and the work of the Stockholm School of Economics, European Institute of Japanese Studies (e.g. Luvsanjamts & Söderberg, 2005) are exceptions.

There have, of course, been numerous critiques of consultants from within academia in donor countries. Nevertheless, just as Western consultants may often have their own rather than consultees' interests uppermost, can we discern a parallel self-interest in academics?

Who and What are we Teaching and Why?

With the rapid growth of higher education in many societies, for reasons of broadening skills bases, raising levels of education and thus adding value to social and cultural capital, the moral standpoint of universities in some 'developed' countries appears to have shifted in the past couple of decades. Given that measures of (un)employment are often (simplistically) cited as an indicator of the success or otherwise of government (economic and social) policy, some administrations appear to have moved from regarding the function of higher education as being necessary for encouraging a more enlightened approach to life, for sustaining professional leadership and for thinking about issues and problems in an analytical way, to being a mechanistic process for facilitating skills acquisition for employment purposes and for attracting foreign income. Universities have been drawn inescapably into a social and ideological environment whereby their moral foundation has been undermined. The necessary skills of original thinking, questioning and being equipped to offer well argued alternative answers to myriad issues seem to have been supplanted by notions of employability and fitness for joining a compliant workforce, under whatever kind of political regime.

Complementing this – if far from complimentary – has been a burgeoning of the number of (UK) universities, almost unattainable government targets for higher education participation rates and a consequent squeezing of many individual universities' budgets. Bottom lines, triple or otherwise, have dominated mission statements and the lure of (long-haul) overseas students' fees has been irresistible. At the time of writing, the list of UK organizations licensed to sponsor migrants within the 'Tier 4 students' category ran to some 119 online pages (UK Border Agency, 2009). Indeed, on the very day this piece was being rounded off, the UK government announced a drastic reduction in the higher education budget for 2010/11, indicating that there should be an emphasis on shorter, two-year and flexible degree programmes, with employability a high priority objective (e.g. Curtis, 2009). This only acted to reinforce the fact that UK universities have increasingly come to rely on the fee income of non-EU international students.

Unfortunately, a number of universities that have been put onto the global street to attract new clients have been young (some reportedly under age) and inexperienced – often addicted to their new-found status after perhaps having languished for so long in a back alley as a local college – and have been found unable to perform adequately for their customers.

In what has been referred to as a 'quasi-colonial' process of recruiting high fee-paying students from 'developing' countries (Tysome, 2006), in 2008 there were some 230,000 non-EU international students enrolled in the UK (a 96% increase in 10 years), generating £2.5bn per annum in fees, or more than 8% of universities' total income. Reflecting the volatile competitive market that is international higher education, in 2005 applications from potential students in China (hitherto a rapidly developed major market) to the UK began to drop. Increasing strength of domestic capacity in the People's Republic, including the setting up of foreign (not least UK) university campuses and teaching within China, a shift from placing a large premium on having a foreign degree to having local work experience, the Chinese media publishing scare stories of decadent student life in the West, and increasing obstacles in accessing (UK) visas all contributed to this trend (Anon, 2005).

A number of universities subsequently, and somewhat belatedly, improved overseas student support mechanisms and increased their assistance for visa application processes, including employing anti-forgery devices in letters of acceptance, to raise credibility with suspicious consular officials. Many now provide excellent support and high quality teaching and learning environments. With a depreciation of the pound sterling forecast at 20% against the US dollar for 2009 and 13% against the Indian rupee, the British Council was looking to a potential doubling in the number of Indian students enrolled in the UK, from 30,000 in 2010 to 60,000 by 2015 (Lipsett, 2008).

But in relation to Hendry's (2004) bi-moral principles, is it too simplistic to suggest that universities have shifted from the first – predominantly altruistic – to the second of these principles in their self-interested policy of giving high priority to high-fee international student recruitment? Those countries with arguably relatively high ethical standards, notably Finland and Sweden, have long maintained free higher education for all at the point of delivery, although even this is now set to change for non-EU student applicants (Helsingin Sanomat, 2008; Parafianowicz, 2009).

The What

The practice of tourism teaching, supervision, research and publication implicitly embraces responsibilities associated with nurturing next generations – of leaders, stewards, practitioners, thinkers, drop-outs – both out in the wider world and within academia.

Debates in the literature on the nature, purpose and application of tourism education content have existed for as long as teaching in this area has been recognized (e.g. Collins, 2002). Tensions between the conceptual and the practical, disciplinary and multidisciplinary, have characterized such discussion. Goeldner and Ritchie (2008), for example, suggest that more than 20 constituent disciplines inform the collective knowledge of tourism.

For Coles *et al.* (2006: 293) 'the study of tourism sits curiously within the contrived academic division of labour'. Such a division underlies a basic tension between the emergence of a 'more flexible, problem-focused outlook and what we might term conservative "old problem, old outlook" discipline-bound straightjackets of the political institutions that now evaluate and rank research performance across numerous national jurisdictions.... Such issues lie at the heart of debate over the relevance of universities and the knowledges that they generate' (p. 313).

These debates have also embraced the nature of delivery and learning style preferences. In Australia and the UK, research has shown that a majority of sampled students had strong learning style preferences that presented some challenges to educators and the planning of learning experiences in higher education (Lashley & Barron, 2006). Students typically preferred learning styles that were concrete rather than abstract, and active rather than reflective. This was seen to present substantial teaching and learning barriers for educators wanting to develop reflective graduates and practitioners. However, Lashley and Barron (2006) also found that students from Confucian heritage backgrounds displayed learning style preferences which were not consonant with this pattern. These groups were more likely to respond positively to abstract and reflective approaches but negatively to concrete and active teaching approaches.

For Chinese international students on tourism-related courses at UK universities, Huang (2005) found that the introduction of problem-based learning (PBL) – encouraging the development of such skills as communication, report writing, teamwork, problem-solving and self-directed learning (Reynolds, 1997) and better retention of knowledge acquired (Blake *et al.*, 2000) – faced a major constraint. Students found it difficult to debate subjects with their tutors, apparently reflecting a Confucian attitude to authority relationships (Dimmock & Walker, 1998). As a consequence, in Huang's (2005) sample, only a small group of the students thought PBL to be a more effective way for them to learn.

In China itself, Lu and Adler (2010), researching into the expectations of undergraduate students in tourism and hospitality, found that opportunities for personal development and high salaries were the most important goals for students to pursue after graduation, perhaps reflecting that country's new-found materialism. They had high expectations of the ability of their programmes to prepare them for applying business and management skills, and were generally optimistic about the tourism industry.

This may reinforce, from a trans-cultural perspective, the literature that argues for the embedding of more appropriate employability skills in tourism curricula, and the need to create a hub and interface between the tourism industries and institutions of higher education (Zehrer & Mössenlechner, 2009). Sheldon *et al.* (2008: 61) speak about 'how to build the capacity for tourism students to lead the industry into the future as it faces increasing pressures for responsibility and stewardship'. Yet the implication is almost for students to be trained to withstand such pressures rather than to acknowledge them and thereby to strive to be more ethically responsible.

In other words, who is tourism education and training ultimately for? The students? The tourism industry? The education profession? National/international employment statisticians? Tourists and 'hosts' of tourism destination areas? All such 'stakeholders'?

Time and again critics have pointed to the apparent incompatibility between business requirements and the ideals of 'responsible' tourism. And while, from our 'safe' position within academia, we may point the finger at 'entrepreneurs' and 'corporations', we can be certain in the knowledge that they will rarely read what we have to say. But what about our direct (and indirect) impact on those who do (have to) listen to us and read our words, and perhaps implement our recommendations?

Hudson and Miller (2005; 2006), sampling students in Australia, Canada and the UK on ethical dilemmas facing tourism managers and students, have found that prior ethical training has no influence on ethical decision-making, suggesting that current curricular methods may not be working in this area. This is despite the long discussed embedding of ethical concepts and practice in the tourism curriculum, including an 'ethical tourism practicum' (e.g. Tribe, 2002).

The How

We assume that well-positioned academics possess a consciously-held collective responsibility in nurturing, supporting, stimulating and working with students, researchers and young colleagues, especially those from societies or sections of societies that lack the attributes of initial advantage from which others have benefited.

Many departments do this very well, as reflected in, for example, 'Southern' postgraduate students and young researchers working in 'Northern' institutions publishing papers as single authors under their own name or as the first named author in collaboration with a supervisor or other senior colleague.

Yet this probe was triggered by the fact that its authors were struck by the apparent dominance of white 'Northern', mostly male, contributors to tourism Recreation Research (*TRR*)'s anniversary issues: particularly ironic given this journal's provenance. As recently re-emphasized by McKercher (2009), academic leadership in most disciplines remains concentrated within a disproportionately small number of individuals (Jackson, 2004; Tschannen-Moran *et al.*, 2000), and this feature is a particular characteristic of the tourism academy (Jogaratnam *et al.*, 2005; McKercher, 2007). While one might have expected such a level of concentration to have diminished over time, McKercher (2009: 136–137) claims that this is not the case in tourism: 'in fact, the opposite is occurring…almost one in four papers published in 25 journals can be attributed to a handful of prolific authors' (see also Ryan, 2005; Zhao & Ritchie, 2007). More optimistically, almost half of all works published in *Tourism Management* now originate from non-native English speaking contributors (Ryan, 2009), although there may well be an element of concentration here too.

Does this suggest that (some) prominent tourism academics are not doing as much as they could to foster and promote work by up-and-coming newcomers (from the 'South' and elsewhere)? Are they being hindered by policy priorities in at least some (UK) universities?

Final Thoughts

This piece will be seen by some to perpetuate a superficial approach to issues of ethics in tourism. Our intention has been simply to turn the focus of 'ethical tourism' debate away, if only briefly, from the concentration on the products offered and experienced by tourists (e.g. Butcher's (2009) 'anti-modern' ecotourism) towards our own actions within academia in reducing, perpetuating or reinforcing moral inequities, through the teaching, supervisory, training, publishing and consultancy processes we may involve ourselves in.

Just as we (especially those of us in positions of privilege in 'developed' societies) need to make much greater personal sacrifices than currently envisaged if human-induced climate change is to be meaningfully mitigated, so we need to adapt our academic behaviour away from individual and institutional self-interest.

Collectively, we may spend much time and effort, as well as many printed words and congenially-located conference presentations, pontificating about the practice of responsible and ethical tourism. But, if such terms are to mean anything, then they need to apply to those of us who preach it in the way we undertake our work. Tourism educators need to think more deeply about their own ethical position within national and international circuits of information.

It seems to us that there are a number of parallels between the pursuit of overseas student recruitment and of consultancy fees. At their worst both can appear to involve a self-interested, almost immoral, form of neo-imperialism, whereby academic endeavour is not an end in itself but a means of repatriating aid money. Both processes can also exhibit an almost cavalier disregard for the sensitivities of local cultures and of the appropriateness of services rendered (e.g. ill-suited consultancy business models, poorly contextualized postgraduate taught courses).

Is it too simplistic and naïve to suggest that the tourism academy needs to return to a more traditional, ethically-based educational morality, away from the 'entrepreneurial' ethics of the marketplace. Indeed, would we now be allowed to do so?

9.2

Academic Capitalism, Academic Responsibility and Tourism Academics: Or, the Silence of the Lambs?[1]

C. Michael Hall

This short response to '"Post Colonialism", Responsibility and Tourism Academics: Where's the Connection?' finds several points of common perspective, but also several of difference. Any analysis of academic responsibility related to broader debates on global issues of importance must, of necessity, deal with ethical issues. But just as significantly we need to understand positionality and the location of individual academics, as well as academic institutions, within broader webs of power, values and interests.

The study of tourism does not occur in isolation from wider trends in scientific and academic discourse nor of the society of which we are a part. Despite 'ivory tower' accusations to the contrary, academic life 'is not a closed system but rather is open to the influences and commands of the wider society which encompasses it' (Johnston, 1991: 1). Discipline development, and the how, where and why of what we do and do not study, 'is an investigation of the sociology of a community, of its debates, deliberations and decisions as well as its findings' (Johnston, 1991: 11). Unfortunately, there is little overt discussion from tourism academics as to sociology of the tourism studies community, the reasons why they study certain topics and the positions that they take although relevant literature is gradually growing (e.g. Hall, 2005; Page, 2005; Coles & Hall, 2006; Coles *et al.*, 2006; Tribe, 2009).

The contents of an area of study at any one time and location, therefore 'reflects the response of the individuals involved to external circumstances and influences, within the context of their intellectual socialization' (Johnston, 1983: 4). Grano (1981) developed a model of external influences and internal change within geography that Hall and Page (2006) and Hall (2004) used as a framework with which to examine the field of tourism studies. According to Grano (1981) the relationship of academic space to external influence can be divided into three interrelated areas:

- knowledge – the content of tourism studies;
- action – tourism research within the context of research praxis; and
- culture – academics and students within the context of the research community and the wider society.

Tourist academics are a sub-community of the social science community within the broader community of academics, scientists and intellectuals which is itself a subset of wider society. That society has a culture, including a scientific/academic sub-culture, within which the subject matter of tourism is developed. Action is predicated on the structure of society and its knowledge base: research praxis is part of that programme of action, and includes tourism research (Hall & Page, 2006). The community of tourism academics is, therefore, an 'institutionalizing social group' (Grano, 1981: 26), a context within which individual tourism academics are socialized and which helps define the internal goals of their subject area in the context of the external structures within which they operate. The content of a subject area or discipline must, in turn, be linked to its milieu, such as changes as to how universities and research are funded and major issues of the day. It is, therefore, in this context that we may better understand notions of ethics and interest, especially because they operate at multiple scales.

Teaching and research in tourism has become highly globalized – as evidenced by the internationalization of journals, books, internet communities and scholarly meetings. This has encouraged the development of international debates on themes such as post colonialism and sustainability – as well as competitiveness, human resources, quality and yield maximization – to name just a few issues. Yet this has occurred as part of a broader process of the globalization of higher education marked by international sets of university and journal rankings, increased competition for the international student market and the opening of new campuses as part of the export of education. Although significant in themselves as indicators of a global 'academic market' (and labour force) which is part of a 'higher education industry', they are indicative of the spread of neoliberal ideas with respect to governance and education policy – ideas that emphasize the role of market forces, deregulation, the role of the state in encouraging the market as a mechanism for distributing goods and services and a reduced social welfare role for the state. In the case of the latter, this also means, in many countries, encouraging universities to search for new sources of income to replace declining government spending for higher education in real terms.

Such shifts in the idea of higher education and the institutions and individuals within them has been discussed under the rubric of concepts such as 'academic capitalism' (Rhoades & Slaughter, 1997; Slaughter & Leslie, 1997, 2001; Slaughter & Rhoades, 2004; Paasi, 2005), 'the entrepreneurial university' (Clark, 1998) and 'new managerialism' (Deem, 1998). New managerialism is a concept that refers both to ideologies about the application of techniques, values and practices derived from the private sector of the economy to the management of organizations concerned with the provision of public services, and to the actual use of those techniques and practices in publicly-funded organizations (Deem, 2001). This is perhaps best reflected in higher education by the use of teaching and research quality audits (Hall, 2005; Page, 2005), increasingly marked by simplistic notions of quality, and the international transfer of ideas and policies that surround them.

The concept of 'the entrepreneurial university' describes the way in which tertiary institutions are 'pushed and pulled by enlarging, interacting streams of demand, [and] universities are pressured to change their curricula, alter their faculties, and modernize their increasingly expensive physical plant and equipment' (Clark, 1998: xiii). Such pressures come from both government policies as well as economic globalization. The notion of academic capitalism 'moves beyond thinking of the student as a consumer to considering the institution as marketer' (Slaughter & Rhoades, 2004: 1), with academics acting 'as capitalists from within the public sector; they are state-subsidized entrepreneurs' (Slaughter & Leslie, 1997: 9). Academic capitalism therefore refers to the way in which academic staff of publicly-funded universities deploy their academic capital to generate external revenues to the institution via the pursuit of market and market-like activities. This means that the 'encroachment of the profit motive into the academy' identified by Slaughter and Leslie (1997: 210), has now become the norm. This has led to a situation in which profit-oriented activities are presently embedded, 'as a point of reorganization (and new investment) by higher education institutions to develop their own capacity (and to hire new types of professionals), to market products created by faculty and develop commercializable products outside of (though connected to) conventional academic structures and individual faculty members' (Slaughter & Rhoades, 2004: 11).

The precepts of 'academic capitalism' and 'the entrepreneurial university' are not evenly adopted across the globe, primarily because of more localized factors affecting higher education institutions, or even across universities, as some faculties and departments which are seen by university administrations as offering commercial and income-generating potential and are more susceptible than others to such reorientation. However, tourism studies, straddling as it does commercial and non-commercial research interests, is a clear subject for the further development of entrepreneurial self-interest. Nevertheless, to adapt academic behaviours away from such self-interest is likely to be extremely difficult given that such self-interest is now very much part of the institutions and structures within which the tourism academia is embedded. If your employment is subject to growth or maintenance of student numbers, attaining a certain number of publications in a determined set of journals, and attracting *x* amount of external income, then it is very hard to behave otherwise. As Hall (2004: 151) stated:

> I have great frustration with much of the research and scholarship undertaken in tourism. Often competently done, but without reflexion and thought as to whose interests are being served – which is normally those from business and government with access to power. For all the talk of sustainable and alternative tourism, few alternatives have really shown up which explore the potential for other spaces and places which reflexivity may provide. In my more sanguine moments I believe that this is because researchers often take the easier path in tourism research because within current academic structures that is what provides the rewards. And who am I to talk? As much as my present position stresses and frustrates me I still sit in a highly privileged position. I have a good salary by New Zealand standards, I have reasonable security of tenure, I can pay off the mortgage and I can travel. Many cannot.

Ethical tourism in its fullest sense is undoubtedly a major issue for tourism academia. Ethics is often given lip-service in many tourism degrees, especially those in business schools where, if it is considered at all, it is usually discussed in terms of a narrow narrative of complying with legal requirements of undertaking business, rather than trying to understand the meta-discourses of business within a capitalist political-economic system and its implications for equity, ecology and the public good. Of course, the reality is that many actually support such a discourse and the role of universities and the contribution they make to 'competitiveness' within it. What some see as unethical self-interest others may regard as 'enlightened' self-interest for the common good. Some actually embrace 'the entrepreneurial university' and the discourse of competition and see nothing wrong with it or the assumptions on which it lies (Hall, 2007). This means, for example, that some academics see nothing wrong in attending conferences in countries with poor human rights records and may believe that being able to give their paper (even if you are not easily able to Google words such as democracy, Tiananmen Square or Tibet) and 'helping' tourism education and research in those countries is more important than the tacit support of such regimes that attending or even running such conferences brings.

Ethics is arguably integral to the development of life-long learning skills as well as greater cultural sensitivity as it encourages students to think in different ways. Pressures for immediately transferable skills without such considerations means that the role of a university as somewhere that teaches you how to think and communicate is being lost. But of course thinking, rather than meeting a set list of metrics that you have to tick off to show you are 'qualified', is dangerous – at least to the powers that be – as it means considering ways of doing other. This applies both to students as well as to the academy itself. Indeed, if there is much discussion of the 'tough choices' that universities must make, there is, amidst the neoliberal discourse of efficiency, productivity and wealth creation, little discussion of alternative choices. 'The academy currently lacks a sustained debate about the economic interests that higher education should serve. It also lacks a sustained debate about the other interests it serves, its role in addressing the critical social, political, and cultural challenges of the day. The "tough choices" being made today are narrowly circumscribed by the material and ideological structures of supply-side policy' (Rhoades & Slaughter, 1997: 32–33).

In the contemporary tourism academy there is still some space to think. Although arguably that space is being increasingly constrained by such matters as conference and journal rankings, the lack of credit being given to books and chapters in research quality exercises, and other constraints on publishing. For what good is it to think unless you can also communicate in a manner in which the communication has 'credit' as being of a certain quality? Yet much of the debate is internal. We are talking to ourselves. And, as important as this is, I am left with a lingering doubt. To be concerned with ethics in tourism, to be critical of interests and structures and the openness of academic space, requires more than just attending the appropriately labelled conference or producing another critique of the tourism system and the need for 'alternative' (often qualitative) methodologies. It actually requires the structures, discourses and institutional arrangements of the contemporary academy to be challenged. It requires not only thinking other but doing other. It requires you to say no.

9.3

Towards an Ethics of Responsibility in Tourism Education

Erik Cohen

This thoughtful but somewhat incoherent probe questions the ethics of contemporary tourism education from a purportedly post colonial position. But the gist of the argument is actually that tourism departments at (UK) universities impart to their students an individualistic, selfish (profit-oriented) contemporary business ethics, to the detriment of traditional humanistic values (which are supposedly the ethical foundation of academic teaching).

Coming from a disciplinary department, I had no personal experience in teaching in a tourism school. But I believe that the need for an ethical reorientation discussed by the authors is similar to that found in other professional schools, which have been introduced into Western universities in the course of the last half century, and which had a profound impact on the ambience, aims and ethics of university teaching. To be sure, there were professional schools in the old academia: medicine to train physicians, law to train jurists (and theology to train priests and pastors). But these professions developed over time a strong code of ethics, which has been imparted to their practitioners in the course of their studies, and over time served as the basis for the formation of formidable professional organizations, on the national as well as international level, to direct and monitor the conduct of its members, who did not always live up to their professional ethics. The newly academized professions mostly lack such a code or, if it exists, it is less fully institutionalized and compelling.

The most important new professional schools in universities are the business schools. And it was here that the ethics of unhindered profit maximization were imparted to the students. The MBAs were the iconic product of these schools, and their practices of unrestrained pursuit of profits were to no small extent a factor which created the conditions for the recent global financial melt-down.

Tourism departments frequently constitute part of the business schools but even if they do not, they partake of the latter's orientation. Strictly speaking, such departments are not about tourism as a distinct field of human endeavour. They are about the application of business approaches to a rapidly expanding service sector. That their subject matter is tourism is of relevance only in so far as they need to adapt the general principles of business administration to the peculiarities of touristic situations. So why should their ethics be different from that of the business schools?

However, the current critique of the business schools' deficiencies led to proposals to reform business education towards a better understanding of the eventual consequences of aggressive business activities, a broadening of students' perspectives, and of a greater sensitivity to the environmental, cultural and social context of business, as prerequisites of its own sustainability. These proposals for change in the orientation of business education, relevant to tourism departments as they are to other professional schools, offer a window of opportunity for a reorientation towards ethics in tourism education.

This should not come about by introducing 'ethics' to the curriculum. Just as it is impossible to teach students to be religious or musical, it is impossible to teach them to be ethical; but people can be made aware or sensitive to ethical dilemmas in real situations. Ethical conduct can be learned from the personal conduct of educators and from practical training, conducted in the spirit of the reorientation of the aims of professional education. The emphasis in such training should in my view be on an ethics of responsibility (rather than on the conservative 'traditional values'). As they handle problems in practice, students should learn to take responsibility for the consequences of their decisions, even if in many complex cross-cultural touristic situations it might be hard to follow the exact path by which these consequences come about.

Individual responsibility, however, cannot stand alone; it needs guidance, during and especially after completion of studies. Tourism, as an emerging cluster of professions, does not have a widely accepted code of professional ethics, or powerful national, and especially international, organs able to enforce it. Tourism departments could play an important role in taking the initiative in the formulation and promotion of such a code, even as they train their students in the spirit of its emergent principles. In these endeavours it ought to be emphasized that responsible restraint of self-interest (as profit-making), enshrined in the clauses of such a code, rather than detrimental to business interests, is an important precondition for the long-range sustainability of the tourism industry.

Concluding Remarks

An intention of the core paper's authors was to turn the focus of 'ethical tourism' debate away from a concentration on tourism products and tourist experiences towards tourism academics' own actions. And while differing in some respects, both subsequent responses have largely agreed with the thrust of the proposition. Putting it into practice, however, is a different matter.

C. Michael Hall has discussed the broader process of the globalization of higher education and a concomitant spread of neo-liberal ideas, at least with respect to governance and education policy. He briefly reviewed the 'new managerialism' of 'academic capitalism' and recognized that the precepts of the latter have not been universally or evenly adopted. But, where they have been adopted, he suggests that it is likely to be difficult to adapt academic behaviours away from the self-interest that has been embedded within the structures and institutions of higher education in several parts of the world as a core element of 'academic capitalism'.

For Cohen, reform of professional schools' curricula through the employment of real life situations presenting ethical dilemmas as case studies, is seen as a practical way forward to shoehorn the ethics of responsibility into a reoriented tourism education programme. Researchers such as Fennell (2006; see also the Tourism and Welfare chapter in this book) might add that there would need to be a clear analysis, understanding and application of 'deep ethics' within this process, otherwise such objectives are in danger of remaining superficial and marginalized.

And, as Cohen rightly emphasizes, individual efforts may be in vain when tourism does not have a widely accepted professional code of ethics nor powerful international organs able to enforce any such code. In such a vacuum, he points to the 'responsible restraint of self-interest' as an important precondition for long-term sustainability of the tourism 'industry'.

Certainly, in this chapter, it has been emphasized that tourism educators need to think more deeply about their own ethical position within national and international circuits of information.

Having done that, the consensus of views expressed in this chapter appears to be that we should indeed be challenging the structures, discourses and institutional arrangements of the contemporary academy. But given the bread and circuses that many academics still enjoy – travel, interpersonal as well as virtual relations with interesting people and our global colleagues, good remuneration and (potential) promotion prospects – will our own personal (and family) self-interest even allow us to contemplate wanting to rock the boat? Would universities, funding councils and professional associations allow us space to do so? And, even if we could challenge, is it not too late?

Discussion Questions

(1) Identify and evaluate the main characteristics of the globalization of higher education.
(2) Attempt to summarize the ethical responsibilities of contemporary tourism educators.
(3) Explain why tourism should have a professional code of ethics.
(4) What are the key elements of a 'deep ethics' approach to tourism?
(5) Why should the structures, discourses and institutional arrangements of the contemporary academy be challenged?
(6) What is wrong with the entrepreneurial ethics of the marketplace in tourism education?
(7) Who does tourism education and training ultimately benefit?

Note

(1) ©C. M. Hall

References

Anon (2005). 'China chipped.' *The Economist*, 3 December: 374(8417): 56.

Anon (2006). 'Consultants take much of foreign aid, UK group claims.' *Community Action*, 21 August.

Anon (2009). 'PNG aid has been misspent: Rudd.' *Sydney Morning Herald,* 28 April. Online at: http://news.smh.com.au/breaking-news-national/png-aid-has-been-misspent-rudd-20090428-alh9.html. Accessed 3 October 2009.

Aziz, K. (2009). 'Foreign aid comes with foreign consultants.' *Chowrangi: Pakistan Politics, Current Affairs, Business and Lifestyle*, 8 May. Online at: http://www.chowrangi.com/foreign-aid-comes-with-foreign-consultants.htm. Accessed 3 October 2009.

Blake, R.K., Hosokawa, M.C. and Riley, S.L. (2000). Licensing examination following implementation of a PBL curriculum. *Academic Medicine* 75: 66–70.

Bramwell, B. and Lane, B. (2009). Editorial: Economic cycles, times of change and sustainable tourism. *Journal of Sustainable Tourism* 17(1): 1–4.

Butcher, J. (2009). Against 'ethical tourism'. In J. Tribe (ed) *Philosophical Issues in Tourism* (pp. 244–260). Bristol: Channel View Publications.

Clark, B.R. (1998). *Creating Entrepreneurial Universities: Organisational Pathways of Transformation*. New York: Elsevier.

Cockburn, P. (2009). 'Kabul's new elite live high on West's largesse.' *The Independent,* 1 May. Online at: http://www.independent.co.uk/news/world/asia/kabuls-new-elite-live-high-on-wests-largesse-1677116.html. Accessed 27 September 2009.

Coles, T. and Hall, C.M. (2006). The geography of tourism is dead: Long live geographies of tourism and mobility. *Current Issues in Tourism* 9(4): 289–292.

Coles, T., Hall, C.M. and Duval, D.T. (2006). Tourism and post-disciplinary enquiry. *Current Issues in Tourism* 9 (4&5): 293–319.

Collins, A.B. (2002). Are we teaching what we should? Dilemmas and problems in tourism and hotel management education. *Tourism Analysis* 7(2): 151–163.

Curtis, P. (2009). 'Academics and vice-chancellors oppose Mandelson's universities move.' *The Guardian*, 23 December. Online at: http://www.guardian.co.uk/

politics/2009/dec/23/academics-vice-chancellors-universities. Accessed 23 December 2009.

Dahal, M.K. (2009). 'Foreign aid in Nepal: The changing context.' *Telegraph Nepal,* 12 July. Online at: http://www.telegraphnepal.com/news_det.php?news_id=3417. Accessed 27 September 2009.

Deem, R. (1998). New managerialism in higher education: The management of performances and cultures in universities. *International Studies in the Sociology of Education* 8(1): 47–70.

Deem, R. (2001). Globalisation, new managerialism, academic capitalism and entrepreneurialism in universities: Is the local dimension still important? *Comparative Education* 37(1): 7–20.

Dimmock, C. and Walker, A. (1998). Transforming Hong Kong's schools: Trends and emerging issues. *Journal of Educational Administration* 36(5): 476–491.

Fennell, D.A. (2006). *Tourism Ethics*. Clevedon, UK: Channel View.

Fennell, D.A. (2008). Tourism ethics needs more than a surface approach. *Tourism Recreation Research* 33(2): 223–224.

Fox, L. (2009). 'PNG supports Rudd's aid spending remarks.' *ABC News,* 29 April. Online at: http://www.abc.net.au/news/stories/2009/04/29/2556473.htm. Accessed 27 September 2009.

Goeldner, C.R. and Ritchie, J.R.B. (2008). *Tourism, Principles, Practices, Philosophies* (10th edn). Toronto: John Wiley and Sons.

Grano, O. (1981). External influence and internal change in the development of geography. In D. Stoddart (ed) *Geography, Ideology and Social Concern* (pp. 17–36). Oxford: Blackwell.

Greenhill, R. (2006). *Real Aid 2: Making Technical Assistance Work*. London: ActionAid International. Online at: http://actionaid.org.uk/doc_lib/real_aid2.pdf. Accessed 17 October 2009.

Hall, C.M. (2004). Reflexivity and tourism research: Situating myself and/with others. In J. Phillimore and L. Goodson (eds) *Qualitative Research in Tourism: Ontologies, Epistemologies and Methodologies* (pp. 137–155). London: Routledge.

Hall, C.M. (2005). Systems of surveillance and control: Commentary on 'an analysis of institutional contributors to three major academic tourism journals: 1992–2001'. *Tourism Management* 26(5): 653–656.

Hall, C.M. (2007). Tourism and regional competitiveness. In J. Tribe and D. Airey (eds) *Advances in Tourism Research: New Directions, Challenges and Applications* (pp. 217–230). Oxford: Elsevier.

Hall, C.M. and Page, S. (2006). *The Geography of Tourism and Recreation* (3rd edn). London: Routledge.

Helsingin Sanomat (2008). 'Tuition fees for foreign students to be introduced on a trial basis in 2010.' *Helsingin Sanomat,* 15 August. Online at: http://www.hs.fi/english/1135238644147. Accessed 17 October 2009.

Hendry, J. (2004). *Between Enterprise and Ethics: Business and Management in a Bimoral Society*. Oxford and New York: Oxford University Press.

Huang, R. (2005). Chinese international students' perceptions of the problem-based learning experience. *Journal of Hospitality, Leisure, Sport and Tourism Education* 4(2): 36–43.

Hudson, S. and Miller, G. (2005). *Ethical Awareness and Orientation of Tourism Students*. Guildford, UK: University of Surrey School of Management. Online at: http://epubs.surrey.ac.uk/tourism/48. Accessed 27 September 2009.

Hudson, S. and Miller, G. (2006). Knowing the difference between right and wrong: The response of tourism students to ethical dilemmas. *Journal of Teaching in Travel and Tourism* 6(2): 41–59.

Jackson, E. (2004). Individual and institutional concentration of leisure research in North America. *Leisure Sciences* 26(4): 323–348.

Jogaratnam, G., McCleary, K., Mena, M. and Yoo, J. (2005). An analysis of hospitality and tourism research: Institutional contributions. *Journal of Hospitality and Tourism Research* 29(3): 356–371.

Johnston, R.J. (1983). On geography and the history of geography. *History of Geography Newsletter* 3: 1–7.

Johnston, R.J. (1991). *Geography and Geographers: Anglo-American Human Geography Since 1945* (4th edn). London: Edward Arnold.

Lashley, C. and Barron, P. (2006). The learning style preferences of hospitality and tourism students: Observations from an international and cross-cultural study. *International Journal of Hospitality Management* 25(4): 552–569.

Lipsett, A. (2008). 'Weaker pound could attract more overseas students.' *The Guardian,* 4 December. Online at: http://www.guardian.co.uk/education/2008/dec/04/overseas-students-council. Accessed 27 September 2009.

Lu, T.Y. and Adler, H. (2010). Career goals and expectations of hospitality and tourism students in China. *Journal of Teaching in Travel and Tourism* 9(1&2): 63–80.

Luvsanjamts, L. and Söderberg, M. (2005). *Mongolia – Heaven for Foreign Consultants*. Stockholm: European Institute of Japanese Studies. *Working Paper 215*.

McKercher, B. (2007). An analysis of prolific authors. *Journal of Hospitality and Tourism Education* 19(2): 23–30.

McKercher, B. (2009). The state of tourism research: A personal reflection. *Tourism Recreation Research* 34(2): 135–142.

Paasi, A. (2005). Globalisation, academic capitalism and the uneven geographies of international journal publishing spaces. *Environment and Planning A* 37: 769–789.

Page, S.J. (2005). Academic ranking exercises: Do they achieve anything meaningful? A personal view. *Tourism Management* 26(5): 663–666.

Parafianowicz, L. (2009). 'Foreign student fees delayed until 2011.' *The Local: Sweden's News in English,* 12 May. Online at: http://www.thelocal.se/article.php?ID=19410. Accessed 16 October 2009.

Reynolds, F. (1997). Studying psychology at degree level: Would problem-based learning enhance students' experience? *Studies in Higher Education* 22(3): 263–275.

Rhoades, G. and Slaughter, S. (1997). Academic capitalism, managed professionals, and supply-side higher education. *Social Text* 51: 9–38.

Ryan, C. (2005). The ranking and rating of academics and journals in tourism research. *Tourism Management* 26(5): 657–662.

Ryan, C. (2009). Editorial: Thirty years of tourism management. *Tourism Management* 30(1): 1–2.

Sheldon, P., Fesenmaier, D., Woeber, K., Cooper, C. and Antonioni, M. (2008). Tourism education futures, 2010–2030: Building the capacity to lead. *Journal of Teaching in Travel and Tourism* 7(3): 61–68.

Simpson, F. and Roberts, L. (2000). Help or hindrance? Sustainable approaches to tourism consultancy in Central and Eastern Europe. *Journal of Sustainable Tourism* 8(6): 491–509.

Slaughter, S. and Leslie, L. (1997). *Academic Capitalism: Politics, Policies and the Entrepreneurial University*. Baltimore: Johns Hopkins University Press.

Slaughter, S. and Leslie, L. (2001). Expanding and elaborating the concept of academic capitalism. *Organization* 8(2): 154–161.

Slaughter, S. and Rhoades, G. (2004). *Academic Capitalism and the New Economy: Markets, State and Higher Education*. Baltimore: Johns Hopkins University Press.

Tribe, J. (2002). Education for ethical tourism action. *Journal of Sustainable Tourism* 10(4): 309–324.

Tribe, J. (ed) (2009). *Philosophical Issues in Tourism*. Bristol: Channel View Publications.

Tschannen-Moran, M., Firestone, W., Hoy, W. and Johnson, S. (2000). The write stuff: A study of productive scholars in educational research. *Educational Administration Quarterly* 36(3): 258–290.

Tysome, T. (2006). 'In the UK, you find you're on your own.' *The Times Higher Education Supplement,* 14 July: 18–19.

UK Border Agency (2009). *Register of Sponsors (Tier 4 General & Child): Register of Sponsors Licensed under the Points-based System*. London: Home Office. Online at: http://www.ukba.homeoffice.gov.uk/studyingintheuk/. Accessed 16 October 2009.

Watt, P. and Greenhill, R. (2005). *Real Aid: An Agenda for Making Aid Work*. London: ActionAid International. Online at: http://actionaid.org.uk/doc_lib/69_1_real_aid.pdf. Accessed 17 October 2009.

Wedel, J.R. (1999a). *Collision and Collusion: The Strange Case of Western Aid to Eastern Europe, 1989–1998*. New York: St Martin's Press.

Wedel, J.R. (1999b). Testimony before the committee on international relations. Washington DC: US House of Representatives, 4 August. Online at: http://janinewedel.info/testimony_HouseInt99.html. Accessed 3 October 2009.

Zehrer, A. and Mössenlechner, C. (2009). Key competencies of tourism graduates: The employer's point of view. *Journal of Teaching in Travel and Tourism* 9(3&4): 266–287.

Zhao, W. and Ritchie, J.R.B. (2007). An investigation of academic leadership in tourism research: 1985–2004. *Tourism Management* 28: 476–490.

Further Reading

Akhimie, P. (2009). Travel, drama and domesticity: Colonial huswifery in John Fletcher and Philip Massinger's 'The Sea Voyage'. *Studies in Travel Writing* 13(2): 153–166.

Osagie, I. and Buzinde, C.N. (2011). Culture and postcolonial resistance: Antigua in Kincaid's 'A Small Place'. *Annals of Tourism Research* 38(1): 210–230.

Palacios, C.M. (2010). Volunteer tourism, development and education in a postcolonial world: Conceiving global connections beyond aid. *Journal of Sustainable Tourism* 18(7): 861–878.

Chapter 10
The Dilemma of Authenticity and Inauthenticity

Erik Cohen, Kjell Olsen and Philip L. Pearce

Context

This probe by Cohen, and commentators Pearce and Olsen, deals with the conceptualizations, permutations, usages, limitations and future prospects of the concept of 'authenticity' in tourism studies. Cohen discusses the trajectory of the concept following its implicit introduction in tourism discourse, by way of MacCannell's 'staged authenticity' thesis. He distinguishes the principal senses in which 'objective' authenticity has been used, the introduction of a constructivist approach to authenticity and the ongoing transition of the discourse to 'subjective' (existential) authenticity. Pearce points out the importance of 'mundane authenticity' (unrecognized in tourism studies), and deals with the questions of who uses the term 'authenticity', what are its linkages with other terms (such as 'mindfulness') and how to measure 'authenticity'. Olsen argues that authenticity could be seen as a Western 'grande idée', points out its inapplicability to non-Western cultures and raises questions regarding the interface of authenticity and power in tourism.

10.1

'Authenticity' in Tourism Studies: *Aprés la Lutte*[1]

Erik Cohen

The discourse of 'authenticity' was introduced into tourism studies by Dean MacCannell (1973), and became the key concept in an emergent sociological paradigm for the study of tourism. MacCannell's work elicited considerable controversy and generated a rich body of conceptual and theoretical analyses, as well as some empirical work. The theoretical discourse of authenticity, however, seems by now to have exhausted itself, even as the concept became empirically less relevant to the study of postmodern tourism. A third of a century after its introduction into the field, it is therefore appropriate to enquire into the paths of change the concept underwent after MacCannell introduced it into the discourse of tourism, explore its contemporary usages, and gauge its fate in an increasingly postmodern world.

MacCannell's (1973) initial concern was less with authenticity as such, but rather with 'staged authenticity', the covert staging by the hosts of sights, sites, objects and events, in order to make them appear genuine to tourists. He left the concept of authenticity itself unexplicated (Cohen, 2004: 103). By relating the alleged tourist's quest for authenticity to modernity, however, MacCannell inserted the study of tourism into mainstream sociology. The concept of authenticity held a promise to become the basic concept in a sociological paradigm for the study of tourism. However, it is a 'heteroglot' (Olsen, 2002: 163) or polysemic concept, with many connotations (e.g. Bruner, 2005: 149–150). Though 'authenticity' soon became 'an important theme in tourism literature' and generated 'numerous and diverse discussions' (Reisinger & Steiner, 2006: 66), these did not lead to a commonly accepted definition of the concept. Rather, diverse connotations vied with each other, leading Reisinger and Steiner to reach the pessimistic conclusion, in the wake of an extensive review of the literature (Reisinger & Steiner, 2006: 67), that 'the concept should be abandoned' (Reisinger & Steiner, 2006: 66), because it is 'too unstable to claim the paradigmatic status of a [basic theoretical] concept' (Reisinger & Steiner, 2006: 66). The authors proposed that the term 'authenticity' 'should be replaced by more explicit, less pretentious terms like genuine, actual, real and true' (Reisinger & Steiner, 2006: 66). The authors, however, do not deal with the problem of possible alternative connotations of those terms.

Somewhat paradoxically, having recommended the abandonment of the concept of authenticity, Reisinger and Steiner proceeded to reconceptualize it.[2] They accomplished that in two articles, devoted, respectively, to object authenticity and existential authenticity (Reisinger & Steiner, 2006; Steiner & Reisinger, 2006). They justified the separation by arguing that, 'existential authenticity is not just a wider take on the authenticity of objects, but another concept altogether' (Reisinger & Steiner, 2006: 68), each demanding distinct treatment. Their reconceptualization, however, is framed in highly abstract and somewhat obscure Heideggerian terms, whose relevance for the theoretical, and especially the empirical, study of tourism has yet to be established.

Steiner and Reisinger (2006: 299) point out that the term 'authenticity' is 'often used in two different senses: Authenticity as genuineness or realness of artifacts or events, and also as a human attribute signifying being one's true self or being true to one's essential nature'. Their separate treatment of the two concepts of authenticity reflects an already existing split in the discourse of authenticity, along two principal lines. Both sub-discourses are rooted in MacCannell's work, but relate to phenomenologically distinct levels of reality: as a quality of the lived-in world or of an experience. Brown (1996: 39) has distinguished between a tourist's 'quest for the authentic Other' and a quest 'for the authentic Self'. He says that the two quests 'push in opposite directions' (Brown, 1996: 39), exemplified by the 'tension between, say, "sex, sea and sand" tourists and their "museum-cathedral" circuit counterparts' (Brown, 1996: 39). Brown's distinction resembles, but is not identical to, Tom Selwyn's (1996: 21–28) distinction between 'hot' and 'cold' authenticity. However, Wang (1999; 2000) has rendered the first systematic account of the two diverging sub-discourses, that of 'objective' and that of 'existential' (subjective) authenticity. (The third category in Wang's typology, 'constructed' authenticity, combines elements of both discourses, as it refers to the subjective criteria used by people to determine the 'objective' authenticity of sights, objects, sites and events.)

The two sub-discourses focused on different issues, and generated considerable efforts of conceptual sophistication and distinction but, unfortunately, there was little cross-fertilization between them. Crucial questions such as the 'objective' conditions in the lived-in world for experiences of existential authenticity, or, contrariwise, the subjective conditions for the perception of a site, object or event in the lived-in world as 'objectively' authentic, have rarely even been raised, except for statements that 'existential' authenticity can be experienced in 'objectively' unauthentic surroundings.

Objective Authenticity

A polysemic concept

Authenticity is not an intrinsic quality of sights, sites, objects or events. As Olsen points out in his critique of MacCannell, the latter tends to ascribe 'the authentic to objects…as [if] it was an essential feature of these and not an idea in contemporary Western culture' (Olsen, 2002: 161). But though Olsen's view seems to be at present generally accepted, there exists little agreement regarding the precise nature of that idea. The concept of 'objective' authenticity has been deployed in tourism studies in several overlapping senses:

(1) Authenticity as 'origins'; this usage is borrowed from the language and practices of museumology (Trilling, 1972; Cohen, 2004: 105; Reisinger & Steiner, 2006: 67), where its antonym is 'falsification'. The judgments of authenticity in this sense are based on such criteria as antiquity, traditional or customary ways of production and usage, pedigree and authoritative certification (Bruner, 2005: 150). Authors in tourism studies have frequently used the term, at least implicitly, in this sense. Its use involves the tacit assumption that the tourist's quest resembles that of the ethnographer or anthropologist (van den Abbeele, 1980: 12), eliciting claims of critics, that 'tourists are not ethnologists' (Nettekoven, 1973; Cohen, 2004: 106).
(2) Authenticity as 'genuineness' ('echt' in German), particularly of a product, as in 'real coffee'. According to Theobald (1998: 411), 'Authenticity means genuine, unadulterated, or the real thing' (quoted in Reisinger & Steiner, 2006: 68). Its antonym is 'surrogate'.
(3) Authenticity as 'pristinity' (Cohen, 2004: 105), an unadulterated state, particularly of nature, such as in 'pristine tropical paradise'; its antonym is 'despoliation'.
(4) Authenticity as 'sincerity', particularly in human relationships, like in the expression of feelings, as in the 'sincere welcome' extended to guests or in 'sincerely yours…' in the conventional ending of a letter. Its antonym is disingenuousness. Taylor (2001) seeks to distinguish between the concepts of sincerity and authenticity, though he considers sincerity to be 'a philosophical cousin of authenticity' (Taylor, 2001: 8).
(5) Authenticity as 'creativity', particularly in cultural production, as in the work of artists, musicians and dancers (Daniel, 1996); its antonym is 'copy', when the imitation is overt, and 'fake,' if it is covert.
(6) Authenticity as 'flow of life,' not interfered with by the 'framing' of sights, sites, objects and events for touristic purposes, by various overt or covert markers. According to this usage, the quintessential authentic

> sight is one, which is utterly unmarked (*cf.* Cohen, 2004: 134); it is authentic, precisely because it is *not* an attraction. In contrast to the other senses of the term, the sight does not have to feature any intrinsic properties: it does not need to be 'pristine'. Indeed, many unmarked sights in the contemporary world are far from pristine. This sense of the term appears to me to be preferable for both theoretical and empirical reasons.

Theoretically, this usage continues to treat authenticity in the spirit of MacCannell's approach, but expands the scope of the concept and adapts it to contemporary realities. Indeed, MacCannell in his original work (1973, 1976), appears to have implicitly meant it in that sense, as unframed reality; the derivative concept of 'staged authenticity', in which he was mainly interested, being a form of covert 'framing'. While MacCannell based that concept on Erwin Goffman's early work, the proposed usage is based on Goffman's later work (Goffman, 1974).

This usage preserves, and in fact makes explicit, the principal paradox inherent in tourism, according to MacCannell's approach: namely, that any sight ceases to be fully authentic once it is marked as such (i.e. 'framed'). Van den Abbeele (1980: 7) comments in his article on MacCannell's *The Tourist* that: 'Once a sight is marked as authentic,...it is by the very fact of its being marked no longer quite authentic...'. Authenticity turns out to be an unstable quality: its very disclosure weakens it. But the boundary between authentic and non-authentic is not a crisp gradient – it varies according to the extent to which a marked-off sight, site, object or event is excised from the local flow of life, for touristic purposes. Mere marking of a sight, which otherwise constitutes part of the local flow of life, as 'authentic', may impair, but does not annul its authenticity. An old church, marked as an attraction, but still in use, can be said to be more 'objectively' authentic than one which was turned into a museum for tourists. A stealthily framed sight, without explicit markers, may appear to be part of the local flow of life, though it is not; it represents an instance of 'staged authenticity'.

Perhaps more important is the empirical reason for preferring this usage of the concept: it is adapted to the contemporary, postmodern situation, from which 'originals' have, according to some authors, vanished (Baudrillard, 1988), while the territorial distinctiveness of ethnic, cultural, religious and linguistic groups has been obliterated by processes of globalization (Hughes, 1995: 783; Cohen, 2007). Authenticity, in the proposed sense, also comprises ongoing situations and events, like a local uprising, revolution or even natural disaster, which a tourist may happen to witness – though he or she may not have bargained for this kind of authenticity.

Indeed, the principal implication of the proposed usage is that exactly the unplanned or unexpected sights and events on the tourist's trip will be the most authentic, and will become the most memorable experiences of the trip. The completely unexpected arrival of the huge tsunami waves, which hit the beaches of southern Thailand in December 2004, were probably the most authentic, though disastrous, sight, which the vacationers had encountered during their visit (Cohen, 2005: 89).

The proposed usage comes quite close to Reisinger and Steiner's Heideggerian view of 'object authenticity', but with an important caveat. These authors assert that, 'If Heidegger [had] used a term like authentic to apply to things, whatever appears would be authentic' (Reisinger & Steiner, 2006: 78). Note that this usage expresses a postmodern perspective: it accepts the surface appearance of things, without delving into their depth: origins, genuineness, etc., are not part of its connotation. My usage would agree with this view, as long as 'what appears' has not been 'framed' for touristic purposes.

Finally, the proposed usage helps to resolve the sticky problem of the relationship between cultural change and authenticity. As cultural change penetrated ever more remote corners of the world, curators, eager to salvage the cultural products of the past, tightened their definition of 'authenticity'. Thus the director of the Museum of Mankind in London in the 1970s defined genuine African art as '...any piece made from traditional materials by a native craftsman for acquisition and use by members of local society... that is made and used with no thought that it ultimately may be disposed of for gain to Europeans or other aliens' (McLeod, 1976: 31). Close to a third of a century later, with the changes in technology and the expansion of the market for African craft products, this narrow definition will automatically designate all contemporary products as inauthentic. In the proposed usage, they will remain authentic, even if produced for the tourist market, as long as they are not marked as 'authentic' or 'original'. Production for the tourist market became the principal occupation of craftspeople around the world; it is thus, an integral part of the local flow of life. The falsification would be in the marking, not in the product itself.

Authentic reproductions and genuine fakes

One of the distinguishing marks of postmodern situations is the blurring of boundaries between different 'limited provinces of meaning' (Schuetz, 1973). The 'simplicity of the distinction between reality and representation' has been threatened (Hughes, 1995: 782). Thus, '...theatre has been taken out of dedicated buildings and out to the streets, painting taken off canvas, out of the frame and out of the gallery, and the definition of sculpture rendered ambiguous by removing the plinth...' (Hughes, 1995: 782). Such blurring is reflected in the discourse of 'objective' authenticity, in the hybridization of the concept of 'authenticity' with its counter concepts, in such expressions as 'authentic reproduction' (Bruner, 2005: 145–168) and 'genuine fakes' (Brown, 1996).

Though written as a 'critique of post-modernism', Bruner's article proposes two senses of authenticity, which have a distinctly postmodern resonance: authenticity as 'verisimilitude' or as 'simulation' (Bruner, 2005: 149). Bruner says that a reproduction may be considered 'authentic' either if it resembles in a 'credible and convincing' manner a historical site or if it simulates such a site completely and immaculately (Bruner, 2005: 149). Regarding the latter, Hughes (1995: 782) remarks that, 'in post-modern constructions, the equivocal relation between reality and its representation [i.e. between the unmarked "flow of life" and "framed" provinces of meaning] has been extended. The simulations became so "authentic" that they achieve a state of hyper-reality'. This makes possible the emergence of 'genuine fakes' (Brown, 1996), in which the '..."completely real" [i.e. the "hyper-real"] becomes identified with the "completely faked". It is reproduced with such attention to detail that it is evidently a fabrication' (Brown, 1996: 35). Brown, however, stresses that 'The genuine fake is not thus the object itself but the relationship between visitors and presenters which the object mediates' (Brown, 1996: 33–34). He illustrates this assertion by the example of 'airport art': 'To the extent that it is not the artifact it is presented to be, it is evidently a fake. To the extent that it is bought knowingly as such, however, it is genuine' (Brown, 1996: 46n). In other words: seen as 'framed', it is false; once 'de-framed' and taken simply for what it is, it is authentic. This leads us to the issue of 'constructed authenticity'.

Constructed Authenticity

Much of the recent discourse of authenticity focused on the issue of 'constructed authenticity' (e.g. Wang, 1999, 2000; Halewood & Hannam, 2001; Olsen, 2002; Mehmetoglu & Olsen, 2003). Rather than considering it as a separate line of discourse, 'constructed authenticity' can best be conceived as a hybrid of both main sub-discourses of authenticity, stressing the subjective practices by which people construct the 'objective' authenticity of a sight, object, site or event.

In ordinary usage, there is at least an implicit assumption, that the right to determine what is 'objectively' authentic is the preserve of various authorities: curators, archeologists, art historians, ethnographers or anthropologists, who are believed to have the knowledge, and hence the authority, to declare a sight, object, site or event as authentic, according to their professional criteria. From a constructivist perspective, however, all judgments are in a sense socially constructed, even those of scientists, experts and professionals. In that respect, the professional criteria of curators or anthropologists for judgments of 'objective' authenticity are as constructed as those of ordinary tourists; the difference is that they are believed to be authorized, due to their professional knowledge, to define those criteria and deploy them in judgments of authenticity. It is a form of intellectual hegemony.[3]

The use of professional definitions of 'objective' authenticity has been criticized as facilitating an authoritarian exercise of power (e.g. Gable & Handler, 1996). The concept of 'constructed authenticity', as it developed in tourism studies, involves a 'democratization' of the right to pronounce a sight, object, site or event as 'authentic'. It implies a loosening of the professional authority and the destabilization of the criteria by which authenticity is judged. In essence, everyone becomes entitled to his or her own set of criteria and thus to decide which aspects of a sight, object, site or event make it 'authentic' (Cohen, 2004: 104–107; Littrell *et al.*, 1993; Bruner, 2005: 145–168).

Several researchers deployed a constructivist approach to the empirical study of authenticity, focusing on the tourists' perceptions of authenticity and rejecting any a priori, authoritative definition of the concept. As Mehmetoglu and Olsen (2003: 151) have pointed out: 'Rather than assuming the responsibility of judgment...of what is authentic or not, it is necessary to understand why some people claim to experience something as sincere or authentic and others do not. To do so, one needs to ascertain the emic view of the actors (i.e. tourists) themselves and to examine...the concrete contexts in which such experiences occur'. This approach is particularly common in studies of tourism to heritage sites (e.g. Chhabra *et al.,* 2003; Halewood & Hannam, 2001; McIntosh & Prentice, 1999; Waitt, 2000). Most of those studies, however, tend to be 'psychologistic': they detail an array of individual perceptions and attitudes but are not concerned with the collective processes, by which individual perceptions coagulate into a *socially* constructed recognition of the authenticity of a sight, site, object or event. This is the problem of 'emergent authenticity' (Cohen, 2004: 109–110): the gradually emerging social consensus that something, which might hitherto have been considered 'inauthentic', is authentic.

In terms of the usage offered above, 'emergent authenticity' can be conceived of as a process of 'de-framing': something which at the outset has been seen as 'framed' for tourism, becoming part of the local flow of life. For example, some Native American crafts products, such as Eskimo soapstone carvings (Graburn, 1976), initially produced specifically for travellers or early tourists, came gradually to be recognized as authentic native products and became collectors' items (Cohen, 2004: 110). Similarly, Disneyland, the initially quintessential, overtly 'framed' tourist attraction, has over time been increasingly recognized as part of contemporary American culture and, as such, as 'authentic'.

Subjective Authenticity

Towards the end of the 20th century, the attention of tourism researchers gravitated gradually away from the discourse of 'objective authenticity' towards that of 'subjective authenticity' (Uriely, 2005: 206–7). A shift in focus has taken place '...from the displayed objects provided by the industry' to the 'subjective negotiation of meanings as a determinant of the experience' (Uriely, 2005: 200).

This move has been partly instigated by the theoretical difficulties revealed in the discourse of 'objective' authenticity discussed above, and perhaps even more strongly by the penetration of tourist studies by postmodern ideas, such as those of Baudrillard (1988), regarding the disappearance or depreciation of 'originals' and the rise of simulacra in the contemporary world. Others stressed that the forces of globalization have demolished the basis on which the concept of objective authenticity rested. Hughes, for example, in the wake of his conclusion that, 'The notion of a territorially insulated place centered culture, is clearly unsupportable in the face of globalization' (Hughes, 1995: 799), asserts that, '...this need not spell the demise of authenticity. By resurrecting a more existential perspective, it is possible to find manifestations of authenticity through individual's assertions of personal identity' (Hughes, 1995: 799).

It is important to note that the discourse of authenticity in tourism began with the consideration of the authenticity of tourist settings (MacCannell, 1973). In the past researchers at least tacitly assumed that objects, which appear to be authentic, would elicit heightened subjective experiences: tourists will be 'exalted' in the presence of a pristine natural attraction, or deeply 'touched' by a great work of art. As long as that assumption held, there was no urge to separate the discourses of objective and subjective authenticity. The insight that there is no necessary connection between the authenticity of the setting experienced by the tourist and the quality of his or her experience led to the abandonment of this assumption and the separation of the two discourses. Researchers asserted that tourists might have authentic experiences, without being in the presence of authentic sights. On vacation they tend to experience a sense of 'self-discovery' (Cary, 2004: 63), a sensation of being 'true to oneself' (Steiner & Reisinger, 2006: 299), of 'really living', and of an immediate relationship with others, culminating in a feeling of communal belonging (Cary, 2004: 63). Such subjective sensations of authenticity can be seen as one instance of Maslow's concept of 'peak experiences', in which the individual senses an undifferentiated 'flow' (Csikszentmihalyi & Csikszentmihalyi, 1988; Cary, 2004: 68). Such sensations occur in supreme moments of love, in mystical experiences, in the mastery of danger (e.g. in free rock climbing) or in care-free vacationing situations, in which the tourists feel 'out of time and place' (Wagner, 1977). Wang (1999, 2000) and, following him, Steiner and Reisinger (2006) relate to such experiences as instances of 'existential' authenticity.

The various ways in which researchers seek to describe 'existential' authenticity are not presented as alternative interpretations (as in the case of 'objective' authenticity). Rather, they are all seen as aspects of the same, exalted experience. However, they are all predominantly psychological concepts. Steiner and Reisinger (2006) recently attempted to provide a more profound, philosophical basis to the concept, in terms of the German existentialist philosopher, Martin Heidegger. The authors claim that '... many of the diverse ideas about existential authenticity [in tourism studies] echo [Heidegger's] view of human authenticity...Heidegger uses the term "authenticity" to indicate that someone is being themselves existentially...' (Steiner & Reisinger, 2006: 303). The authors stress that 'this is deeper than being oneself behaviourally or psychologically. To be oneself existentially means to exist according to one's nature or essence, which transcends day-to-day behaviour or activities or thinking about self' (Steiner & Reisinger, 2006: 303). However, '...the existential self is transient, not enduring...It changes from moment to moment. There is no authentic self. One can only momentarily be authentic in different situations. Thus, there are no authentic and inauthentic tourists, as much as researchers might like there to be such handy categories' (Steiner & Reisinger, 2006: 303).

Steiner and Reisinger's conception of 'existential' authenticity as a momentary state agrees with Cary's idea of 'The tourist moment' (2004: 64), which '...represents a spontaneous experience of self-discovery and communal belonging'.

Experiences of 'existential' authenticity are said to be revelatory/epiphanic (Cary, 2004: 64), akin to sacred or numinous (Cary, 2004: 67), and hence, like the latter, not fully explicable or representable in words. They occur serendipitously (Cary, 2004: 64), in a 'liminal moment', in which Cary says, citing Ryan (1991: 35), '...the 'tourist has ceased to be a tourist', having 'become entirely subsumed' in the ongoing activity (Cary, 2004: 64).

However, Cary's account of the 'tourist moment' comes with a twist, absent in other descriptions of 'existential' authenticity: while its occurrence appears to enable the individual to transcend momentarily his or her tourist role – the tourist's 'wonderful feeling of [communal] belonging', serendipitously experienced on a trip, is soon terminated by a reflective reminder of being 'alone in a foreign place. This discordant note of exteriority signals a mis-recognition of belonging' (Cary, 2004: 69). The tourist's feeling of communitas is spoiled by a 'sense of exteriority' (Cary, 2004: 69).

Cary's account of the ephemeral character of the experience of 'existential' authenticity parallels the instability of 'objective' authenticity, which, according to MacCannell and van den Abeele, is spoiled by the very act of being marked. Reflectivity on the subjective level is thus homologous to marking on the objective level; both make authenticity an essentially transient state. Indeed, Cary makes the point that serendipity can strike only once: 'If serendipity appears to strike twice in the same situation, the tourist experience may shift into the as-of-yet-unexplored realm of staged serendipity' (Cary, 2004: 66) – just as an authentic sight can be discovered only once, re-discovering an 'undiscovered' sight is obviously a case of 'staged authenticity'.

It is important to point out that serendipitous moments of 'existential' authenticity in tourism are not always pleasant ones. In the tsunami disaster on the Andaman coast of southern Thailand, the vacationing tourists were alerted from their placid existence by the sudden recession of the sea (prior to the onrush of the huge wave), and rushed forward excited, to have a look at the extraordinary natural sight. This was probably the most exciting or exalting experience they had on their trip, but, alas, for many it was also the last one, as the onrush of the crest of the wave overwhelmed them (Cohen, 2005: 89). In the chaos following the sudden, destructive impact of the wave, there was a moment resembling communitas, the survivors helping one another indiscriminately, tourists ceasing momentarily being 'tourists'; but they were soon reverted to their role, being given preferential treatment in the rescue and evacuation operation following the disaster (Cohen, 2005: 90–91). As in war, the experiences of the tourists, though certainly 'existential', proved overwhelming and, like many a traumatic war experience, these were experiences, which the survivors preferred to forget rather than treasure.

The experience of 'existential' authenticity is believed to be unique in several respects. According to Heidegger, it is a rare experience, '…in which people find themselves in their unique place in the world, in a unique situation in relation to the connectedness around them' (Steiner & Reisinger, 2006: 307). Experiences of 'existential' authenticity typically emerge serendipitously in unexpected situations and cannot be repeated at will (Cary, 2004: 66). They constitute a 'special state of Being in which one is true to oneself and acts this way as opposed to becoming lost in public roles and public spheres' (Steiner & Reisinger, 2006: 301, citing Berger, 1973).

In tourism people are believed 'to distance themselves from their norms and look at their lives from a different perspective'; they are allowed to 'escape from role playing' by engaging in tourism activities, deemed to be 'non-ordinary, free from constraints of daily life' (Steiner & Reisinger, 2006: 306). Tourism, thus, seems to be eminently suited to attain unique experiences of 'existential' authenticity.

However, such a view disregards the realities of contemporary mass tourism and the influence of the tourist industry, and of the popular media on the institutionalization of the tourists' roles and on the homogenization of their expectations. Rather than create opportunities for 'existential' experiences, the institutionalization of all forms of tourism, including such initially 'counter-cultural' ones as drifting and backpacking (Cohen, 2004: 49–63; Richards & Wilson, 2004), in reality stifles them. Tourists can in reality achieve such experiences only incidentally, particularly when they escape, not just their everyday roles, but the tourist role itself. The very promise of unique experiences in the promotional literature denies that uniqueness. This is the reason that the most exalted sights, thoroughly institutionalized and heavily promoted, like the Mona Lisa, the Egyptian pyramids or even Machu Picchu in the Peruvian Andes, do not (any more) elicit the most exalted experiences. These are encountered on the receding liminal margins of contemporary tourism.

Subjective experiences in tourism come in different degrees of intensity. Just as I asserted in an earlier formulation, that conventional recreational experiences reflect on a superficial level a deeper theme, a quest for personal renewal ('re-creation') (Cohen, 2004: 70), so another conventional tourist experience, having an 'authentic "Good Time"' (Brown, 1996: 37), is one manifestation of the sense of 'real living', and, as such, one of the, perhaps less exalted, modes of experience of 'existential' authenticity.

Discussion

Several shifts in the study of authenticity occurred in the third of a century since it was introduced in the discourse of tourism. The principal shift was from the initial discourse of the much contested concept of 'objective' authenticity to the more recent discourse of the concept of subjective, in particular 'existential', authenticity. This shift in a way salvaged the centrality of the concept, in a world from which 'authenticity', in its rigorous sense of originality and genuineness, is rapidly vanishing.

Another important shift took place at the level at which authenticity has been discussed. In the earlier literature the discussion centered on the type of tourists and the typical experiences they are seeking. Thus, in an early article on modes of touristic experiences, I talked of tourists travelling in an 'existential' mode (Cohen, 2004: 77–80). More recent work shifted the discourse to the level of discrete experiences, which are said to vary in the course of a trip (Uriely *et al.*, 2002). Rather than characterizing a personal quest throughout a trip, crucial 'existential' experiences are seen as spontaneous, fleeting and rare occurrences, coming to pass in what Cary (2004) calls a 'tourist moment', and Steiner and Reisinger (2006: 306) an 'existential moment'.

Finally, some authors called on the touristic discourse to go 'beyond authenticity' altogether (Russell, 1992; Lacy & Douglass, 2002). Lacy and Douglass (2002: 17) pointed out that in the Basque region of Spain, 'There is "traditional" (Indians of Europe[4]) and "cosmopolitan" (coastal casinos and the Guggeheim [Museum]) tourism...', both of which are aspects of contemporary Basque identity. The authors assert that their study of Basque cultural tourism attempted to go '...beyond a concern with the cultural authenticity of a particular tourist site, in the belief that the anthropology of tourism...is a...prime context within which to view the dynamics of cultural generation and representation' (Lacy & Douglass, 2002: 18). The latter, being innovations, are in the understanding of the authors, 'beyond authenticity'.

While Lacy and Douglass seek primarily to expand the discourse of tourism to encompass what are not considered 'authentic' aspects of an ethnic cultural identity, other authors asserted more radically that the concept of authenticity itself is under threat, as the public is becoming 'openly skeptical about authenticity...as a foundational value' (Gable &

Handler, 1996: 569). America, in the view of Gable and Handler, is entering a '"post-authentic" age' (Gable & Handler, 1996: 569).

Conclusion

What is the viability of the concept of authenticity in the study of postmodern tourism? The answer depends on how the concept will be conceived in the future. Postmodern tourism makes the paradigmatic approach based on MacCannell's beginnings increasingly less relevant, as sophisticated post-tourists seek the enjoyments of distinctive, but familiar, experiences away from home, and the less sophisticated ones, the thrills of fantasy, supplied by high-tech theme parks and similar contrived attractions (Cohen, 2012). Contemporary tourism may appear to be moving into the 'post-authentic' age, but authenticity is lurking beneath the surfaces of postmodern attractions, though in an inverted, and in the eyes of some, perverted guise. The future of tourism may well be presaged by the extraordinary success of such an 'emblem of post-modernism' (Douglass & Raento, 2004: 7) as Las Vegas, which thrives at being the embodiment of all that is opposed to authenticity. However, in their celebratory essay on Las Vegas's 'tradition of invention', Douglass and Raento argue that 'the city's critical massing of simulacra constitutes its own authentic reality' (Douglass & Raento, 2004: 7). Imaginative, jarringly discordant innovations, bordering on the fantastic, may well in the postmodern future replace the longing for the wholeness of the pre-modern other, as the principal socially accepted manifestations of a novel kind of authenticity.

10.2

Staged Authenticity: A Grande Idée?

Kjell Olsen

With reference to Susanne Langer's book '*Philosophy in a New Key*' (1942), Clifford Geertz writes about *Grande idée* in social sciences. Such *Grande idée*(s) often:

> *...resolve so many fundamental problems at once that they seem also to promise that they will resolve all fundamental problems, clarify all obscure issues. Everyone snaps them up as the open sesame of some new positive science, the conceptual center-point around which a comprehensive system of analysis can be built*. (Geertz, 1973: 3)

The destiny of a *Grande idée* often is that they become metaphorical and metonymically extended in such a way that their weaknesses, constraints and limitations become obvious. The result is that they are either thrown away, or, as Geertz suggested in the case of the concept of culture, scholars start; '...cutting the [...] concept down to size' (1973: 4). Which one of these two destinies will MacCannell's concept of staged authenticity end up with, remains to be seen. Subjectively argued, perhaps some insight might be gained by delimiting the concept. A useful approach to cutting authenticity down to size, while remaining in tune with MacCannell, is essentially to regard the concept as an important *idée* in Western modernity that can be analyzed specifically through concrete studies. The ideas embedded in the concept of authenticity are typically concerned with the relationship between the copy and the original; about the moral value of certain forms of relationships in modern societies and; the difference between societies labelled modern and those regarded as traditional (MacCannell, 1999: 3, 93–97). The application of the framework of authenticity in tourism studies, particularly, may enable the development of a better understanding of people's experiences in contemporary times.

The reason for such a position is embedded in the myriad possibilities that the concept of authenticity, as envisaged by MacCannell, opens up for tourism research. Essentially, the possibilities can be summarized in the following three ways: (1) the use of staged authenticity as a *Grande idée* binds tourism research to the core of many disciplines in social science and humanities through the analysis of the development of modern society as a whole, (2) in using Peircean semiotics, the concept opens up for an analysis of the idea of authenticity both as a structural feature of modern societies, and as an experience at the individual level, and (3) the concept has a closeness to an empirical reality found in literature, advertising, newspapers, magazines and many people's own talk about experiences and expectations when engaging in tourism.

To commence with the last point (3 above), considering MacCannell's (1999: 3) claim that 'For moderns, reality and authenticity are thought to be elsewhere: in other historical periods and other cultures', it seems reasonable to argue that as an empirical description for some aspects of modern tourism, it may hold some truth. At least this is the case for much of the discourse on tourism. For instance, in an analysis of travelogues, Dann (1999) demonstrated that the discourse, on the relationship between travellers and tourists, pivots around ideas connected to MacCannell's use of authenticity.

Similar focus on values concerned with authenticity can be found in the continually growing amounts of writings, TV programmes, advertising and tourists' own talk on their experiences, thereby building up on ideas of authenticity. Thus there appear to be similarities between the everyday discourse on tourism and the scholarly discourse that seem to be derived from MacCannell's work. This might be a result of tourism research not being '...just works "about" social processes, but materials which in some part constitute them' (Giddens, 1991: 2), or being a result of our work which is '...coauthored, not simply because informants contribute data to the text, but because, [...], ethnographer and informant come to share the same narratives' (Bruner, 1986: 148). Hence, it is this congruency between discourses on the object of investigation that is maintained in the use of the concept.

It is claimed that ideas of authenticity are important, regardless of other ideas that may be found in the discourses on modern Western society. Therefore delimiting MacCannell's concept of staged authenticity down to size enables a focus on individuals' experience, alongside paying attention to the grand scale processes that shape and are shaped by the individual. This lends to the second point (2 above). MacCannell's use of Peircian semiotics (attraction = marker/sight/tourist) connects the concept of staged authenticity to an individual's (tourist) experience, as well as to macro scale processes, sights and the production of markers that shape the individual experience. In my opinion, Cohen's (1988) work on authenticity as emerging in social processes is a first step in analyzing such macro scale processes. While places and objects are attached to the values of authenticity over time, there is an urgent need to understand the social processes that elevate specific aspects and relegate others in tourism research.

In regard to authenticity as values held by individuals in modern society, but shaped in macro processes, this raises the question of whether the current dichotomy between object-related and existential authenticity could be transcended. According to Wang (2000), the concept of existential authenticity refers to personal or intersubjective feelings whereby:

> *...people feel that they are* themselves *much more authentic and more freely self-expressed than they are in everyday life, not because the toured objects are authentic, but rather because they are engaging in non-everyday activities, free from the constraints of daily life*. (2000: 49–50)

Without denying that people have such feelings during their holidays, I will still argue that these feelings are the result of macro scale processes of 'authentification' (Cohen, 1988: 379), thereby ascribing values which can be analyzed, regardless of whether they relate to objects or personal relationships. As MacCannell (1999: 94) wrote about 30 years ago: '...In our society, intimacy and closeness are accorded much importance. These qualities are seen as the core of social solidarity and they are also thought by some to be morally superior to rationality and distance in social relationships, and more "real"'. In spite of interesting studies utilizing the concept of 'existential authenticity' (Kim & Jamal, 2007; Steiner & Reisinger, 2006), the concept continues to retain some of the essentialism that Cohen (1988) and Bruner (2005) had attempted to eliminate. To interpret authenticity as values and ideas strongly tied to a Western modernity, and free from its previous essentialist connotations, is particularly important at a time when tourism researchers take an interest in the practices of other cultural traditions, only to subjugate them to the tourism label. Authenticity as a concept is strongly connected to a Western modern tradition (Bendix, 1997), and might become a fallacy when applied to non-Western modernities that are now engaging in practices and performances being catered for by a tourist industry. In such cases, the concept of authenticity, when applied to other modernities, might well show up to be among those *Grande idées* that becomes too metaphorically and metonymically extended and therefore should be thrown away.

Cohen (in this chapter) seems to call for a track of research that appropriately addresses the theme of authentic experiences, object related or not, in collective social processes. In support of this call, it may be suggested that its perspective be extended to appropriately include the economic, social and cultural power relations at work in the discourse on authenticity in tourism. This relates to those processes where political interests play a definite part, such as the branding of nation states, the traditionalization of ethnic minorities, the heritagification that occurs in many Western societies, to identify just a few. This implies the re-contextualization, rather than de-contextualization, of the concept of staged authenticity in empirical studies. This suggestion resonates with Bruner's argument that, '...because disembodied decontextualized narratives have no politics, as there are no persons or interests involved' (2005: 128). The same applies to the growing importance of labels such as peak experiences, experience economy and self-discovery, so prevalent in a global Western tourism discourse to the extent that these labels have an impact on how people make meaning out of what they do during their holidays, even when performances are divergent. Also, one should pay more attention to the power relations in those social and cultural processes that shape people's experiences of 'existential authenticity' to the extent that people deem it more 'true to themselves' in some social contexts than in others. This refers in particular to questioning what these changes may entail for the modern, autonomous, self-maintained individual – sometimes called the liberal self or 'the detached self'.

Finally there is an urgent need to gain an understanding of the power relations that are so obvious in tourism. Traditional disciplines, like folklore

(Bendix, 1997), have already traced the changes in the meanings offered by this variation in the concept. In such cases, the concept of authenticity has played a prominent role for both scholars and their object of study. However, this alternative context relates to our object of study, as it also pertains to the scholarly use of the concept in a way that lodges tourism studies within, or near about, the core of the parent disciplines.

10.3

Authenticity Matters: Meanings and Further Studies in Tourism

Philip L. Pearce

While preparing this analysis of the contributions of several scholars to the topic of authenticity in tourism research, an obituary appeared in an Australian newspaper (Butler, 2007). The newspaper column reflected on the achievements and the influence of Jean Baudrillard (1929–2007) whose work can certainly be read as pivotal to the history of the authenticity debate. Together with a number of colleagues, Baudrillard asserted the value of a postmodern perspective and effectively challenged the view that we seek authenticity (Baudrillard, 1983, 1988). It now seems particularly timely, following the passing of one of the masters of the postmodern perspective, to ask: is the original concept of authenticity, which was developed initially in pre-modern times, something of enduring value? Alternatively, should we adopt the view, as Olsen (2007) hints, that it is a once grand idea, now thoroughly reworked by postmodern appraisals and ready to be discarded? The assertion of the analysis presented here is that authenticity still matters. The pathway to reach this conclusion will be pursued in this commentary by asking some fundamental questions which help assess the continuing value of authenticity as a research concept. The questions are as follows:

What Does the Term Authenticity Cover?

Erik Cohen's (2007) comprehensive analysis identifies multiple contemporary meanings of the concept of authenticity. Cohen suggests that there is now a fairly widespread view that the notion of a simple, physical or material quality to authenticity is no longer viable. He suggests that most researchers have sought other senses of the term 'authenticity' to characterize the settings they examine. Cohen offers six such senses: authenticity as customary practice or long usage; authenticity as genuineness in the sense of an unaltered product; authenticity as pristinity, an unadulterated state of nature such as in pristine tropical paradise; authenticity as sincerity when applied to relationships; authenticity as creativity with special relevance to cultural performances including dance and music; and authenticity as the flow of life in the sense that there is no interference with the setting by the tourism industry or other managers. All of these senses of the term may be labelled as aspects of objective authenticity because they all contain the promise that the situation may be appraised using etic or external criteria. Cohen favours the uninterrupted flow of life approach to authenticity where the inhabitants or sites are not framed for tourism. He suggests this sense best fits MacCannell's (1976) original intention and adds that this usage is appropriate to the unplanned and unexpected sights of travel often described in everyday language as authentic.

By way of contrast, 'subjective authenticity' derives more directly from postmodern perspectives on social and cultural life. It can be understood as a negotiated and socially constructed personal view. The term 'existential authenticity' is used by some researchers as a special variant of 'subjective authenticity' (Cary, 2004; Steiner & Reisinger, 2006). This expression may be conveniently reserved for the most uplifting dramatic and numinous traveller experiences where there are distortions of time and heightened meanings attached to the event. Such experiences may have consequences and insights for the travellers' views of their everyday world. Presumably, but not directly labelled by any commentators, there is also a 'mundane subjective authenticity'; a term which might recognize that everyday activities such as noting cultural and location differences and similarities as well as standing in line, losing luggage and walking long distances, are real and genuine travel experiences with a different kind of authenticity value and import. To confirm this point, it can be suggested here that the term 'mundane authenticity' be added to the lexicon of terms as a way of identifying non-transforming but still worthwhile insights into local sights and cultures which are mindfully processed by the embodied tourist and become an important part of their travel storytelling and reminiscences.

The instances where subjective authenticity and objective authenticity are linked gives rise to a further term – 'constructed authenticity', where this concept explores the practices, judgments and evaluations individuals use to assess some or all of the senses of objective authenticity. One additional term completes the family lexicon. 'Emergent authenticity' relates to framed events and spaces acquiring their own cultural value so that they too fit into the regular flow of life. In this sense cultural spaces and events become famous in themselves and are seen as genuine cultural icons. Disneyland is a notable example.

This brief traverse of the array of meanings of authenticity alerts us to the possibility that accepting or rejecting the social science usefulness of the term is not a simple one dimensional decision. The usefulness and use of such variants to the term 'authenticity' awaits attention to further questions.

Who Uses the Term 'Authenticity'?

The value of academic conceptual schemes and terms is partly realized through their frequency of use and their penetration through the invisible colleges of scholarship. 'Authenticity' fares quite well on these accounts. Its core home in tourism lies in the many publications using the expression in *Annals of Tourism Research*. By 2012 and some 40 years after MacCannell's book was first published, more than 45 papers are indexed in this source alone as including 'authenticity' as a keyword. The pursuit of the ideas surrounding authenticity seems to attract sociologists, anthropologists, geographers and psychologists but leisure study researchers and cultural commentators are also more than occasional visitors to authenticity territory (Harris, 2005). This spread of the term has consequences of import beyond the value of the term itself. As Jafari (2005) stresses, an important status marker and achievement for tourism studies is to export knowledge to other fields and in these terms and despite its complexities, authenticity is a leading commodity. In this sense to discard authenticity might reduce the potential trade amongst disciplines – a trade which Jafari and other commentators assert is important to the future of sophisticated tourism study.

The use of the term 'authenticity' is not restricted to academic analysts and researchers. It is widespread in at least two other audiences – tourism promoters and amongst tourists themselves. In the promotional sphere it can be suggested that the earliest use of the term in the sense of objective authenticity often prevails with frequent descriptions of ethnic groups and cultural performances as imbued with the magical material or physical essence of authenticity. Thus a promotional piece on taking a four- wheel-drive day tour visiting central Sardinia offers:

> the chance to see and meet with the real people of the region whose authentic way of life is right in front of the traveller at every turn of the road. (Sardinia Tipico, 2004)

Tourists too, through their travel diaries, through their survey responses and increasingly through blogs represent and describe their travels as authentic (Woodside *et al.*, 2007). This use of the term seems to be closer to that of existential authenticity since it is applied to and captures the highlights and the peaks of travel experience rather than being used to characterize the everyday (but presumably still authentic) mundane activities of quietly observing the world and coping with travelling (Cary, 2004). In particular, the term 'authenticity' as used by tourists often resembles the marking of an insightful moment or a revelation in the passage of their travels and is linked to what Pearce (1988) described as the spontaneous and gift-like quality of the incident. Cohen (2007) also addresses this theme, although in a horrific rather than joyful sense, in his use of the 2004 tsunami experience for travellers in South East Asia.

There are some other people using the term 'authenticity' who need to be considered. The heritage managers and the legion of designers, conservationists and preservationists who manage and present many of the world's major tourist attractions also engage with the authenticity lexicon. This application is linked to one of the two uses of the term 'constructed authenticity' – the first being applicable to the places and performances which are professionally and elaborately created, while the second use of the term refers to the psychological process of visitors selecting criteria to assess places for their acceptability on an authentic–fake dimension.

A recent addition to the users of terms 'authentic' and 'authenticity' can also be identified amongst the new coalition of researchers known as positive psychologists. In their appraisal and conceptualization, authenticity is a character strength defined as speaking the truth and identifying oneself in a genuine way (Seligman *et al.*, 2005; Peterson & Seligman, 2004). The link to tourism studies with this use of the term is not yet a strong one, although it can be noted that Sofield (1991) identified an important extension to the original objective authenticity usage with a focus on the behaviour of the performers and participants in addition to the settings and stages which had preoccupied MacCannell's vision. The new use of the term by the large number of adherents to positive psychology research is a further instance of the growing diaspora of people using or in some ways linked to the concepts of authenticity (Seligman, 2002). These multiple uses and users may best be seen as a positive argument for continuing to employ the term. In addition to the pragmatic perspective that such a commonly used term is unlikely to fade away, the subtleties inherent in the concept offer a bridge for dialogue amongst different users. This in itself is good for tourism and for the wider effort of making social science matter (Flyvbjerg, 2001) even if there remain struggles to be clear about the meanings different discussants bring to the conversations.

How is Authenticity Linked to Other Concepts in Tourism Studies?

A concept which is set apart from others occupies a potentially vulnerable space. It can be seen as a product of its time and further, viewed as failing to offer researchers continuing insights as new ideas are developed. The fate of such ideas is that they appear less in academic publications; they are dropped from the text books and fade into the history of the earlier preoccupations of the discipline area. In tourism an example is that of Plog's allocentric-psychocentric approach to motivation (Mansfeld, 1992; Pearce, 1993; Goeldner & Ritchie, 2002) and if authenticity is not to meet a similar demise its connections to other major ideas in tourism study may need further development. As with many other efforts in appraising authenticity Cohen (2002) has led the way and has provided an account of the links between sustainability and authenticity. Further connections of substance may be envisaged. Few researchers have linked authenticity in detail to the life-cycles of destinations (Butler, 2006) yet many rich possibilities for study can be conceptualized such as how subjective authenticity and constructed authenticity develop and how audiences react to the promoters' accounts of objective authenticity.

Authenticity has been closely linked to the concept of gaze throughout its history. Indeed, there are sometimes confusing or possibly intentional substitutions of the terms sight and site in the research literature (*cf*. Cohen, 2007: 1, for the joint use of sights and sites to describe MacCannell's earliest work). Further links which can be suggested include establishing connections to the work on tourism as an embodied experience. In particular, the roles of dance and music and the performances built on these modalities may possibly provide appropriate sites to study physiologically stimulating and embodied encounters involving more than patterns of gaze (Daniel, 1996; Roberts, 2006).

The production of tourism settings has received attention through the work of Ritzer (1993, 1999) on McDonaldization as well as that of Bryman (2004) and Harris (2005) on Disneyfication. There are several connections to be developed here since the outcomes of these processes with their standardized offerings and in the theme parks their hyperreal sights help to define an international social representation of global leisure culture, capitalism and possibly inauthenticity. The very topic of social representations, which are the collective, shared everyday meanings and mini theories of how parts of life operate, is a potentially pivotal research nexus to be developed. A study of the social representations of the authentic goes beyond the summary of individual views (a perspective Cohen labels psychologistic and by implication inadequate) because a proper Moscovici (1988) inspired account of representations attends to the collective processes by which people acquire information from each other, the media and other information sources (*cf*. Pearce *et al*., 1996).

A further link to other tourism concepts which may better embed authenticity in the growing armoury of tourism research developments is to make a link to the concept of mindfulness and the topic of interpretation. Mindfulness is an active mental state where visitors are attending to the setting, processing information and integrating their experience with pre-existing material (Moscardo, 1999). Mindfulness is stimulated by several key external stimuli including the physical dimensions of the world being viewed as well as the interpretive efforts of guides or advisory stimuli material. The links to authenticity here are varied. The role of information and interpretation in shaping subjective authenticity is one direction of interest. The characteristics of surprise and novelty, which characterize some existential authenticity accounts, are both linked to mindful mental processing. An exploration of when scenes and sites are processed mindfully or mindlessly may hold important cues as to whether authenticity is questioned or simply accepted and, further, when it is questioned, how it is judged. As authenticity becomes better embedded in and linked to more tourism concepts its continuing role in tourism analyses will be fruitful and enduring.

How do you Measure Authenticity?

For researchers interested in pursuing the diverse topics already outlined there are several suggestions which may enhance authenticity studies. First, the ability to choose sites which may be meaningfully compared along some well-organized dimensions – such as length of operation in terms of time or tradition, activity style, audience involvement, mediums of performance – may stimulate comparative authenticity studies. The consideration of multiple sites varying in these dimensions may move research beyond the isolated and often idiosyncratic analyses of case studies. Some creativity in assessing authenticity responses may also be encouraged with direct questions about the authenticity of the overall performances or sites needing to be supplemented with interviews, response to photographs and assessment of unsolicited blogs and analysis of media reports. In seeking these broader inputs to accessing audience responses researchers could also seek to identify anchor points and contrast options, some of which may be partly defined by the processes and products reported in the literature on Disneyfication.

Beyond the Questions

This consideration of issues relating to authenticity in tourism studies suggests that some new developments and possibly a new era of research extending this work are at hand. Through better definitions and partitioning of the concept, through linking to multiple users and uses of the term, and by connecting to other topics in tourism with fresh methods authenticity may have a continuing life in and beyond tourism studies.

Concluding Remarks[5]

Several common threads of inquiry unite the treatment of authenticity in the three papers constituting this chapter. In common with tourism and social science scholars globally, the authors seek to identify the multiple meanings of this well-used and distinctive tourism-derived term. They tackle the task diligently. Cohen's contribution offers a core distinction between objective and subjective authenticity. He argues that objective authenticity, at core a belief that the properties of what we see define authenticity, has to be understood in six ways. The Pearce paper summarizes these same approaches which are reinforced by the comments of Olsen. In the end, Cohen opts for the approach of thinking about objective authenticity as representing the uninterrupted flow of life at a site or for a community where the space and behaviours are not marked off or framed for tourism. As soon as marking and tourist-directed identification occurs, then in this view, objective authenticity no longer exists. It is a stringent but clear approach.

Cohen's contribution also stresses a core distinction between objective and subjective authenticity. A subdivision of subjective authenticity identified is that one can propose the existence of two forms: existential authenticity and mundane authenticity. The former includes a focus on momentary states which emerge serendipitously and are unlikely to be repeated. Mundane authenticity is less dramatic but can be viewed as moments in ordinary tourist activity where distinctive sights and content contribute to powerful memories and iconic stories defining the individual travel experience.

Olsen, Cohen and Pearce all contemplate the future of the authenticity concept. They are optimistic that it can have enduring value but see it evolving from MacCannell's front stage–back stage metaphor emphasizing objectivity concerns to now emphasizing the collective social processes which make the term meaningful. Additionally, future research can continue to tackle our understanding of which individual experiences, episodes, relationships and sights are labelled as authentic.

It can be argued that the common concerns of the authors about the meaning of the term and its future health will be viewed and used differently by researchers working within different research paradigms. For positivist and post-positivist researchers, authenticity may remain a frustrating and slippery concept requiring careful emic approaches to aggregate individual perceptions and establish broader social representations. To those working in constructivist and interpretive paradigms, the nuances of the concept may be more easily appreciated but here the challenge lies in using creative methods to evoke rich and consensually validated responses from those who travel. For researchers operating in a critical realism paradigm the attention to authenticity may move in the direction of asking the question: 'How is authenticity possible? Or who makes it possible?' The resulting answers may help sort out the array of actors contributing to the making of places and experiences. Finally, and certainly not unimportantly, researchers who work with more critical theory approaches and language may seek to ask questions about power and privilege, some of which Olsen also raises. Such attention may expose political and developmental agendas seeking to preserve rather than develop places, cultures and communities. Further some insights might be generated through the critical theory approach into the epistemology of tourists' beliefs in terms of what they have seen and learned. In brief, the chapter offers options and opportunities for the multiple observers of today's tourism.

Discussion Questions

(1) How do the suggested new meanings of authenticity relate to Goffman's front and back stage concepts which were employed in the original MacCannell work?
(2) What kinds of conceptual links can be developed between authenticity and sustainability as pivotal tourism concepts?
(3) Does the concept of authenticity, in the 'non-marked, flow of life' sense, have meaning for non-Western tourists?
(4) Is there a relationship between authenticity and slow tourism?
(5) What is the role of interpretation and one's sense of time in shaping the subjective authenticity of tourists' experiences?
(6) Can authentic moments exist in well planned commercial tourism settings such as theme parks and shopping villages?

Notes

(1) The title paraphrases that of Anthony Giddens' (1976) 'Functionalism: *Aprés la lutte'*, published close to a third of a century ago.
(2) The ambiguous attitude of the authors to the viability of the concept is curiously reflected in the difference between the title of their article, 'Reconceptualizing object authenticity', and the running title on the top of its pages, 'Abandoning Object Authenticity'.
(3) For an interesting study of the way in which the management of a heritage site manipulates the simultaneous use of authoritarian 'objective' and constructed criteria of authenticity, see Gable and Handler (1996).

(4) 'Indians of Europe' is an epithet by which some promotional literature referred to the Basque people in France (Lacy & Douglass, 2002: 11).
(5) Contributed by Philip L. Pearce.

References

Baudrillard, J. (1983). *Simulations*. London: Semiotext.

Baudrillard, J. (1988). *Simulacra and Simulations. Selected Writings* (pp. 166–184). California: Stanford University Press.

Bendix, R. (1997). *In Search of Authenticity: The Formation of Folklore Studies*. Madison: The University of Wisconsin Press.

Berger, P. (1973). Sincerity and authenticity in modern society. *Public Interest* 31: 81–90.

Brown, D. (1996). Genuine fakes. In T. Selwyn (ed) *The Tourist Image* (pp. 33–47). Chichester: John Wiley and Sons.

Bruner, E.M. (1986). Ethnography as narrative. In V.W. Turner and E.M. Bruner (eds) *The Anthropology of Experience* (pp. 139–155). Urbana: University of Illinois Press.

Bruner, E.M. (2005). *Culture on Tour: Ethnographies of Travel*. Chicago: The University of Chicago Press.

Bryman, A. (2004). *The Disneyization of Society*. London: Sage

Butler, R.W. (ed) (2006). *The Tourism Area Life Cycle: Conceptual and Theoretical Issues* (Vol 2). Clevedon: Channel View.

Butler, R.W. (2007). Theorist upset lazy thinking. *The Australian*, 9 March (Arts, p. 16).

Cary, S.H. (2004). The tourist moment. *Annals of Tourism Research* 31(1): 61–77.

Chhabra, D., Healy, R., and Sills, E. (2003). Staged authenticity and heritage tourism. *Annals of Tourism Research* 30(3): 702–719.

Cohen, E. (1988). Authenticity and commoditization in tourism. *Annals of Tourism Research* 15(3): 371–386.

Cohen, E. (2002). Authenticity, equity and sustainability in tourism. *Journal of Sustainable Tourism* 10(4): 267–276.

Cohen, E. (2004). *Contemporary Tourism: Diversity and Change*. Amsterdam: Elsevier. (Please see 'Nomads from affluence: Notes on the phenomenon of drifter-tourism' pp. 49–63; 'A phenomenology of tourist experiences' pp. 65–85; 'Authenticity and commoditization in tourism' pp. 101–114; 'Contemporary tourism: Trends and challenges' pp. 131–143.)

Cohen, E. (2005). Tourism and disaster: The tsunami waves in Southern Thailand. In W. Alejziak and R. Winiarski (eds) *Tourism in Scientific Research* (pp. 81–114). Krakow: Academy of Physical Education and Rzeszow, University of Information Technology and Management.

Cohen, E. (2007). Authenticity in tourism studies: *Après la lutte*. *Tourism Recreation Research* 32(2):75–82.

Cohen, E. (2012). Major trends in contemporary tourism. In a book based on papers from the 9th Biennial Conference of the International Academy for the Study of Tourism (Beijing, 2005). (forthcoming).

Csikszentmihalyi, M. and Csikszentmihalyi, I.S. (1988). *Optimal Experience: Psychological Studies of Flow of Consciousness*. Cambridge: Cambridge University Press.

Daniel, Y.P. (1996). Tourism dance performances: Authenticity and creativity. *Annals of Tourism Research* 23(4): 780–797.

Dann, G.M.S. (1999). Writing out the tourist in space and time. *Annals of Tourism Research* 26(1): 159–187.

Douglass, W.A. and Raento, P. (2004). The tradition of invention: Conceiving Las Vegas. *Annals of Tourism Research* 31(1): 7–23.

Flyvbjerg, B. (2001). *Making Social Science Matter*. Cambridge: Cambridge University Press.

Gable, M. and Handler, R. (1996). After authenticity at an American heritage site. *American Anthropologist* 98(3): 568–578.

Geertz, C. (1973). *The Interpretation of Cultures*. New York: Basic Books.

Giddens, A. (1976). Functionalism: *Aprés la Lutte. Social Research* 43(2): 325–366.

Giddens, A. (1991). *Modernity and Self-Identity*. London: Polity Press.

Goeldner, C.R. and Ritchie, J.R.B. (2002). *Tourism: Principles, Practices, Philosophies* (9th edn). New York: Wiley.

Goffman, E. (1974). *Frame Analysis: An Essay on the Organization of Experience*. New York: Harper and Row.

Graburn, N.H.H. (1976). Eskimo art: The Eastern Canadian Arctic. In N.H.H. Graburn (ed) *Ethnic and Tourist Arts* (pp. 39–55). Berkeley: University of California Press.

Halewood, C. and Hannam, K. (2001). Viking heritage tourism: Authenticity and commodification. *Annals of Tourism Research* 28(3): 568–580.

Harris, D. (2005). *Key Concepts in Leisure Studies*. London: Sage.

Hughes, G. (1995). Authenticity in tourism. *Annals of Tourism Research* 22(4): 781–803.

Jafari, J. (2005). Bridging out; nesting afield: Powering a new platform. *Journal of Tourism Studies* 16(2): 1–5.

Kim, H. and Jamal, T. (2007). Touristic quest for existential authenticity. *Annals of Tourism Research* 34(1): 181–201.

Lacy, J.A. and Douglass, W.A. (2002). Beyond authenticity: The meanings and uses of cultural tourism. *Tourist Studies* 2(1): 5–21.

Langer, S. (1942). *Philosophy in a New Key. A Study in the Symbolism of Reason, Rite and the Art.* Cambridge, Massachusetts: Harvard University Press.

Littrell, M.A., Anderson, L.F. and Brown, P.J. (1993). What makes a craft souvenir authentic? *Annals of Tourism Research* 20(1) 197–215.

MacCannell, D. (1973). Staged authenticity: Arrangements of social space in tourist settings. *American Journal of Sociology* 79(3): 589–603.

MacCannell, D. (1976). *The Tourist: A New Theory of the Leisure Class.* New York: Schocken Books.

MacCannell, D. (1999). *The Tourist. A New Theory of the Leisure Class.* Berkeley: University of California Press.

Mansfeld, Y. (1992). From motivation to actual travel. *Annals of Tourism Research* 19: 399–419.

McIntosh, A.J. and Prentice, R.C. (1999). Affirming authenticity: Consuming cultural heritage. *Annals of Tourism Research* 26(3): 589–612.

McLeod, M.D. (1976). Limitations of the genuine. *African Art* 9(3): 31, 48–51.

Mehmetoglu, M. and Olsen, K. (2003). Talking authenticity: What kind of experiences do solitary travelers in the Norwegian Lofoten Islands regard as authentic. *Tourism, Culture and Communication* 4: 137–152.

Moscardo, G. (1999). *Making Visitors Mindful: Principles for Creating Quality Sustainable Visitor Experiences Through Effective Communication*. Champaign, Illinois: Sagamore Publishing.

Moscovici, S. (1988). Notes towards a description of social representations. *European Journal of Social Psychology* 18(3): 211–250.

Nettekoven, L. (1973). Touristen sind eben keine völkerkundler [Tourists are not ethnologists]. *Auslandskurier* 3: 28–29.

Olsen, K. (2002). Authenticity as a concept in tourism research. *Tourist Studies* 2(2): 159–182.

Olsen, K. (2007). Staged authenticity: A *Grand idée*? *Tourism Recreation Research* 32(2): 83–85.

Pearce, P.L. (1988). *The Ulysses Factor: Evaluating Visitors in Tourist Settings*. New York: Springer Verlag.

Pearce, P.L. (1993). Fundamentals of tourist motivation. In D. Pearce and R. Butler (eds) *Tourism Research: Critiques and Challenges* (pp. 85–105). London: Routledge and Kegan Paul.

Pearce, P.L., Moscardo, G. and Ross, G.F. (1996). *Tourism Community Relationships*. Oxford: Pergamon.

Peterson, C. and Seligman, M.E.P. (2004). *Character Strengths and Virtues: A Handbook and Classification*. Washington DC: American Psychological Association.

Reisinger, Y. and Steiner, C.S. (2006). Reconceptualizing object authenticity. *Annals of Tourism Research* 33(1): 65–86.

Richards, G. and Wilson, J. (eds) (2004). *The Global Nomad: Backpacker Travel in Theory and Practice*. Clevedon: Channel View Publications.

Ritzer, G. (1993). *The McDonaldization of Society: An Investigation into the Changing Character of Contemporary Social Life*. London: Sage.

Ritzer, G. (1999) *Enchanting a Disenchanted World: Revolutionising the Means of Consumption*. Thousand Oaks, CA: Pine Forge press.

Roberts, J. (2006). *A Sense of the World*. London: Simon and Schuster.

Russell, C. (1992). Beyond authenticity: The discourse of tourism in ethnographic and experimental film. *Visual Anthropology* 5: 131–141.

Ryan, C. (1991). *Recreational Tourism: A Social Science Perspective*. London: Routledge.

Sardinia Tipico (2004). *Guide to Tours of Central Sardinia*. Cagliari: Sardinia Tourism.

Schuetz, A. (1973). Multiple realities. In A. Schuetz *Collected Papers Vol. 1* (pp. 99–117). The Hague: M. Nijhoff.

Seligman, M.E.P. (2002). *Authentic Happiness: Using the New Positive Psychology to Realize Your Potential for Lasting Fulfillment*. New York: Free Press/Simon and Schuster.

Seligman, M.E.P., Sheen T.A., Park, N. and Peterson, C. (2005). Positive psychology progress: Empirical validation of interventions. *American Psychologist* 60(5): 410–421.

Selwyn, T. (1996). Introduction. In T. Selwyn (ed) *The Tourist Image* (pp. 18–28). Chichester: John Wiley and Sons.

Sofield, T. (1991). Sustainable ethnic tourism in the South Pacific: Some principles. *Journal of Tourism Studies* 2(1): 56–72.

Steiner, C.J. and Reisinger, Y. (2006). Understanding existential authenticity. *Annals of Tourism Research* 33(2): 299–318.

Taylor, J.P. (2001). Authenticity and sincerity in tourism. *Annals of Tourism Research* 28(1): 7–26.

Theobald, W. (1998). *Global Tourism* (2nd edn). Oxford: Butterworth-Heinemann.

Trilling, L. (1972). *Sincerity and Authenticity*. London: Oxford University Press.

Uriely, N. (2005). The tourist experience. *Annals of Tourism Research* 32(1): 199–216.

Uriely, N., Yonay, Y. and Simchai, D. (2002). Backpacking experiences: A type and form analysis. *Annals of Tourism Research* 29: 519–537.

van den Abbeele, G. (1980). Sightseers: The tourist as theorist. *Diacritics* 10: 2–14.

Wagner, U. (1977). Out of time and place: Mass tourism and charter trips. *Ethnos* 42(1–2): 3–52.

Waitt, G. (2000). Consuming heritage: Perceived historical authenticity. *Annals of Tourism Research* 27(4): 835–862.
Wang, N. (1999). Rethinking authenticity in tourism experience. *Annals of Tourism Research* 26: 349–370.
Wang, N. (2000). *Tourism and Modernity: A Sociological Analysis.* Kidlington: Pergamon.
Woodside, A.G., Cruickshank, B.F. and Dehuang, N. (2007). Stories visitors tell about Italian cities as destination icons. *Tourism Management* 28(1): 162–174.

Further Reading

Gilmore, J.H. and Pine, B.J. (2007). *Authenticity: What Consumers Really Want.* Cambridge, MA: Harvard Business School Press.
Honore, C. (2004). *In Praise of Slow*. London: Orion.
Jackson, T. (2009). *Prosperity Without Growth: Economics for a Finite Planet.* London: Earthscan Ltd.
Murphy L.E., Moscardo, G., Benckendorff, P. and Pearce, P.L. (2011). *Tourist Shopping Villages: Forms and Functions.* New York: Routledge.
Nisbett, R.E. (2003). *The Geography of Thought*. Nicholas Brearley: London.
Zimbardo, P. and Boyd, J. (2008). *The Time Paradox.* London: Rider.

Chapter 11

Heritage Tourism: Heritage Tourists

Gregory Ashworth, Brian Wheeller and C. Michael Hall

Context

The core argument was the charge that tourists destroy the heritage they have come to experience and is as old as heritage tourism itself. It is a longstanding, deeply and commonly held view that is expressed overtly, but more often covertly, by managers of heritage resources and sites and by place managers and politicians representing heritage localities. This essay argues both for and against five serious accusations levelled at tourists and the tourism industry, namely:

- tourist heritage is inherently inferior, being trivial, superficial and often just wrong;
- tourists have inappropriate motives and engage in inappropriate behaviour;
- tourists damage heritage sites;
- tourists crowd out and displace other more worthy users of heritage;
- tourists are free-riding economic parasites.

The concluding verdict depends upon the prior establishment of answers to two major questions: the first about the distinctiveness of the tourist and the second about the distinctiveness of the heritage. These arguments were pursued, extended and further exemplified by commentaries from Wheeller and Hall.

11.1

Do Tourists Destroy the Heritage They Have Come to Experience?

Gregory Ashworth

The Charge

The charge that tourists destroy the heritage they have come to experience is as old as heritage tourism itself. It is a long-standing, deeply and commonly-held view that is expressed overtly, but more often covertly, by managers of heritage resources and sites, and by place managers and politicians representing heritage localities. In part, it is just one aspect of an 'anti-grockle' sentiment in which 'locals' express their disdain for tourists, but it is especially sharply felt in heritage tourism, with many feeling profoundly uncomfortable with what, to them, seems an incongruous and even contradictory combination of adjective and noun.

The managers of heritage sites and structures are impaled on the horns of a dilemma, an uncomfortable situation that no doubt increases the vehemence of their reaction. On the one hand, they need the direct financing, but also less direct legitimation, that tourism numbers provide. Even when largely dependent upon public subsidy, visitor numbers are a powerful argument for more state money. On the other hand, 'the golden horde' is both feared and despised. Consequently, opposition to tourism and dislike of the tourist is often unexpressed in public or expressed through various code words. Policy statements freely use the word 'sustainable' nearly always as a synonym for less. The slippery concept of 'carrying capacity' is similarly often misused. Which heritage agency report has concluded that more tourists would be quite sustainable or that the carrying capacity of a heritage attraction far outstrips the current visitation numbers? This is often combined with a search for the 'right sort' of tourist, in which the current tourists are to be replaced, usually in smaller numbers, by tourists who more closely resemble the economic, educational, cultural and social class of the heritage managers. At best, there is a grudging acceptance of the tourist as a necessary evil to be contained and controlled as far as possible.

At heart, many local policy-makers are of the 'stay at home, send the money and we will send the postcards' persuasion.

The Propositions

Tourist heritage is inherently inferior, being trivial, superficial and often just wrong

The prosecution

The tourist experience of local heritage is short, with visits to particular towns measured in hours and to a particular site or exhibit better measured in minutes or even seconds. Because the tourist, who by definition is a cultural outsider, lacks knowledge, contextual background and sensitivity to the heritage being visited, a rigorous selection of heritage highlights is made, usually by the guidebook. Complexity is reduced to simplicity in a sanitized past, lacking depth and context. 'Nottingham becomes the city of Robin Hood and Heidelberg the city of the student prince' (Ashworth & Tunbridge, 1990: 54). The unusual or the spectacular becomes memorable snippets created for easy consumption and instant gratification. The tourist having visited a tower in Pisa notes only that it leans: the name of the architect and period of architecture is unimportant, unremarked and unremembered. The nightmare scenario is 'Disneyfication', in which pasts are condensed into easily consumed, bite-sized pieces lacking any authenticity. The 'Seoul Declaration' of ICOMOS (2005), on managing tourism in historic towns in Asia, expressed its concern about the 'importance of accurate and aesthetic interpretation and presentation of heritage places for tourism'.

The defence

History can be true or false: heritage cannot. As a product of the human creative imagination, heritage has no authenticity of the object or the historical record, only the authenticity of the experience as perceived by the user. It can be well or badly presented, relevant or irrelevant to the visitor, and effectively or ineffectively communicated: it cannot be wrong. As to the idea that the locals have a profounder and more nuanced appreciation of heritage than the superficial and transitory tourist, the reverse is increasingly the case. The rise of special-interest tourism and heritage tourism is only a *pot pourri* of very diverse special interests, which means that often it is the tourist who has the deeper more specialized knowledge, often in a global comparative context, and who discovers and rediscovers local heritages unknown or unappreciated by the locals themselves. Calls for authenticity by the usually self-appointed guardians of 'our' heritage and its assumed values, generally mean their particular authenticity and, above all, their right to be the authenticator of it.

Tourists have inappropriate motives and engage in inappropriate behaviour

The prosecution

Tourists are in search of fun: they are spending their free time and money on entertainment and distraction, motivated by curiosity and the pursuit of pleasure. This can rapidly degenerate into the distasteful and the offensive. Battlefield tourism (Seaton, 2002) or 'Schindler tourism in the "Auschwitzland" product of Krakow' (in essence 'let's have fun with the holocaust'; Ashworth, 1996) are just a few of the many cases of an objectionable heritage tourism, offensive to those with more worthy motives.

Tourists are on holiday freed from the daily mundane disciplines of work and liberated from the social and cultural constraints of their home society. They are thus prone to behave in ways that give offence to locals and to more legitimate and appropriate users of heritage sites. Their dress, demeanour and behaviour are likely to be unacceptable and destructive of the value of the sites and buildings they visit. Spiritual sites, whether churches, mosques, synagogues or just sacred spaces, such as Ayers Rock, suffer especially from the inappropriate and often offensive behaviour of fun-seeking tourists. If permitted at all, tourists must be regulated and controlled through imposed 'behavioural codes' and 'responsible tourism codes' (Malloy & Fennell, 1998) so that they engage in an 'ethical travel' (Pattullo, 2006) that does not give offence to either locals or to other, more worthy, prioritized heritage users. Roowaan (2005) notes that visitors to the *Neue Wache* memorial to the German war dead in Berlin are expected to behave with *Anstand* (respect), and Till (1999) even reports being escorted out of that building for 'loud talking'.

The defence

Who decides which motive is more worthy and which behaviour is more acceptable? There is an assumption that residents are more commendable than visitors, and that education or aesthetic fulfilment is more meritorious a motive than mere pleasure or entertainment and thus should be prioritized. The frequently encountered on-site lists of rules and admonitions, as well as the suggested codes of conduct and behaviour for tourists, make much use of the word 'respect'. It is insisted that the tourist should respect the behaviour and mores of the locals, even presumably if these are distasteful to the tourist. The tourist may encounter political oppression, racial bigotry, child prostitution, animal cruelty and even just repellent food habits, which are supposed to be respected. The asymmetry of this relationship of host and guest is inherent in such thinking. Should not the hosts also respect the mores of the tourists, freely dispensing their discretionary time and money? With trade in many other products, consumers are becoming increasingly ready to use their purchasing power to intervene to remedy unsatisfactory production conditions (such as child labour). Why is it only in tourism services that the reverse position is recommended?

Tourists damage heritage sites

The prosecution

Heritage sites and artefacts, being often old, may be particularly fragile and vulnerable to damage. Tourists unavoidably cause damage in three main ways. First, they physically damage the structures and artefacts they visit through their feet, hands, breath, sweat, digestive and microbiological systems. Secondly, their physical presence in large numbers destroys the ambience of the site. There is little remaining sanctity in a cathedral in which visitors shuffle around a one-way pedestrian flow system, controlled by traffic lights. 'It is difficult to experience much aesthetic pleasure from an Athenian acropolis around which visitors are "crocodiled" in continuously moving unbroken columns along roped channels, shepherded by guards with whistles' (Ashworth & Tunbridge, 1990: 53). Thirdly, tourists require and attract ancillary services facilities. Herds of tourism buses parked outside Notre Dame cathedral, scrums of insistent hawkers blocking the entrance to the temple of Borobodur, tower block modern hotels literally overshadowing Buddhist temples in Bangkok, could be replicated at almost any major heritage tourism attraction. In historic cities, 'the damage from visitors to historic buildings, streets and squares is recognized as the townscape is spoilt by overcrowding and as it is worn down by numbers' (Orbasli, 2000: 160).

The World Monument Fund (WMF) monitors damage to heritage buildings and sites. It identifies three major threats facing heritage sites: namely, political conflict, climate change and tourism. The tourist is thus seen to be as damaging as war or rising sea levels. In the WMF 2008 list of the 100 most endangered monuments in the world, approximately one-third were diagnosed as being 'in danger' mainly from the tourist.

The defence

It cannot be denied that old structures already weakened by the natural forces of decay may be damaged by visitors. However, it is not tourists as a separate and destructive category of humanity that is to blame: it is people.

Much the same phenomenon occurs in a crowded shopping street, a popular sports match or music festival or on busy commuter transport. People *en masse* are liable to cause wear and damage, which in turn can and should be avoided, or at least mitigated, through management. Damage is, thus, just a reflection of bad management. The floor of Anne Hathaway's cottage at Stratford-on-Avon was literally sinking under the weight of tourists' feet until a new floor was suspended above it. Tourists on-site are relatively easy to manage. Their behaviour is predictable and can be anticipated; their lack of local knowledge renders them particularly dependent upon information, which site managers can use as an instrument of control. The managers of sensitive natural areas learned a generation ago that tourists and fragile ecosystems could be combined using signposts rather than barbed wire and by creating alternative honey-pots that offer a better experience to the visitor than the site to be protected. If the tourists to Hadrian's Wall are wearing it away then build a Vindolanda heritage centre, which will siphon off much of the demand from the wall as well as provide a higher quality, and indeed more 'authentic', experience for the visitor.

It is also worth noting that most heritage is located in urban areas, which are inherently physically robust and constructed to accommodate a high density of users. In comparison, tourist damage to eco-systems in more sensitive rural, coastal and natural areas is likely to be far more severe than that caused by urban heritage tourists (see UNEP, 2005).

Tourists crowd out and displace other, more worthy users of heritage

The prosecution

Very little heritage was actually created for tourists or by the tourism industry. Those who created it, or for whom it was created, are often in danger of being crowded out and ultimately displaced by the tourist. Casual fun-seeking holidaymakers compete with those visiting heritage for more serious and socially beneficial, educational, aesthetic or spiritual reasons. The tourist competes with the resident and, in poorer countries, the richer visitor competes with the poorer local resident. Heritage is a non-renewable resource in danger of being depleted or even exhausted by a growing tourism demand. To counter this displacement of the worthy by the unworthy, the former must be prioritized and the latter, if not banned altogether, must at least be restricted.

The defence

In theory, all heritage is created by the user of it and all heritage consumers create their own heritage. Tourists, therefore, produce their own heritage: they do not appropriate someone else's heritage. In practice, of course, these different heritages may be experienced simultaneously at the same site and even use the same resources. Most heritage is multi-user and this can and must be managed using various well-known techniques. Many Polish churches visited by tourists have two entries and subsequent interior spaces, for tourists and worshippers respectively, and language is employed as the simple control mechanism. Differential pricing for locals and visitors is very commonly used to guarantee the rights of locals at many renowned heritage tourism sites and, again, it is often language, or even skin colour and general appearance, that is the differentiating factor.

The charge is predicated on the mistaken idea that heritage resources are in fixed supply: there is a limited and non-renewable, non-replicable number of heritage objects, buildings and spaces. This assumption clearly underlies the 'Seoul Declaration' of ICOMOS (2005), which prescriptively declaims that, 'tourism sector representatives must work with conservation authorities to establish ways to achieve sustainable tourism without exhausting non-renewable cultural resources such as heritage'. This is just a misunderstanding of heritage or perhaps rather a confusion of heritage with history. As heritage is a demand-derived set of contemporary uses constructed as required then the resources of which it is composed have no limits other than the limits of the human creative imagination. There can be no question of resource shortage or depletion: the resource is ubiquitous and can be created according to the demand for it. Over-use, let alone resource exhaustion, can only be a temporary consequence of bad management, capable of being solved by good practice (see the arguments in the case of the heavily visited heritage city of Bruges; Jansen-Verbeke, 1990).

Take the well-known and long-standing perceived crisis in tourism to the lagoon city of Venice, which has long struggled with an apparent mismatch between what is regarded as a limited supply of heritage tourism sites, above all space, and increasing demands for these from tourists. That this mismatch is to a large extent unreal, as much of the supply remains unused much of the time and is anyway capable of considerable extension, is not the point. The managers of Venice have for some time been reacting to this perceived over-use by attempting to control demand, rendered a practical possibility by the restricted causeway access to the lagoon city (Borg, 1998; Graham *et al*., 2000). Users are routinely prioritized on the grounds of the moral worth of their use ('real' art connoisseurs above mere curious tourists) or their economic return (those staying in hotels in the lagoon city as opposed to day excursionists). The alternative strategy, based on a heritage paradigm, would be to expand the product to meet the rising demand. This can be done by utilizing off-peak surplus capacity or simply by developing 'new Venices', using either the largely unused areas in the existing lagoon islands (the *Arsenale* for instance), some of which are almost empty (Torcello for instance) or duplicating the Venice product in the numerous sites around the Adriatic and Eastern Mediterranean (from Capodistria/Koper to Santorini).

Tourists are free-riding economic parasites

The prosecution

Heritage, from the viewpoint of the tourism industry, provides a near ideal supply of potential resources, which are ubiquitous, highly flexible, infinitely renewable and, above all, generally free of cost. The paradox is that heritage has many of the characteristics of a zero-priced, freely accessible public good. In many ways it is. For the tourist visiting the morphological forms of the city or consuming the historic commemorations and associations of public history, the resource is consumed without direct cost to the consumer. However, heritage from the position of the producer, or indeed the place, is not free. To the costs of selection, maintenance, accommodating collections, promotion, interpretation, marketing and consumer site-management, must be added the opportunity costs of development options foregone. The tourist is thus free-riding on heritage experiences paid for by others. Also a zero-priced, freely accessible public good will inevitably be over-used in a 'tragedy of the commons' scenario (Hardin, 1968).

To this argument can be added the misconception of many local resource managers about the windfall gain economic model. This assumes that the extra demands of tourism provide a windfall gain to heritage facilities that serve other users and would continue to function without tourism. Such a free benefit can only be accrued if the extra demand incurs no extra costs and can be accommodated without depleting or changing the product or reducing the benefits accruing to the other prioritized users. These conditions are unlikely to be met in such a Faustian contract.

The defence

With any multi-user resource, costs and benefits may well be asymmetrically allocated and this is often the case with heritage tourism. However, it could be that it is the resident that is free-riding on facilities that would not exist or be uneconomic without the support of tourism demands. The solution to the asymmetry is obvious in theory and well-known in practice. Many of the costs and benefits of heritage tourism are economic externalities; that is, they accrue to individuals or groups outside the internal economic production system. Many benefit, but do not pay: many pay, but do not benefit. The solution, as developed in environmental economics, is to internalize these externalities through fiscal and other compensatory measures (Ashworth, 2009).

The Verdict?

The argument outlined above, and the sharp divergences of viewpoint based upon such opposed assumptions, is not to be resolved here. The charge that heritage tourists destroy the heritage they have come to experience will continue to be made and be rebutted. However, the basis for a resolution may depend upon the prior establishment of answers to two major questions: the first about the distinctiveness of the tourist, and the second about the distinctiveness of the heritage.

The charge is dependent upon a fundamental assumption that tourists, by virtue of just being tourists, are different in significant respects from locals. The differences in knowledge, background, continuity, context, commitment and other attributes lead to the assumptions about irreconcilable differences in motivation, behaviour and, more fundamentally, in worth. However, if most tourism is not an activity that is different from its antithesis, then the tourist cannot be designated a separate species of humanity. Much of the anti-tourist argument is predicated on the assumption that the host/guest difference is also a dichotomy of poor/rich, uneducated/educated, undeveloped/developed, powerless/powerful, in an economic and cultural colonial relationship. In many instances, this may still be the case, but not in all or even most. Most tourism flows are between places and communities that differ little in these respects. Many are reciprocal. Where then is the difference? Secondly, most tourism is some form of special interest tourism. People use their discretionary time while on holiday to continue and extend interests, tastes, predilections and preferences that they already have while not on vacation. Thus, the tourist is only the resident on holiday: the resident is only the tourist between trips.

If the charge depends upon separating tourists from locals, it also distinguishes between the heritage of the locals and that of the visitors, in terms of its type, quality, depth, worth and ownership. At its heart, there is the core belief that the heritage of the local place belongs to the local people, not to the visitors. However, heritage as a contemporary construct is in a constant state of creation and recreation in which different groups and individuals interact. The idea that locals create a heritage for their own needs and then sell this to visitors, whose consumption leaves the heritage unchanged, is a chimera. In many instances around the world, it was the outsiders who 'discovered' the heritage attraction of a place of which the locals were unaware. Until an outsider arrives, appreciates and records a local culture or heritage, the locals will be unaware of its distinctiveness or its value. Equally, the heritage tourist, ostensibly consuming someone else's heritage, cannot other than in reality be consuming a version of their own heritage in another place. There is a creolization process in action in which locals and tourists together extend, adapt, recreate, modify and discard a continuously shifting array of heritages. This heritage creolization model is well illustrated in the case of New Mexico (Ashworth, 1999).

If these arguments are sound, then the failure to distinguish locals from tourists and local heritage from tourist heritage means that the charge cannot be sustained and can be dismissed as conceptually flawed.

11.2

Heritage Tourists: Responsible, (f)or What?

Brian Wheeller

There is the ever-present danger that much of debate in tourism hangs on definitional niceties (Wheeller, 2007), to the extent that analysis becomes academic – inexorably taking discussion ever further away from the world of the tourist into the (separate) realm of tourism academe. 'Heritage' is a case in point: suitably ambiguous, its precise meaning – let alone what it generically represents – open to conjecture. Even if it were possible to draw exact parameters around an acceptable, 'workable' definition, interpretation of that thereby encompassed would still be subjective. The complexities as to what actually constitutes a heritage attraction are perhaps best exemplified in Prentice's (1993) attempt at a 23-category classification.

The emphasis of Ashworth's probe appears to be focused on the effects tourists have on the built heritage/environment. But I would hazard a guess that a layperson's (i.e. tourist's¿) interpretation of 'heritage' would embrace far wider dimensions than this rather narrow perspective. As, indeed, would those of guardians (self-appointed or otherwise) of 'heritage'. Uzzell's (1989) influential *Heritage Interpretation* was sub-titled *The Natural and Built Environment*. In the UK, the recent popularity of the television series *Coast* would suggest that natural heritage and, currently, coastal heritage in particular, constitute significant elements in the nation's psyche…and not only in an idealized, abstract state of mind, but also in a physical, tangible 'product' form. And, in our material world, this is unfortunately crucial. The Tourism Society acknowledged this in awarding their 2007 annual award to *Coast* for 'contribution to tourism'. The programme helped 'sell' the coast; and boosting tourist demand and expanding markets are both high on the Tourism Society's agenda (see Wheeller, 2009). As Busby (2006: 73) informs, 'It is the economic perspective that has led to what Winter and Glasson (1996: 174) identify as "a deepening commodification of heritage" with few resources incapable of being transformed into heritage of some sort.' And, while it may well be that some cultural fabric is supported by tourism, over time, the forces of commodification often take hold, distorting 'heritage' to match tourist demand. But whether this constitutes 'destruction' is a moot point.

'Destroy' is a prerogative word: in the context of heritage interpretation (seen over the passing of time) 'alter', 'erode', 'change', 'modify' are perhaps more appropriate terms. And what of tourists' ability to create 'heritage'? When traditions stop or become eclipsed by the march of time do they metamorphose and transform into 'heritage'? Or should that be cultural heritage? 'Heritage tourism usually refers to the built or natural environment and can be more clearly delineated whereas cultural tourism is a much more ambiguous term' (Busby, 2006: 69). If repetitive patterns of behaviour become traditions and traditions, in turn, become part of heritage, then it seems reasonable to suggest that tourists do, in fact, create our/their own heritage. The 'typically' British holiday camp tradition of the 1940s, 1950s and 1960s – a phenomenon of the time, epitomized by Butlins – has, it is suggested here as an example, become part of working class cultural leisure heritage in the UK.

'Heritage' then is open to conjecture. So too, the meaning and interpretation of 'responsible'. In the context of tourists, responsible for what? Their own behaviour? The behaviour of others around them...who, presumably, should know better? Responsible to whom? Themselves, other tourists, the locals? Responsible for destroying the heritage? Or are they Responsible Tourists? But are Responsible Tourists responsible? Maybe they are – but in the sense of 'responsible for the destruction of...' 'Responsible' suggests admirable intent and behaviour yet, contrary to the continued hype, the green sheen of responsible tourism is tarnishing fast. Replaced by a telling patina of verdigris which, though still superficially deceptively attractive, is beginning to show its age.

Clearly, in such a short paper it is impossible for Ashworth to comprehensively cover all relevant matters...and then to apportion 'blame' accordingly. However, the apparent site-specific perspective adopted – in that the assumption appears to be that the tourist actually goes to visit heritage (however defined) 'in situ' – raises a number of quandaries and possible omissions. None more so than the vexed issue of the role of museums, particularly those displaying an array of artefacts from a variety of heritages, 'garnered' from far-flung shores. If these articles are on permanent display, there is then the question as to whether visitor/tourist demand justifies the presumed institutional plundering involved in 'sourcing' the museums in the first place. One step removed in both time and space from the actual site, but nevertheless are these tourists vicariously encouraging 'destruction' or, at least, the desecration of heritage in its original context? For many tourists it is not so much a matter of seeking their own heritage; often it's more a passing interest in taking a look at that of others. They may be on holiday at a neutral destination continents away from either the artefacts' original source and/or the tourists' home and heritage, and – on a whim – just decide to pop into a museum to have a look around. Are these tourists' impacts on heritage to be ignored or are they to be held responsible, tarred with the same brush and damned to the same degree as those trampling over pristine sites?

It is, of course, not just tourist demand that affects the condition of archaeological and religious sites – demand for artefacts from collectors worldwide fans the flames of plunder and illegal trading. And then there are examples where governments, always seeking to be seen in a favourable light, choose to deliberately distort, or even ignore, certain historical events. Nor is it only human frailties that are responsible for modifying and altering 'heritage'. Natural disasters can wreak devastation: earthquakes are notoriously damaging; witness the flattening of the ancient Iranian city of Bam in 2003 – though, perversely, Pompeii would appear to owe its success as tourism mecca to volcanic eruptions.

With respect to a couple of specific points, Ashworth's comment as to the compression of time spent 'on site' is given some credence by, for example, Isaac's (2008) observation of visitors to cultural sites in Breda, The Netherlands. And anecdotally in Houellebecq's (1999: 80) disturbing, and disturbingly prescient, 'tourism' novel *Platform,* where the enigmatic protagonist, on visiting Ayutthaya, observes: 'It was three o'clock in the afternoon. According to the Michelin Guide, you needed to set aside three days for a complete visit, one day for a quick tour. We had three hours. It was time to get out the camcorders.'

Personally, though, I'm not too convinced by the argument that tourists and locals are always (or even mostly) one and the same. Ashworth pursues the definitional problems of 'tourists' from this interesting time–space perspective, arguing that 'the tourist is only the resident on holiday: the resident is only the tourist while not on vacation'? Axiomatically, a half truth? Maybe I haven't fully grasped his concept but surely any attempt at wider, if not universal, applicability reveals obvious flaws in logic for (globally) there are more residents than there are tourists. Other than at an abstract level I am, therefore, not convinced as to the contention's veracity: maybe 'potential' tourist would (theoretically) be nearer the mark. But there is plenty of literature to suggest that we are different beasts when we escape the rigours of home for the abandonment of holidays: our behavioural pattern changes and we become more liberal and hedonistic (Carr, 2002).

When it comes to heritage (and interpretation thereof) fragile sensitivities are exposed and passions can run high. So what's in a name? Ashworth refers to Ayers Rock rather than Uluru. Making a point with a deliberate, considered gesture? Or simply an oversight? Similarly, there is ambiguity and potential political capital inherent in Burma/Myanmar; Derry/Londonderry; the re-naming of roads (Rhodes?) in South Africa, etc…all aspects of (re) claiming heritage. And all part of the process of presentation and interpretation. And significant, too, in creating destination imagery in the tourist imagination.

And what of religion? While the tangible, physical deterioration damage/effects of tourism on churches, cathedrals, mosques, sacred sites, etc., can be evidenced and possibly measured and monitored, what of the more impalpable ramifications of tourists of whatever creed: be they cultural/heritage or otherwise? 'A Moroccan minister has provoked uproar by suggesting that, to avoid waking tourists, muezzins should make less noise when calling the faithful to prayer at dawn' (Campbell, 2008: 22). You don't have to be classified as a cultural or heritage tourist to be instrumental in disrupting or destroying said culture or heritage.

I commend Ashworth's accessible style and lively, at times provocatively stimulating, content. In particular his novel argument for tourists' 'rights' being especially innovative, interesting and noteworthy. And again on a personal note, I empathize with – and strongly endorse – his sentiments as to the issue of 'worthiness'. These, and similar concerns of tourists' social class, status and perceived hierarchy, underpinned much of my early work and are themes I have developed subsequently.

The writing of this piece coincided with me accompanying students on a field-study tour to Cambodia: it seems appropriate, therefore, to contextualize the discussion by drawing a little on pertinent experience gleaned while there. Particularly notable, and relevant, were the temples of Angkor Wat where cultural/heritage tourism and the trappings of mass tourism are surely synonymous, impossible to disentangle and differentiate.

If I understand the situation correctly, the overgrown remains of the dilapidated edifices were 'discovered' and then restored by the (alien) French, thereby lending support to (Ashworth's) assertion as to it being the outsider who brings with them 'appreciation' to otherwise overlooked and oft-neglected sites. I use the word 'restored': some might argue 'stage-managed'. Writing on the magical Ta Prohm Temple, Freeman and Jacques (2003: 91) report, 'It has been maintained in this condition of apparent neglect', suggesting 'The trees that have grown intertwined among the ruins are especially responsible for Ta Prohm's atmosphere and have prompted more writers to descriptive excess than any other feature of Angkor.'

And, evidently, seduces film directors, too, as it is the location for scenes from Simon West's 2001 *Lara Croft: Tomb Raider*. Now, interestingly, many of the local guides seem to make more of the Croft connection than of the more conventional traditional, historical associations. In drawing visitor attention to the exact location of the filming is a new (and new kind of) heritage being created¿

It would seem that the argument that at vulnerable 'attractions' tourist receipts are ring-fenced, ear-marked for renovation purposes and ploughed back into the site is perhaps questionable. And naive. At Angkor Wat estimates suggest that as little as one dollar of the US$ 20 entry fee is used for this purpose: if that. This, though, may well only be negative hearsay and rumour. So, too, the word that corruption is rife and the 'fact' that Siem Reap and the surrounding region, while receiving the most international tourists into Cambodia, remains the second poorest region in the country. But, no smoke without ire. As a colleague, Theo de Haan, put it, 'If those charged with the responsibility of protecting the sites do not take their responsibilities seriously, then tourists will indeed destroy the heritage'. This, however, assumes not only that some managerial structure is actually in place but furthermore (and crucially) that 'managers' have enough clout and sufficient resources at their disposal to resist the spectre of corruption, take effective action and ensure implementation. Rather a tall order. As we know, hardly surprisingly, globally there is considerable variation as to the degree such prerequisites are met.

Meanwhile, Cambodia's 'coastal heritage' is being ruthlessly sold off. In their devastatingly damning article 'Country for sale', Levy and Scott-Clark (2008) reveal a web of deceit and a litany of dubious dealings that chronicle the cynical machinations behind the recent massive (underhand¿) sale of prime land for tourism development. If the allegations made in the distressing article are true, then it is just another example of the catalogue of self-interest and misuse of 'power' that is rapidly becoming the norm, rather than the exception – a potent cocktail of 'get rich quick', cold capitalism; the usual, increasingly ubiquitous mix of corruption, greed and a 'what's in it for me, now' mentality. And tourists are just another poisonous ingredient in the concoction.

So, if there is a problem, then resting responsibility squarely on the shoulders (feet¿) of the tourists is surely amiss, or at best only partially, apportioning accountability...sort of 'shooting the messenger': even if we broaden the net to include 'tourism' in the blame name game we are still wide of the mark, missing the point. Surely the tourist and tourism are inevitable products of 'our' society and are, as such, symptomatic of wider corrosive factors endemic in that society...epitomized by the driving forces of avarice and selfish, short-termism (see Wheeller, 2005, 2007). The current situation in Cambodia would appear a case in point. Regardless of whether the growing tourist numbers are demand or supply led, if the widespread rumours are to be believed the calculated brutality of property speculation/ development and the apparent culpable connivance of the government are indicative of this wider and deeper malaise.

It is, then, similarly irresponsible to singularly blame a government, however corrupt, for flouting decency and selling off the land for private gain. The buyers (be they Western or Asian) know equally well what the score is and are happy to have their fingers in the pie, irrespective of the concomitant social distress inevitably involved. It takes two to tango. Though here it appears more of a conga, with all stakeholders that possess the requisite power and influence gleefully joining in, grabbing what's going for themselves whilst simultaneously (and knowingly) marginalizing the poor and disenfranchised – the enforced 'wall-flowers' looking on, often from a (displaced) afar, helpless, hapless and forlorn.

Here, the tourist is the harbinger of this wider malaise. But if we are looking for an answer, please don't simply shoot the messenger...after all, you and I may very well be first in the line of fire.

11.3

Tourists and Heritage: All Things Must Come to Pass[1]

C. Michael Hall

Heritage represents the things we want to keep. Therefore, heritage is the things of value which are inherited. If the value is personal we speak of the legacy of family or personal heritage, if the value is communal or national, we speak of 'our' heritage. The World Heritage Convention, often regarded as the pinnacle of international heritage conservation efforts, idealistically aims to conserve places which have universal value for the whole of humankind. The notion of inheritance and the responsibilities that it entails are at the heart of the World Heritage Convention. For example, Article 4 of the Convention states that each party to the Convention recognizes 'the duty of ensuring the identification, protection, conservation, presentation and transmission to future generations of the world's cultural and natural heritage' (World Heritage Centre, 2008).

More often than not, heritage is thought of in terms of acknowledged cultural values. This is significant as it stresses the extent to which heritage is socially constructed and intimately related to issues of memory and identity (Graham & Howard, 2008). For example, an object or building is not usually officially deemed as public heritage unless it can be seen as part of the symbolic property of the wider culture or community, as an element of that culture's or community's identity. The linkage between heritage and identity is crucial to understanding not only the significance of heritage as something to be valued, but also the difficulties managers face in identifying and conserving heritage, and particularly the issue of whether tourists destroy the heritage they have come to experience (Hall & MacArthur, 1998). As the Wellington City Art Gallery (1991) stated with respect to an exhibition on heritage: 'References to heritage typically propose a common cultural heritage. Distinguished old buildings are spoken of as being part of "our" heritage. It is suggested that "we" metaphorically own them and that their preservation is important because they are part of our identity. But who is the we?'

Questions of identity, meaning and values indicate the likelihood of there being conflicting notions of ownership attached to heritage and, therefore, conflicting sets of values and interests with which the managers of heritage sites have to contend, in terms of both maintenance of the physical quality of heritage and its social meanings. Indeed, the emergence of multiple perspectives on heritage has led to an expanded meaning of heritage beyond simply the things we want to keep (Harvey, 2001, 2008).

Tunbridge and Ashworth (1996) identified five different aspects of the expanded meaning of heritage:

- a synonym for any relict physical survival of the past;
- the idea of individual and collective memories in terms of non-physical aspects of the past when viewed from the present;
- all accumulated cultural and artistic productivity;
- the natural environment; and
- a major commercial activity, e.g. the 'heritage industry'.

Undoubtedly, there is significant overlap between these various conceptions of heritage. However, according to Tunbridge and Ashworth (1996: 3), 'there are intrinsic dangers in the rapidly extending uses of the word and in the resulting stretching of the concept to cover so much. Inevitably, precision is lost, but more important is that this, in turn, conceals issues and magnifies problems intrinsic to the creation and management of heritage'. In one sense, the concept of 'heritage' has increasingly assumed a similar mantle to 'sustainability' in that the meaning of the word has been extended in so many different directions that 'the generalized message of sustainability may appear to mean nothing specific at all' (Leman-Stefanovic, 2000: xv). But then, appearances are deceiving. In some cases, definitions are like the proverbial struggle to describe an elephant – difficult to do but you will know one when you see it! – yet in other situations, and particularly in a political or regulatory context, words become extremely important in seeking to define limits and freedoms of individual and organizational action, as well as giving effect to policy.

Ironically, the uncertainty about what constitutes heritage is occurring at a time when heritage has assumed greater importance because of its relationship to identity in a constantly changing world (Hall & Tucker, 2004; Graham & Howard, 2008). As Glasson *et al*. (1995: 12–13) commented: 'One reason why the heritage city is proving such a visitor attraction is that, in easily consumable form, it establishes assurance in a world which is changing rapidly'.

The formulation of what constitutes heritage is related to broader political, social, economic and technological changes which reflect concerns over the end of certainty and the convergence between cultural forms that were once seen as separate aspects of everyday life, e.g. education and tourism or, in a heritage context, marketing and conservation. Much discussion in heritage studies has focused on the recognition of multiple meanings of heritage, particularly with respect to the recognition of other voices or memories of heritage, that are not recognized in the heritage of dominant groups in a society. Yet, while the cultural construction and complexity of heritage is now readily acknowledged by researchers (e.g. Corner & Harvey, 1991; Hooper-Greenhill, 1992; Tunbridge & Ashworth, 1996; Smith, 2006; Frost & Hall, 2009), translation of this understanding into practical approaches for heritage managers who are faced with multiple demands on heritage and maintaining the quality of the experience of heritage has not been readily forthcoming (Hall & MacArthur, 1998).

The nature of heritage means that, in a social constructivist sense, heritage is not only made by tourists, or any other individual or collective for that matter, but may also be destroyed by them. Heritage is, therefore, a resource. To repeat the frequently quoted words of Zimmermann's seminal work on resources, 'Resources are not, they become; they are not static but expand and contract in response to human wants and human actions' (Zimmermann, 1951: 15). From a resource perspective, something only becomes a heritage resource if it is seen as having utility value, and different cultures and nationalities can have different perceptions of the value of the same object. What may be a heritage resource in one culture may be 'neutral stuff' in another. Or, in other words, what may be a heritage tourist attraction in one culture or location may not be recognized as an attraction in another. There is no singular 'heritage gaze'.

A tourist heritage resource is that component of the environment (physical or social) which either attracts the tourist and/or provides the infrastructure necessary for the tourist experience (Hall, 2007: 34). Yet, resources are an entirely subjective, relative and functional concept. What actually constitutes a tourism resource depends on the motivations, desires and interests of the consumer, and the cultural, social, economic and technological context within which those motivations occur. The heritage experience is, therefore, not only co-created or co-produced from 'nothing', but can also be co-destroyed, i.e. over time what was once seen as heritage can become 'nothing'.

Such a perspective is not to deny the significance of physical destruction of heritage, whether it be the activities of fundamentalist religious zealots, or the slower effects of pollution and climate change, yet destruction – in the sense of a quick, immediate action – is actually a rare event. Instead, what is sometimes termed destruction is really only another word for change, which is something to which we all contribute. People, whether we describe them as tourists or otherwise, do contribute to changes to heritage both physically and socially, it is an unavoidable by-product of something being a resource. Even non-use will produce change, especially as vicarious appreciation alone may not be sufficient to retain high heritage values in the longer term. Direct experience appears vital for the assignment of significant heritage value.

It is quite possible to measure the extent to which change is wrought on a physical object or even some of the more intangible elements of a culture. But such changes may in some case not affect the value that people attach to what they consider heritage. Instead, there may perhaps be a need to try and separate the object from the value associated with it. However, such academic concerns may be extremely difficult in those societies, cultures, education systems and television channels in which heritage is presented as something unchanging and consistent, rather than something which is constructed and deconstructed over time. In fact, the co-production of heritage as a resource for consumptive purposes is inherently dependent on nostalgia for fixity and certainty, rather than being presented as a resource that is inherently dynamic.

The problem of heritage and tourists is, therefore, arguably more a problem of definition than it is one of destruction. Tourists – the Other – are an easy target via virtue of visibility and difference. However, rather than deal with the concept of tourists, let's acknowledge that we are dealing with people. The problem is ourselves. In absolute terms there are too many of us and in relative terms there is too little acceptance of the value of diversity and difference and, in some cases, of recognizing the sometimes negative effects of our own consumption.

In a world in which public heritage is being increasingly constructed as corporate and political hegemonic capital, a little bit of heritage destruction may not necessarily be a bad thing. As Smith (2006) noted with respect to the ways in which heritage is used in the present-day, on the one hand heritage is about, 'the promotion of a consensus version of history by state-sanctioned cultural institutions and elites to regulate cultural and social tensions in the present. On the other hand, heritage may also be a resource that is used to challenge and redefine received values and identities by a range of subaltern groups' (Smith, 2006: 4). Personally, I remain in favour of further creative destruction of heritage, though not necessarily the physical environment and objects on which it is based, if it helps promote a greater diversity and acceptance of identities, values and voices and appreciation of the flows of history and change.

Concluding Remarks

Much of the discussion above reverts eventually and inevitably to asking what must be the oldest question ever posed in studies of heritage, namely 'whose heritage is it?' Equally ancient is the answer, 'those who claim it or, for whatever reason, associate with it'. The difficulty with such a gratifyingly inclusive answer is that it does not contribute towards the resolution of multiple conflicting claims. Heritage tourism i.e. travelling to experience and enjoy heritage outside the place of residence is evidently prone to such conflicts. It is logical that the resources from which heritage experiences are created do not pre-exist like some mineral deposit awaiting discovery and exploitation: they are activated by the demand for them and therefore heritage tourists create their own heritage; they do not appropriate someone else's. This may be intellectually satisfying but it is unhelpful to management in specific time and space bound situations.

The complexity, intransigence and much of the frustration contained in the issues raised above stem in essence from two related characteristics of most heritage sites. First, many are intrinsically public or semi-public goods and secondly, heritage tourism generates important externalities, whether positive or negative. From these characteristics stem the economic issues (essentially 'who pays?' and 'who benefits?') but also the problems of management (essentially the lack of knowledge and the lack of control intrinsic to the consumption of public goods).

However heritage tourism is by no means unique in this respect and this characteristic of conflicting claims upon space is a management issue shared with many other human activities. All that can be concluded from the insights and examples presented above, and elsewhere in a copious literature, is that no one identifiable group or activity has an automatic, self-evident, universal, absolute priority. Residents have no prior claim and certainly no monopoly of moral worth but neither do visitors, however they may be motivated. The tasks for management are unsurprisingly the same as managing people in many other similar situations, namely, the identification of stakeholders and interests, the establishment of probably multiple goals, the implementation of a range of on and off-site policies to achieve those goals and the continuous monitoring of results. Both discussants of the argument above concluded that however tempting it may be to pillory others for being less aware, less beneficent and less commendable than us, we are, as individuals and as a wider society, both the problem and, fortunately, the solution to it.

Discussion Questions

(1) Is it possible and useful to attempt to calculate a 'carrying capacity' for a heritage tourism site?
(2) Can we build more Venices?
(3) If a particular heritage site is being over-used by tourists, what on-site management responses are available?
(4) If much heritage is freely accessible, who is, or should be, paying to maintain it?
(5) Are 'responsible tourists' responsible?
(6) Why should tourists respect the cultural values and norms of the people whose places they visit?
(7) Should local residents be prioritized in the use of heritage sites? How could this be done?

Note

(1) ©C. M. Hall

References

Ashworth, G.J. (1996). Holocaust tourism and Jewish culture: The lessons of Krakow-Kazimierz. In M. Robinson, N. Evans and P. Callaghan (eds) *Tourism and Cultural Change* (pp. 1–12). Newcastle: Centre for Travel and Tourism.
Ashworth, G.J. (1999). Tourism in the communication of senses of place or displacement in New Mexico. *Tourism, Culture and Communication* 2: 115–128.
Ashworth, G.J. (2009). Heritage and economy. In R. Kitchen and N. Thrift (eds) *Encyclopaedia of Human Geography Volume 5*. (pp. 104–108). London: Elsevier.
Ashworth, G.J. and Tunbridge, J.E. (1990). *The Tourist-Historic City*. London: Belhaven.
Borg, J. van der (1998). Tourism management in Venice or how to deal with success. In D. Tyler, Y. Guerrier and M. Robertson (eds) *Managing Tourism in Cities* (pp. 125–135). London: Wiley.
Busby, G. (2006). The Cornish church heritage as a tourist attraction: The visitor experience. PhD thesis (unpublished), University of Exeter.
Campbell, M. (2008). 'Sshh, you'll wake the tourists.' *Sunday Times*, 20 April: 22.
Carr, N. (2002). Going with the flow. *Tourism Geographies* 4(2): 115–134.
Corner, J. and Harvey, S. (eds) (1991). *Enterprise and Heritage: Crosscurrents of National Culture*. London: Routledge.
Freeman, M. and Jacques, C. (2003). *Ancient Angkor*. Bangkok: River Books.
Frost, W. and Hall, C.M. (eds) (2009). *Tourism and National Parks: International Perspectives on Development, Histories and Change*. London: Routledge.

Glasson, J., Godfrey, K., Goodey, B., Absalom, H. and van der Borg, J. (1995). *Towards Visitor Impact Management: Visitor Impacts, Carrying Capacity and Management Responses in Europe's Historic Towns and Cities*. Aldershot: Avebury.

Graham, B. and Howard, P. (eds) (2008). *The Ashgate Research Companion to Heritage and Identity*. Aldershot: Ashgate.

Graham, B., Ashworth, G.J. and Tunbridge, J.E. (2000). *A Geography of Heritage: Power, Culture and Economy*. London: Arnold.

Hall, C.M. (2007). *Introduction to Tourism in Australia: Development, Issues and Change* (5th edn). South Melbourne: Pearson Education.

Hall, C.M. and MacArthur, S. (1998). *Integrated Heritage Management*. London: The Stationery Office.

Hall, C.M. and Tucker, H. (eds) (2004). *Tourism and Postcolonialism*. London: Routledge.

Hardin, G. (1968). The tragedy of the commons. *Science* 162: 1243–1248.

Harvey, D.C. (2001). Heritage pasts and heritage presents: Temporality, meaning and the scope of heritage studies. *International Journal of Heritage Studies* 7(4): 319–38.

Harvey, D.C. (2008). The history of heritage. In B. Graham and P. Howard (eds) *The Ashgate Research Companion to Heritage and Identity* (pp. 19–36). Aldershot: Ashgate.

Hooper-Greenhill, E. (1992). *Museums and the Shaping of Knowledge*. London: Routledge.

Houellebecq, M. (1999). *Platform*. London: William Heinemann.

ICOMOS (2005). *Managing Tourism in Historic Towns in Asia.* Seoul Declaration (Recommendation 3). ICOMOS.

Isaac, R. (2008). Understanding the behaviour of cultural tourists. PhD thesis (unpublished), Groningen University, The Netherlands.

Jansen-Verbeke, M. (1990). Toerisme in De Binnenstad Van Brugge Een Planologische Visie. *Nijmeegse Planologische Cahiers* 35 K.U. Nijmegen.

Leman-Stefanovic, I. (2000). *Safeguarding Our Common Future: Rethinking Sustainable Development.* Albany: Suny Press.

Levy, A. and Scott-Clark, C. (2008). Country for sale. *Guardian Weekend*, 26 April: 30–41.

Malloy, D.C. and Fennell, D.A. (1998). Codes of ethics and tourism: An exploratory content analysis. *Tourism Management* 19(5): 453–461.

Orbasli, A. (2000). *Tourism in Historic Towns: Urban Conservation and Heritage Management.* London: Spon.

Pattullo, P. (2006). *Ethical Travel Guide*. London: Earthscan.

Prentice, R. (1993). *Tourism and Heritage Attractions.* London: Routledge.

Roowaan, R. (2005). The *Neue Wache* in Berlin. In G.J. Ashworth and R. Hartmann (eds) *Horror and Human Tragedy Revisited: The Management of Sites of Atrocity for Tourism* (pp. 149–162). New York: Cognizant.

Seaton, A.V. (2002). Another weekend away looking for dead bodies: Battlefield tourism on the Somme and in Flanders. *Tourism Recreation Research* 25(3): 63–78.

Smith, L. (2006). *Uses of Heritage*. London: Routledge.

Till, K.E. (1999). Staging the past: Landscape designs, cultural identity and *Erinneringspolitik* at Berlin's *Neue Wache. Ecumene* 6: 251–283.

Tunbridge, J.E. and Ashworth, G.J. (1996). *Dissonant Heritage: The Management of the Past as a Resource in Conflict*. Chichester: John Wiley and Sons.

UNEP (2005). *Making Tourism More Sustainable: A Guide for Policy Makers.* Nairobi: UNEP.

Uzzell, D. (1989). *Heritage Interpretation: The Natural and Built Environment*. London: Belhaven Press.

Wellington City Art Gallery (1991). *Inheritance: Art, Heritage and the Past*. Wellington: Wellington City Art Gallery.
Wheeller, B. (2005). The truth, the hole truth and everything but the truth. *Current Issues in Tourism* 6: 467–477.
Wheeller, B. (2007). Sustainable mass tourism: More smudge than nudge – The canard continues. *Tourism Recreation Research* 32(3): 73–75.
Wheeller, B. (2009). Tourism and the arts. In J. Tribe (ed) *Philosophical Issues in Tourism* (pp. 191–208). Bristol: Channel View Publications.
Winter, M. & Gasson, R. (1996) Pilgrimage and Tourism: Cathedral visiting in contemporary England. *International Journal of Heritage Studies* 2(3): 172-182.
World Heritage Centre (2008). *Convention Concerning the Protection of the World Cultural and Natural Heritage*. Online at: http://whc.unesco.org/en/conventiontext/. Accessed 1 April 2008.
Zimmermann, E.W. (1951). *World Resources and Industries* (revised edition). New York: Harper and Brothers.

Further Reading

Ashworth G.J. and Tunbridge, J.E. (2000). *Prospect and Retrospect on The Tourist-Historic City.* London: Elsevier.
Graham, B.G. and Howard, P. (eds) (2008). *The Ashgate Research Companion to Heritage and Identity.* Aldershot: Ashgate.
Prentice, R. (1993). *Tourism and Heritage Attractions.* London: Routledge.
Smith, L. (2006). *Uses of Heritage.* London: Routledge.
Timothy, D.J. (ed) (2007). *Managing Heritage and Cultural Tourism Resources:* Critical Essays (Vol 1). Aldershot: Ashgate.
Timothy, D.J. (2011). *Cultural Heritage and Tourism: An Introduction.* Bristol: Channel View Publications.
Tunbridge, J.E. and Ashworth, G.J. (1996). *Dissonant Heritage: The Management of the Past as a Resource in Conflict.* Chichester: Wiley.

Chapter 12
Nature-Based Tourism: There's a Lot in a Name

David Fennell, Ralf Buckley and Alexandra Coghlan and Betty Weiler

Context

The paper 'What's in a Name? Conceptualizing Natural Resource-Based Tourism' was conceived at a time when a number of new tourism types were gaining traction both in theory and practice. The 1990s and up until the early new millennium was a time dedicated to establishing clarity around these types – nature-based tourism, adventure tourism, ecotourism, wildlife tourism, green tourism, and so on. The aim of the paper was to provide a few dimensions that could provide added meaning in the effort to differentiate some of these related types. These are in no way definitive, and researchers continue to investigate ways by which to examine these types. Having said this, there is still a degree of overlap between certain forms such as wildlife tourism and ecotourism. What stands in the way of universal understanding of the differences is a large degree of insular thinking. Tourism scholars continue to ignore interdisciplinary knowledge (sociology, ethics and biology) that would help provide needed clarity in understanding dynamic human–human and human–environment relationships in tourism.

How is it, for example, that wildlife tourism or ecotourism can be deemed non-consumptive if they support the viewing of captive animals (Fennell, 2011)? In tourism, we usually view these issues through a very restricted lens.

Buckley, Coghlan and Weiler have been helpful in redefining the nomenclature that is so important in matching supply with demand for these environment-dependent forms of tourism. If the goal is to balance the needs of the environment with the needs of an increasingly specialized and expanding base of nature-based tourists, managers need this knowledge to formulate good policies and guidelines, which in turn lead to good practice.

12.1

What's in a Name? Conceptualizing Natural Resource-Based Tourism

David Fennell

Responsible travel that conserves the environment and sustains the well-being of local people.
Responsible travel to natural areas that conserves the environment and improves the welfare of local people.

The definitions above describe two different concepts: the first, ecotourism (e.g. birding), and the second, nature-based or nature tourism (e.g. backpacking, boat tours, cycling, environmental education, farms, fishing, hunting and ecotourism). Increasingly, governmental agencies and tourism associations have found it either necessary or convenient to view ecotourism and nature-based tourism synonymously. To many ecotourism researchers, however, the two concepts differ significantly (Brandon, 1996; Goodwin, 1996). In the words of Burton (1998), 'it is clear that strict ecotourism is a far cry from the everyday tourism activities that occur in a natural environment and that are known as nature-based tourism'. This paper argues for a clearer distinction between ecotourism and other types of natural resource-based tourism, and maintains that a solid conceptual footing for these two forms of tourism does not exist as a result of definitions and policies which are simply too passive (Orams, 1995), and which have done little to insulate ecotourism from the tentacles of a more consumptive *'nature-based'* umbrella.

Historical Linkages

Early research in this area used *nature tourism* as a label to refer specifically to what we view today as ecotourism. For example, Laarman and Durst (1987: 5) defined nature tourism as tourism in which the 'traveller is drawn to a destination because of his or her interest in one or more features of that destination's natural history' (Halbertsma, 1988). Laarman and Durst described a typical nature tourism (ecotourism) experience as follows:

> First they would tour the Galapagos Islands and hear a lecture or two on Darwin, the Beagle, and the evolution of species. Then they would return to the mainland to visit a mid-elevation biological station known for its variety of birdlife. (Laarman & Durst, 1987)

Somewhere along the line, however, nature-based tourism grew to encompass many 'outdoor recreational-based' tourism experiences, including ecotourism. Although difficult to trace, this phenomenon may be attributed to the efforts of governmental agencies that were intent on organizing the various nature-based tourism types under one roof. The reason for doing so may relate to territoriality, economics and/or marketing (in the sense of trying to sensationalize the growth of the sector). Nevertheless, problems still exist in differentiating ecotourism from nature-based tourism, with significant overlap in meaning and definition.

While it is apparent from the work of Goodwin (1996) and Burton (1998), and others, that ecotourism and nature-based tourism are indeed different, some suggest the need for a different nature-based tourism taxonomy. Ewert and Shultis (1997) use 'resource based tourism' in describing what is essentially nature-based tourism. The 'resource' label provides a needed description to activities which are inherently dependent upon the natural resource base. Garrod and Fyall (1998) caution, however, that researchers should be clear that the word 'resources' means more than simply natural resources. They suggest that human-made and socio-cultural resources are just as important in the context of tourism. Secondly, Fennell (1999) argues for a change in terminology on the basis of the historical evolution of the ecotourism term, as outlined above, and the difficulty in equating 'nature-based tourism' with activities such as hunting, which are more consumptive. Consequently, clarification may be achieved in uniting the resource-based tourism label (forms of tourism that rely on the *natural* resource base), with the nature-based tourism label (avoiding the assumption that all other more consumptive forms of nature-based tourism are ecotourism) into Natural Resource-Based Tourism (NRBT) or Natural Resource Tourism (NRT).

Conceptualizing NRBT

Laarman and Durst (1987), Lindberg (1991) and Orams (1995) are among those who have developed ecotourist typologies. Typologies of natural resource-based tourism, however, are virtually non-existent. The conceptual frameworks to follow were created for the purpose of differentiating any number of NRBT activities, including ecotourism. Principally, four variables have been identified to accomplish this end (acknowledging that other variables may be used), including impact and natural resource values (Figure 12.1), and skill and learning (Figure 12.2).

As identified in Figure 12.1, an activity is judged to be either consumptive or non-consumptive depending upon the extent of its ecological impact on the natural world. The second variable, natural resource values, underscores the belief that differences do exist between preservationism and conservationism. Gifford Pinchot (in Hays, 1959) viewed conservation as the use of natural resources for the greatest good of the greatest number for the longest time. While this perspective referred to using resources wisely, conservation today is viewed as representing different degrees of management, from preservation and protection to multiple use and extractive use (Nelson *et al.,* 1985). Passmore (1974) saw a fundamental difference between the two concepts in writing that conservation was the saving of resources *for* use, while preservation was the saving of resources *from* use. Consequently, natural resource-based tourists will differ in their preservation and conservation beliefs on the basis of their chosen pursuits and what and how they wish to gain from their experiences.

Reliance on technical skills (e.g. whitewater skills, belaying, navigation) is a variable that has not been used to describe the behaviour of ecotourists.

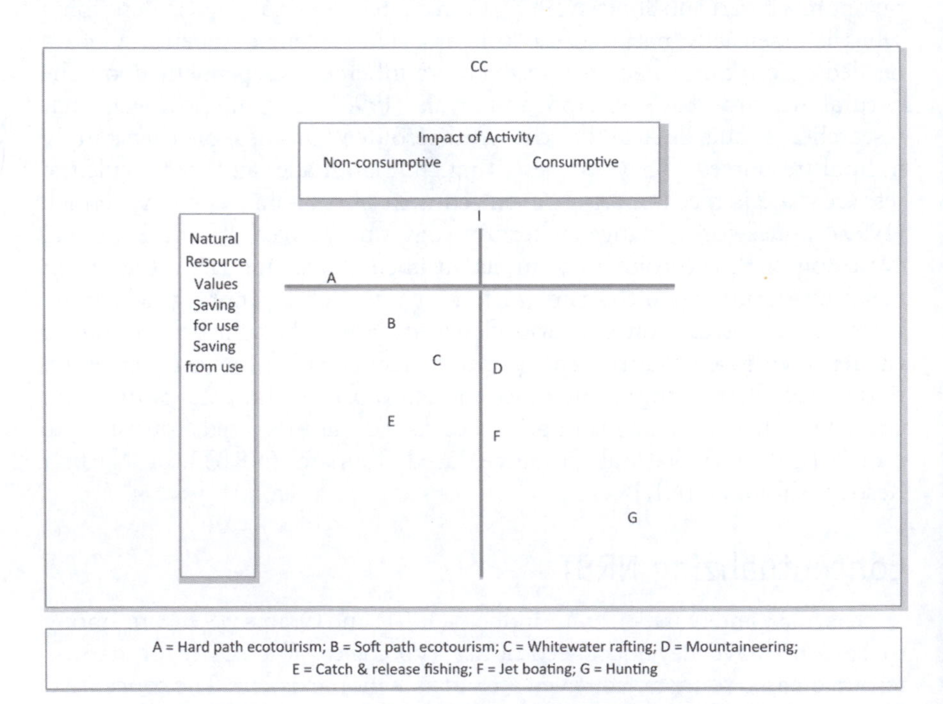

Figure 12.1 NRBT impact and values dimensions (hypothetical)

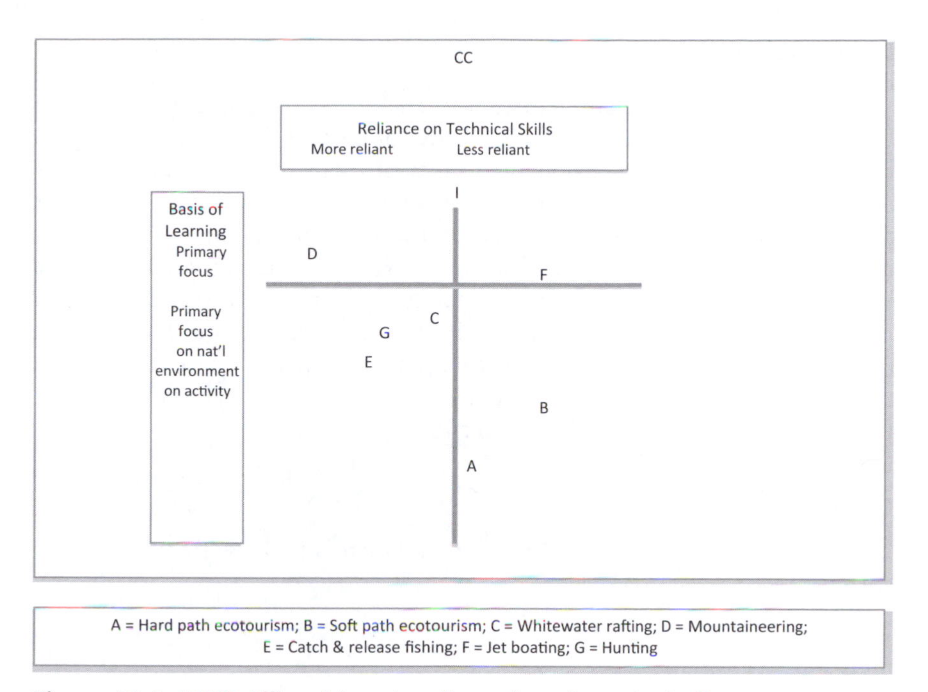

A = Hard path ecotourism; B = Soft path ecotourism; C = Whitewater rafting; D = Mountaineering; E = Catch & release fishing; F = Jet boating; G = Hunting

Figure 12.2 NRBT skill and learning dimensions (hypothetical)

Nevertheless, it may be used to set apart NRBT travellers – particularly adventure tourists – who rely on varying levels of skill to safely accomplish their ends (Figure 12.2). While it may be wrong to suggest that some or all types of adventure tourists do not learn from their experiences, their basis of learning is *primarily* activity-based. For other natural resource-based tourists, e.g. ecotourists, the learning focus is ecologically-based or centred *primarily* on the natural environment.

Figures 12.1 and 12.2 show the hypothetical positions of seven NRBT activities, which have been placed according to their perceived degrees of (i) impact, (ii) natural resource values, (iii) reliance on technical skills, and (iv) basis of learning. In Figure 12.1, ecotourism ('A' and 'B') are situated in the 'non-consumptive' and 'saving from use' cell; whereas, jet boating ('F') and hunting ('G') are considered to be more 'consumptive'. In reality, the place of these activities in such a framework would be subject to many influences, including regional policies, settings and the practices of individual operators and tourists in the field. Intuitively, one may feel that an activity like whitewater rafting is, for example, less ecologically damaging than jet boating. Strict regulations in the jet boating industry, and a lack thereof in the whitewater rafting industry, may alter the place of these two activities according to the variables suggested in the framework.

Conclusion

It is indeed unfortunate that some of the most oft-quoted definitions of ecotourism are those that are most narrow or passive. Such is the case with the first definition identified at the outset of this paper, which is even *less* restrictive than its nature-based counterpart. It is also unfortunate that these passive definitions have been instrumental in helping to formulate government and industry policies on NRBT. These directives have done little to shield ecotourism from being marketed as, or subsumed by, other more consumptive 'nature-based' activities. The framework developed in this paper is meant to act as a starting point from which to further explore types of NRBT, and would benefit from empirical research in helping to further clarify relationships between these types.

Acknowledgement

Many thanks to John Larsen, of Brock University, for his comments on this paper.

12.2

Nature-Based Tourism in Breadth and Depth

Ralf Buckley and Alexandra Coghlan

Revisiting Fennell's (2000) taxonomy of natural resource-based tourism after a decade, we can see that whilst his conclusions remain valid and his illustrations illuminating, recent research has added both greater breadth and greater depth.

In retrospect, three of the seven types of tourism in his two figures, namely whitewater rafting, mountaineering and jet boating, would now be treated as adventure tourism (Buckley, 2010a). Two, namely hunting and fishing, would be considered as consumptive nature-based tourism. Even catch-and-release fishing kills some of the fish caught, so it is consumptive;

and if ecotourism is considered to include an ethical dimension, as argued by Fennell (2003), then catch-and-release also fails the ethical test. The remaining two categories are listed only as 'hard-path' and 'soft-path' ecotourism, with the terms taken, perhaps, from American outdoor recreation vernacular.

There have been many analyses of the distinctions between ecotourism and related terms (Buckley, 2009). Some terms, such as nature-based tourism, are defined by features of the product or setting. Others, such as responsible tourism, are defined by measures taken to improve either social or environmental outcomes. Several authors, including Fennell (2003), have also proposed hybrid terms which recognize that commercial tourism products as actually sold in the retail market do not necessarily fall neatly into single analytical categories.

In 2012, therefore, the increasing breadth of terminology in the broad outdoor tourism sector is well-known to tourism researchers, but we can still see how the seven types examined by Fennell (2000) fit into this modern taxonomy of outdoor tourism.

More interesting at present, perhaps, is to examine how our understanding of nature-based tourism has grown in depth, over the past decade. To address this issue we draw on a recent review by Coghlan and Buckley (2012).

Nature-based tourism includes all forms of tourism where relatively undisturbed natural environments form the primary attraction or setting (Buckley, 2009; Newsome *et al.*, 2002). It can include consumptive and adventurous as well as non-consumptive contemplative activities, which in turn can include ecotourism (Buckley, 2009; Fennell, 2003; Weaver, 2008) and conservation tourism (Buckley, 2010b). It is a significant component of the global tourism industry, with estimates of worldwide economic scale ranging from hundreds of billions up to one trillion dollars annually (Buckley, 2009). It is also a very diverse sector, and Coghlan and Buckley (2012) argue that we should recognize the diversity and variety amongst both the products supplied, and in the customers who demand and buy them.

Individual nature-based tourists commonly have widely differing origins, interests, motivations and behaviours (Mehmetoglu, 2007; Silverberg *et al.*, 1996; Strasdas, 2006). Vespestad and Lindberg (2010) argue that nature-based tourism experiences can be divided into four categories: a search for self, a form of entertainment, a state of being and a form of social affiliation. Taking a somewhat different approach, Arnegger *et al.* (2010) use a matrix of travel service arrangements vs. travel motivations in order to characterize tourists. They categorized service arrangements as standardized, customized, *a la carte* or independent, and travel motivations as sports and adventure, hedonism, nature experience and nature protection. This approach leads to a 'pick and mix' view (Coghlan & Buckley, 2012), under which tourists can select different levels of engagement with nature on different tours.

Human relationships with natural environments change over time and differ between countries and demographic groups. In the West, nature has a relatively short history as a tourist attraction (Buckley, 2000; Meyer-Arendt, 2004). It has been promoted both as a development opportunity in rural or remote areas (Rinne & Saastamoinen, 2005), and as an alternative to mass tourism (Fennell, 2003). Interest in nature-based tourism may also reflect the growth in environmental activism (Weaver, 2008), and a so-called post-Fordist trend towards individual choice and flexibility (Saarinen, 2005).

Nature-based tourists of Caucasian ethnic origin, from Anglophone Western nations, are not necessarily representative of those from other continents, countries, linguistic traditions and ethnicities, whether Western (Gössling & Hultman, 2006; Lopez *et al.,* 2005; Priskin & Sarrasin, 2010) or Eastern (Buckley *et al.,* 2008; Ma *et al.,* 2009; Su *et al.,* 2007). There has been little research to date on domestic nature-based tourists and their preferences in the BRIC nations. These nations, however, will be critical to the continuing development of nature-based tourism worldwide.

If nature-based tourism is considered in the context of the 'experience economy', then the critical issue arises that the tour operators who provide, choreograph and manage their client experiences cannot control nature in the same way as other components of their products. Nature is a powerful attraction on which operators can capitalize, but an uncertain one, so tour operators focus instead on service aspects more amenable to control (Tonge & Moore, 2007). These can include emotional components (Bigne *et al.,* 2005; Meyer-Arendt, 2004; Zins, 2002), ranging from contemplative quasi-religious experiences (Heintzman, 2010; Vespestad & Lindberg, 2010), to adrenalin-inducing performative experiences (Cater & Cloke, 2007). These can temporarily displace or outweigh anxieties which clients suffer in their everyday lives.

Individual nature-based tourism experiences thus depend on individual perceptions or constructions of nature (Coghlan & Buckley, 2012); though with no guarantee that such experiences will be transformative (Budeanu, 2007; Lee & Moscardo, 2005; Powell *et al.,* 2009). This diversity in individual constructions of nature also makes the management of visitor behaviour, impacts and satisfaction particularly complex and difficult (Buckley, 2009; Fletcher & Fletcher, 2003; McCool, 2009; Manning, 2011).

Looking to the future, Coghlan and Buckley (2012) call for continuing international comparative research on: the factors that influence the satisfaction and dissatisfaction of nature-based tourists; the degree to which these factors are controllable either by land managers or by tour operators; and the ways in which these may influence individual behaviour and impacts. That is, they argue for a research programme in nature-based tourism and its management, which focuses at least as much on the social as the natural sciences.

12.3

Ecotourism and Nature-Based Tourism: What's Beyond the Names and Labels?

Betty Weiler

In his research probe entitled 'What's in a name?', Fennell (2000) sets out to articulate a clearer distinction between ecotourism and other types of nature-based tourism. In his rationale, he cites misconceptions and misuses of the labels ecotourism and nature-based tourism, a problem that few would disagree with, including not only scholars and researchers but also many in government, non-government and the commercial tourism sectors. His frustration is palpable and I empathize with his desire to set the record straight, as it were, but do not see the solution to the problem as introducing new terminology and a new taxonomy. This is confirmed by Buckley and Coghlan's observations, a decade later, that Fennell's taxonomy has not stood the test of time.

This short probe firstly presents the reasons why Fennell's solution – presentation of a new typology of nature-based tourism (or what he calls natural resource-based tourism) in two-dimensional space – is a non-solution to the problems he is trying to address. I do this mainly to steer other authors away from the temptation to engage in similar exercises. This is followed by a reframing of the problems posed by Fennell that, together with Buckley and Coghlan's probe, opens up new solutions.

Fennell is correct in saying that ecotourism definitions are often ill-conceived. Academics, policy-makers and tourism operators have struggled for many years to define ecotourism in such a way that it captures the breadth of experiences and products on offer, while distinguishing it from other nature-based tourism activities. Most authors agree that ecotourism is nature-based, learning-centred and conservation-oriented (Nowaczek & Smale, 2010; Orams, 2001). One of the more rigorous approaches to delineating what ecotourism is meant to achieve is Edwards *et al.*'s (2003) analysis of ecotourism definitions used in government policy documents. A stated purpose of their study was to determine how government agencies in the Americas define ecotourism. In their study, ecotourism policy included 'actions governments are engaged in as they carry out ecotourism policy..., policy outputs that governments or their partnerships have developed..., and identifiable organizational mechanisms that address ecotourism...' (Edwards *et al.*, 2003: 293). Their analysis was based on a census survey of the policies of the federal governments of the US and Canada, 50 states in the US, all 10 provinces and the two territories of Canada, as well as more than 50 nations in Latin America and the Caribbean. As a result of their qualitative analysis of the 42 ecotourism definitions identified by the 119 governmental tourism agencies contacted, together with triangulation with ecotourism definitions identified in the literature and cross-checking of the findings with practitioners, the following key elements of ecotourism were identified:

> *Ecotourism, regardless of its scale, ought to produce at least three objectively verifiable outcomes: (1) it is a positive force for conservation, emphasizing protection and perpetuation of the very landscapes and features that attract the tourists, (2) it benefits host communities economically and ensures that the people who must endure the social and environmental impacts of tourism development also share in its rewards, and (3) it promulgates environmental awareness both among tourists and local communities.* (Edwards *et al.*, 2003: 305)

I return to these outcomes later in this probe, but the point I wish to make here is that Edwards *et al.* (and many others) conclude from their study that ecotourism is distinguishable more by its ideological orientation than by any attributes it exhibits. This, of course, has been both the boon and the bane of ecotourism scholars and practitioners alike. One can hardly take issue with any government body or tourism operator that sets out to achieve these three outcomes, but it is the *labelling* of an experience or product as ecotourism rather than the *striving to be* ecotourism that is contentious.

Turning again to Fennell's probe, and if we accept Edwards *et al.*'s conclusion that ecotourism is an ideology, then Fennell's frameworks are not necessarily flawed so much as they are misplaced as solutions to the problems he poses. Figure 12.1 attempts to position two types of ecotourism (hard and soft) together with other nature-based products along two dimensions which are not dissimilar to Edwards *et al.*'s outcome 1. However, there are a range of activities and experiences that could be characterized as hard ecotourism and soft ecotourism. More importantly, the framework ignores other outcomes and attributes that, based on the literature, distinguish ecotourism from other forms of nature-based tourism. As a result, the framework falls short of answering the question posed by Fennell in both the introduction and conclusion of his paper: how does one define ecotourism and distinguish it from other types of nature-based tourism? Given the ideological nature of ecotourism, I would suggest that it makes more sense to conceptually position specific ecotourism products or experiences in three-dimensional space using the three outcomes identified by Edwards *et al.* If we imagine three axes where one end of the axis represents a high level of success and the other end a low level of success in achieving the outcome, then government bodies (and indeed consumers) can make their own judgments as to how far along the continuum of each axis a product or experience needs to be.

Of course, assessing individual products and experiences on each of the three outcomes is much more easily said than done, as many certification and accreditation schemes have discovered when trying to evaluate ecotourism businesses and products against sets of criteria, standards or indicators. The label 'ecotourism' has been used by governments and accreditation bodies to distinguish products that seek to conserve natural and cultural resources, benefit host communities and promote environmental awareness (Edwards *et al.*'s three outcomes), and by commercial operators to convince consumers that their business subscribes to the ideology of ecotourism. However, for reasons discussed by Buckley and Coghlan and many other scholars, such efforts often descend into greenwashing rather than genuinely reflecting actual business practices and products.

Fennell's Figure 12.2 strays even further from the problem he poses, as it attempts to model *ecotourists* as opposed to ecotourism, although this is not reflected in the title he has given to his graph. Buckley and Coghlan, in their reflections on Fennell (2000), also acknowledge the need to accommodate the heterogeneity of ecotourists. A similar observation by Wight (2001) some 10 years earlier partly explains the development of a number of ecotourist typologies or classification schemes, which Hvenegaard (2002) notes as being both researcher-driven and respondent-driven. However, as with other tourist typologies, Hvenegaard cautions that individuals are not necessarily confined to a single classification, and may even cross boundaries during a single trip. Nonetheless, it has been suggested and in some cases empirically demonstrated that ecotourists can be differentiated and therefore potentially marketed to and catered for based on a number of dimensions including motivation and/or commitment to be environmental responsible (Acott *et al.,* 1998; Lindberg, 1991; Weaver, 2001; Weiler & Richins, 1995; Wight, 2001), motivation to learn and/or cognitively and emotionally engage with the natural environment (Weiler & Richins, 1995; Wight, 2001), desire for and/or acceptance of physical challenge or difficulty (Laarman & Durst, 1987; Weaver, 2001; Weiler & Richins, 1995), degree of specialization in or focus on ecotourism activities (Laarman & Durst, 1987; Lindberg, 1991; Wight, 2001) and other dimensions. While the first two dimensions, degree of environmental responsibility and learning-centredness, can be argued to be closely aligned with the definitions and outcomes of ecotourism, the other dimensions have little to do with the goals or outcomes of ecotourism. Nonetheless, these dimensions do assist in flagging the need for a diversity of products to meet the range of motivations, expectations and capacities of ecotourists.

In an effort to develop a typology for nature-based tourism that marries the demand (ecotourist) and supply (ecotourism product) sides, Arnegger *et al.* (2010) present a four by four (16-cell) matrix, with one dimension being the degree to which nature is the primary motivation for travel, and the other dimension being the extent to which service arrangements are independently made. However, as with Fennell's framework, this is another two-dimensional response to a multi-dimensional phenomenon, which relies on a single dimension for demand and a single dimension for supply. Thus, despite arguing that their matrix will help managers to better meet tourists' needs and demands, I would argue that it is unlikely to do so.

In response to this issue, I am very much in agreement with Buckley and Coghlan's redirecting of Fennell's 'what is the essence of ecotourism' issue to a different and probably more useful focus, that of the experience and the satisfaction of nature-based tourists. Providing products that 'conserve the environment and sustain the well-being of local people' (Fennell, 2000: 97) or achieve Edwards *et al.*'s three outcomes of conservation of resources, benefits to host communities and enhancement of environmental awareness may well be consistent with the ideology of ecotourism, but in fact there is little evidence that achieving these outcomes contributes to enhancing the visitor experience or visitor satisfaction. Tourist typologies go some way to tackling this issue, but by focusing only on tourist motivations, capacities and activities, they somewhat miss the point: how can we better ensure that ecotourism delivers on its ideological outcomes and at the same time provide satisfying and enriching experiences for the visitor? And in today's highly competitive tourism environment with instant word-of-mouth and word-of-internet feedback, failure to focus on the visitor experience and satisfaction is a recipe for business failure.

A substantial body of literature is developing on the tourist experience, including some very good monographs and seminal journal articles (Jennings & Nickerson, 2006; Ritchie & Hudson, 2009; Ryan, 2002). It is thus timely for researchers and scholars to turn their attention to applying and refining the theory and findings regarding the tourist experience to ecotourism and nature-based tourism contexts. One of perhaps many ways of enhancing the alignment between the desired outcomes of ecotourism and the motives and capacities of the various types of ecotourists and nature-based tourists, at least in the context of guided ecotourism and nature-based tourism, is to focus on the role of the ecoguide. As the broker of experiences, a knowledgeable, personable, skilled and motivated guide can play a critical role in managing expectations, assessing capacities, challenging and protecting visitors, and delivering engaging and memorable experiences, while at the same time delivering on the conservation, host community, natural and cultural learning and awareness-building outcomes that form the ideological pillars of ecotourism. Regarding the latter, the contribution of guides to sustainability outcomes is discussed more fully in Weiler and Kim (2011). Regarding the former, the results of many studies of tour guides in many contexts reveal the pivotal role of guides in facilitating quality experiential outcomes. Aspects of nature, whether that is encounters with wildlife, spectacular ocean views or simply the quiet and solitude of a rainforest setting, are often cited by tourists as being key motivators as well as key factors in facilitating satisfaction. However, in the context of a guided experience, nature takes a distinct back seat to the quality of the guiding and interpretation in what visitors identify as the most satisfying or most memorable aspects of their experience (Ham & Weiler, 2007; Hughes, 1991; Weiler & Yu, 2007; Wong, 2001). Perhaps more importantly, these same tourists identify as key dissatisfiers experiential elements that have little to do with the tenets of nature-based or ecotourism, but can potentially lead to negative word-of-mouth advertising, for example, poor tour management, poor guiding, issues with service in the transportation, accommodation and restaurant sectors, and issues with the wider destination context that are beyond the control of the tourism industry (Weiler & Yu, 2007). Buckley and Coghlan refer to these elements as being beyond the control of the operator, and also see them as important. Thus, a second focus for research and practice is how ecotourism operators and governments can better manage expectations of both the natural and the many other elements of the tourist experience that are beyond their control, and indeed influence policy and practice in these areas. This may go some way to addressing the ongoing criticism and cynicism of Wheeller (2007) and others regarding the narrow perspective with which ecotourism and nature-based tourism are judged to be sustainable.

In conclusion, I support Buckley and Coghlan's call to shift the focus of discussion, research and management of ecotourism and nature-based tourism beyond terminology and taxonomies to what's delivered on the ground, and particularly on the tourist experience. Clearly the visitor experience begins and ends – temporally, geographically, physically and psychologically – well beyond the confines of the products and activities included in Fennell's and most other ecotourism taxonomies. Focusing on the transportation, accommodation, hospitality, services and non-tourism components of the experience may be less sexy and fun than researching tourism and tourists in natural areas, but it is badly needed.

Concluding Remarks

The aim of this probe was to distinguish between different forms of natural resource-based or nature-based tourism. Four dimensions were selected (not the only dimensions to consider), for the purpose of stretching our thinking beyond the conventional mindset. Space prevents delving into the responses offered by Buckley and Coghlan, and Weiler. It is sufficient to point out that their responses adhere closely to their specific research agendas on management and product delivery or 'what's delivered on the ground'. Their arguments point to the belief that we ought to be more concerned about how ecotourism products are delivered in securing positive tourism experiences, than on conceptualizing ecotourism. What they call for is well underway in the literature, but the state of affairs was far different when the paper was conceived in 1999.

Without the space to fully address criticisms, I direct my thoughts towards their focus on visitor experience and the practice of ecotourism. As far as trying to 'steer other authors away from the temptation to engage in similar exercises' (Weiler in reference to typologies), I believe strongly that this is advice that we should not follow – especially in the absence of an acceptable alternative. Empirical, theoretical and conceptual research is vital to any field's progression. This view reminds me of the discussion on responsible tourism (RT), i.e. that RT is not a type of tourism or product but rather a way of doing tourism. But how do we 'do' tourism in the absence of good theories and approaches that establish boundaries between and within? You simply cannot have one without the other. I'm not sure the extent to which Buckley and Coghlan, and Weiler, subscribe to this mindset, but it is one that even the pragmatist school of philosophy would shy away from. The point is, we need to establish a dynamic tension between theory and practice – each informing the other. I offer Figure 12.3 as an example of my position. Good definitions lead to good policies and standards, which in turn lead to good practices. The feedback loop demonstrates that we have to continually revisit our received ideas about what constitutes good practice, policy and the conceptual bases that drive these. The dashed line leaves open the possibility of refining the way we think about ecotourism. For me, 'the visitor experience begins and ends', only when practitioners – and scholars – have made these connections.

I would be remiss in not pointing out that I was mindful that the probe was only part of a more broadly based set of issues corresponding to supply and demand, theory and application, and management and experience, all orbiting outside the bounds of the probe itself. These thoughts culminated in the publication of *Ecotourism Programme Planning* (Fennell, 2002), which is a comprehensive statement on 'what's delivered

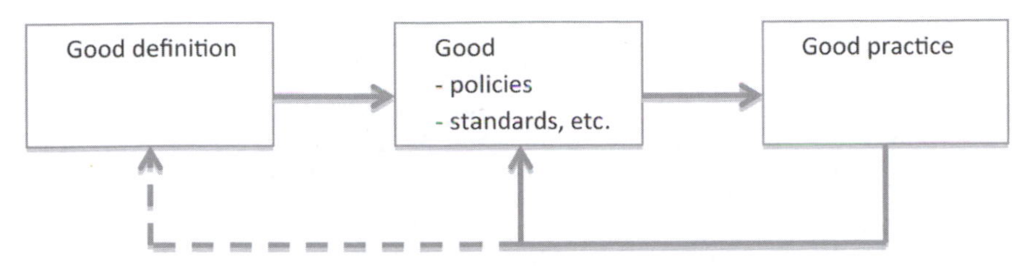

Figure 12.3 The definition, policy, practice loop

on the ground'. To date I have not seen another work in ecotourism that speaks so directly to what Buckley and Coghlan, and especially Weiler, are calling for, above. It is important that ecotourism scholars recognize the value of an applied side approach, but I could never imagine such a field void of a spirited tension between 'doing' and 'thinking'. We need to move forward, not backward.

Discussion Questions

(1) How does hard-path ecotourism differ from other types of natural resource-based tourism?
(2) What's the difference between hard-path ecotourism and soft-path ecotourism?
(3) What is the chief criticism of Fennell's work?
(4) How does Fennell counter the criticisms waged against his work?
(5) What is the relationship between definition, policy and practice?

References

Acott, T.G., Latrobe, H.L. and Howard, S.H. (1998). An evaluation of deep ecotourism and shallow ecotourism. *Journal of Sustainable Tourism* 6(3): 238–253.

Arnegger, J., Woltering, M. and Job, H. (2010). Toward a product-based typology for nature-based tourism: A conceptual framework. *Journal of Sustainable Tourism* 18(7): 915–928.

Bigne, J.E., Andreu, L. and Gnoth, J. (2005). The theme park experience: An analysis of pleasure arousal and satisfaction. *Tourism Management* 26: 833–844.

Brandon, K. (1996). *Ecotourism and Conservation: A Review of Key Issues*. Environment Department Paper No. 23, The World Bank, Washington, D.C.

Buckley, R.C. (2000). NEAT trends: Current issues in nature, eco and adventure tourism. *International Journal of Tourism Research* 2: 437–444.

Buckley, R.C. (2009). *Ecotourism: Principles and Practices* (p. 368). Wallingford: CAB International.

Buckley, R.C. (2010a). *Adventure Tourism Management* (p. 268). Oxford: Elsevier.

Buckley, R.C. (2010b). *Conservation Tourism* (p. 214). Wallingford: CAB International.

Buckley, R., Zhong, L-S., Cater, C. and Chen, T. (2008). Shengtai Luyou: Cross-cultural comparison in ecotourism. *Annals of Tourism Research* 35: 945–968.

Budeanu, A. (2007). Sustainable tourist behaviour – A discussion of opportunities for change. *International Journal of Tourism Studies* 31: 499–508.

Burton, F. (1998). Can ecotourism objectives be achieved? *Annals of Tourism Research* 25(3): 755–758.

Cater, C. and Cloke, P. (2007). Bodies in action: The performativity of adventure tourism. *Anthropology Today* 23: 13–16.

Coghlan, A. and Buckley, R.C. (2012). Nature-based tourism. In A. Holden and D. Fennell (eds) *A Handbook of Tourism and the Environment*. London: Routledge.

Edwards, S., McLaughlin, W.J. and Ham, S.H. (2003). A regional look at ecotourism policy in the Americas. In D. Fennell and R. Dowling (eds) *Ecotourism: Policy and Planning (pp. 293*–307). Wallingford, UK: CAB International.

Ewert, A. and Shultis, J. (1997). Resource-based tourism: An emerging trend in tourism experiences. *Parks & Recreation* September: 94–103.

Fennell, D. (1999). *Ecotourism: An Introduction*. London: Routledge.

Fennell, D. (2000). What's in a name? Conceptualizing natural resource-based tourism. *Tourism Recreation Research* 25: 97–100.

Fennell, D. (2002). *Ecotourism Programme Planning*. Wallingford: CABI.

Fennell, D. (2003). *Ecotourism*. London: Routledge.

Fennell, D. (2011). *Tourism and Animal Ethics*. London: Routledge.

Fletcher, D. and Fletcher, H. (2003). Manageable predictors of park visitor satisfaction: Maintenance and personnel. *Journal of Park and Recreation Administration* 21: 21–37.

Garrod, B. and Fyall, A. (1998). Beyond the rhetoric of sustainable tourism? *Tourism Management* 19(3): 199–212.

Goodwin, H. (1996). In pursuit of ecotourism. *Biodiversity and Conservation* 5(3): 277–291.

Gössling, S. and Hultman, J. (2006). *Ecotourism in Scandinavia: Lessons in Theory and Practice*. Wallingford: CAB International.

Halbertsma, N.F. (1988). Proper management is a must. *Naturopa* 59: 23–24.

Ham, S.H. and Weiler, B. (2007). Isolating the role of on-site interpretation in a satisfying experience. *Journal of Interpretation Research* 12(2): 5–24.

Hays, S.P. (1959). *Conservation and Gospel of Efficiency* (p. 263). Cambridge, MA: Harvard University Press.

Heintzman, P. (2010). Nature-based recreation and spirituality: A complex relationship. *Leisure Sciences* 32: 72–89.

Hughes, K. (1991). Tourist satisfaction: A guided "cultural" tour in North Queensland. *Australian Psychologist* 26(3): 166–171.

Hvenegaard, G.T. (2002). Using tourist typologies for ecotourism research. *Journal of Ecotourism* 1(1): 7–18.

Jennings, G. and Nickerson, N. (eds) (2006). *Quality Tourism Experiences*. Oxford: Elsevier Butterworth-Heinemann.

Laarman, J.G. and Durst, P.B. (1987). Nature travel in the tropics. *Journal of Forestry* 85(5): 43–46.

Lee, W.H. and Moscardo, G. (2005). Understanding the impact of ecotourism resort experiences on tourists' environmental attitudes and behavioural intentions. *Journal of Sustainable Tourism* 13: 546–565.

Lindberg, K. (1991). *Policies for Maximizing Nature Tourism's Ecological and Economic Benefits*. International Conservation Financing Project Working Paper. Washington D.C. World Resources Institute: 1–37.

Lopez, R., Lopez, A., Wilkins, R.N., Torres, C., Valdez, R., Teer, J. and Bowser, G. (2005). Changing Hispanic demographics: Challenges in natural resource management. *Wildlife Society Bulletin* 33(2): 553–564.

Ma, X., Ryan, C. and Bao, J. (2009). Chinese national parks: Differences, resource use and tourism product portfolios. *Tourism Management* 30: 21–30.

Manning, R.E. (2011). *Studies in Outdoor Recreation: Search and Research for Satisfaction* (3rd edn). Corvallis, OR: Oregon State University Press.

McCool, S.F. (2009). Constructing partnerships for protected area tourism planning in an era of change and messiness. *Journal of Sustainable Tourism* 17: 133–148.

Mehmetoglu, M. (2007). Nature-based tourism: A contrast to everyday life. *Journal of Ecotourism* 6: 111–126.

Meyer-Arendt, K. (2004). Tourism and the natural environment. In A. Lew, M. Hall and A. Williams (eds) *A Companion to Tourism* (pp. 425–437). Oxford: Blackwell Publishing.

Nelson, J.G., Smith, P.G.R. and Theberge, J.B. (1985). Environmentally significant areas (ESA) in the Northwest Territories, Canada: Their role, identification, designation and implementation. *Environments* 17: 93–109.

Newsome, D., Moore, S.A. and Dowling, R.K. (2002). *Natural Area Tourism: Ecology, Impacts and Management*. Clevedon: Channel View Publications.

Nowaczek, A. and Smale, B. (2010). Exploring the disposition of travellers to qualify as ecotourists: The ecotourist predisposition scale. *Journal of Ecotourism* 9(1): 45–61.

Orams, M.B. (1995). Towards a more desirable form of ecotourism. *Tourism Management* 16(1): 3–8.

Orams, M.B. (2001). Types of ecotourism. In D.B. Weaver (ed) *The Encyclopedia of Ecotourism* (pp. 23–36). New York: CABI.

Passmore, J. (1974). *Man's Responsibility for Nature*. New York: Scribner.

Powell, R., Kellert, S. and Ham, S. (2009). Interactional theory and the sustainable nature-based tourism experience. *Society and Natural Resources* 22: 761–776.

Priskin, J. and Sarrasin, B. (2010). France and Francophone nations. In R.C. Buckley (ed) *Conservation Tourism* (pp. 110–124). Wallingford: CAB International.

Rinne, P. and Saastamoinen, O. (2005). Local economic role of nature-based tourism in Kuhmo Municipality, Eastern Finland. *Scandinavian Journal of Hospitality and Tourism* 5: 89–101.

Ritchie, J.R.B. and Hudson, S. (2009). Understanding and meeting the challenges of consumer / tourist experience research. *International Journal of Tourism Research* 11(2): 111–126.

Ryan, C. (ed) (2002). *The Tourist Experience*. London: Thomson Learning.

Saarinen, J. (2005). Tourism in the northern wildernesses: Wilderness discourses and the development of nature-based tourism in Northern Finland. In C.M. Hall and S. Boyd (eds) *Nature-Based Tourism in Peripheral Areas – Development or Disaster?* (pp. 36–49) Clevedon: Channel View Publications.

Silverberg, K.E., Backman, S. and Backman, K. (1996). A preliminary investigation into the psychographics of nature-based travelers to the Southeastern United States. *Journal of Travel Research* 35: 19–28.

Strasdas, W. (2006). The global market for nature-based tourism. In H. Job and J. Li (eds) *Natural Heritage, Ecotourism and Sustainable Development* (pp. 55–64). Kallmunz: Lassleben.

Su, D., Wall, G. and Eagles, P. (2007). Emerging governance approaches for tourism in the protected areas of China. *Environmental Management* 39: 749–759.

Tonge, J. and Moore, S.A. (2007). Importance-satisfaction analysis for Marine-Park Hinterlands: A Western Australian case study. *Tourism Management* 28: 768–776.
Vespestad, M. and Lindberg, F. (2010). Understanding nature-based tourist experiences: An ontological approach. *Current Issues in Tourism* iFirst article: 1–18. DOI: 10.1080/13683500.2010.513730.
Weaver, D.B. (2001). Ecotourism in the context of other tourism types. In D.B. Weaver (ed) *The Encyclopedia of Ecotourism* (pp. 73–83). New York: CABI.
Weaver, D.B. (2008). *Ecotourism* (p. 348) (2nd edn). Brisbane: John Wiley and Sons.
Weiler, B. and Kim, A.K. (2011). Tour guides as agents of sustainability: Rhetoric, reality and implications for research. *Tourism Recreation Research* 36(2): 113–125.
Weiler, B. and Richins, H. (1995). Extreme, extravagant and elite: A profile of ecotourists on Earthwatch expeditions. *Tourism Recreation Research* 20(1): 29–36.
Weiler, B. and Yu, X. (2007). Dimensions of cultural mediation in guiding Chinese tour groups: Implications for interpretation. *Tourism Recreation Research* 32(3): 13–22.
Wheeller, B. (2007). Sustainable mass tourism: More smudge than nudge – The canard continues. *Tourism Recreation Research* 32(3): 73–75.
Wight, P.A. (2001). Ecotourists: Not a homogenous market segment. In D.B. Weaver (ed) *The Encyclopedia of Ecotourism* (pp. 37–62). New York: CABI.
Wong, A. (2001). Satisfaction with local tour guides in Hong Kong. *Pacific Tourism Review* 5(1): 59–67.
Zins, A.H. (2002). Consumption emotions, experience quality and satisfaction: A structural analysis for complainers and non-complainers. *Journal of Travel and Tourism Marketing* 12: 3–18.

Further Reading

Balmford, A., Beresford, J., Green, J., Naidoo, R., Walpole, M. and Manica, A. (2009). A global perspective on trends in nature-based tourism. *PLoS Biol* 7(6): e1000144.
Fennell, D. (2011). *Tourism and Animal Ethics*. London: Routledge.
Pröbstl, U., Wirth, V., Elands, B. and Bell, S. (eds.) (2010). *Management of Recreation and Nature Based Tourism in European Forest*. London: Springer.

Chapter 13
Tourism and Climate Change: A Need for Critical Analysis

C. Michael Hall, Susanne Becken, Ralf Buckley and Daniel Scott

Context

The relationship between tourism and climate change is emerging as one of the key issues facing tourism in the 21st century. However, there is increasing recognition that there are a number of different approaches to climate change and the manner in which the tourism industry as well as destinations should adapt and/or mitigate. The four contributing probes in this chapter highlight some of the issues surrounding this debate. The lead probe piece by Hall provides a thematic snapshot of climate change and tourism knowledge with reference to the wider work of bodies such as the IPCC as well as the academic literature. The paper notes the unevenness of tourism and climate change research both with respect to topics as well as locations. The first rejoinder by Becken suggests a different approach to tourism and climate change that focuses on integration across disciplines and partnerships between researchers and tourism organizations. The second rejoinder by Buckley outlines a number of research priorities, the most significant of which is the social and environmental consequences of extreme weather events at destinations and the flow-on effects for tourism. The final contribution by Scott draws on the work of all of the other contributions and emphasizes the need for more critical reflection on climate change, a point that is also taken up in the conclusion to the collection of papers.

13.1

Tourism and Climate Change: Knowledge Gaps and Issues[1]

C. Michael Hall

The relationship between tourism and climate change has become one of the fastest growing areas of tourism research since the turn of the century. In one sense this should not be surprising. There is widespread public concern over the topic and it has become highly politicized especially as different stakeholders, including academics, have sought to advance their own economic and personal interests. As noted in the following chapters, despite tourism being recognized as a sector that is vulnerable to climate change different discipline areas and research traditions treat the tourism and climate change relationships in different ways. This means that the assessment of tourism and climate change is extremely uneven with respect to focus as well as approach, with particular methods being dominant in some fields and reports (Scott *et al.*, 2012).

The fact that there are differences in approaches to climate change is nothing new within academic literature. However, for the wider public, given that climate change science is often portrayed as 'objective' science, the fact that the climate change 'problem' could be portrayed in different ways may come as something of a surprise. Indeed, one of the major difficulties in managing climate change knowledge is that, 'to speak of the "science" of economic damage assessment as if it were a separate domain exchanging independently generated ideas with policymakers is to conceal the shared commitments that define them as part of a single cultural and political order' (Demeritt & Rothman, 1999: 406). The choice of valuation procedure is part of the social construction of knowledge and is ultimately personal and political rather than objective and rational (Demeritt & Rothman, 1999; Hall, 2012). For example, most assessments of climate change ignore intrinsic value, adopt a narrow decisionist framework (Wynne, 1997) and ignore the exchange relations and structural imperatives of the capitalist economy that actually drives emissions (Wainwright, 2010). All of this has meant that anthropogenic climate change has primarily been defined in environmental rather than political terms, or one that requires a more fundamental re-framing of the contemporary capitalist economic system. Indeed, Demeritt (2001: 316) argues that treating the objective physical properties of greenhouse gases in isolation from their social, economic and political relations has served 'to conceal, normalize, and thereby reproduce those unequal social relations'.

Demeritt (2001: 309) suggests that instead of accepting an idealized objective vision of scientific truth and denying the socially and politically situated and contingent nature of scientific knowledge, 'the proper response to it is to develop a more reflexive understanding of science as a situated and ongoing social practice'. Indeed, the portrayal of climate change research as being suddenly politicized is fundamentally misleading. Science has always been political in the sense that its outputs and outcomes, if not the process itself, lead to winners and losers. Indeed, given the lack of a post-Kyoto agreement and the attention given to 'climategate' in the media, Demeritt and Rothman (1999: 405–6) presciently observed that it is enormously tempting for politicians to argue that climate change policy must be based upon an objective scientific assessment of the economic costs and benefits, thereby absolving themselves of any responsibility to exercise discretion and leadership. Because this division of labor enhances their power and prestige, it is also attractive to the IPCC participants. But this is a dangerous strategy. It provides neither a very democratic nor an especially effective basis for crafting a political response to global climate change. It enshrines in apparent scientific objectivity the particular values embodied by the IPCC assessment. When, as is almost inevitable, these political presumptions are publicly exposed and deconstructed, there is the danger that the resulting acrimony will tend to harden positions rather than make negotiation and mutual understanding easier.

Those who write of the social construction of climate change are not arguing that it does not exist, rather they are drawing attention to the ways in which the climate change problem, and hence its solutions, is constructed and communicated and how this meets some interests and world views and not others (Winter & Koger, 2004). Similarly, Wainwright (2010) suggests that although climate scientists engage in debates about the meaning of their results, they rarely reopen the 'black boxes' that are taken for granted in their research and provide an example of how carbon may be considered by physical and social scientists.

> Two physical scientists might engage in heady debate about the precise role of CO_2 or CH_4 in forcing a certain atmospheric process, but it is hard to imagine that carbon's basic qualities – its atomic number or weight, chemical properties, and so on – would be called into question. By contrast, two social scientists discussing, say, the hegemony of carbon emissions markets in climate policy discourse ... would need to agree on the meaning of hegemony, markets, climate policy, discourse, and so on.

> This turns out to be no mean feat, because distinct interpretations of these and related concepts reflect different conceptions of the world… there is no metalanguage that lies outside of social life with which to objectively calibrate these concepts. Consequently, debates over the meaning of the building-block concepts for social thought are, by necessity, complex and interminable. (Wainwright, 2010: 984)

The following debate on tourism and climate change is therefore part of the important process of opening up the black box in which research appears so as to make the process not only more transparent but also to highlight the way in which knowledge gap and research agendas are constructed, and developed.

The Policy Context

The results of academic research are not value free, perhaps nowhere more so in the case of such a politically charged area as climate change where research is an important element of policy determination. The reasons for the highly charged policy environment primarily relate to the implications of the costs of any mitigation and adaptation strategies (Monbiot, 2006; Randerson, 2006). For example, application of the 'polluter pays' principle to an area such as aviation would have significant impacts on costs of airline passenger and freight costs and therefore levels of demand. Furthermore, climate change is now part of wider political debates and appears to be a factor in electoral campaigns. For example, Australia's non-ratification of the Kyoto agreement under the conservative coalition government of Prime Minister John Howard was an issue that emerged a number of times during Australia's 2007 federal election (e.g. BBC News, 2007). Similarly, political parties in the UK were seeking to position themselves with respect to climate change in the run-up to the next election, with the Labour government announcing an amended climate change bill including adding emissions from the aviation and shipping industry in the UK's greenhouse gas (GHG) emissions targets (Ryan & Stewart, 2007). The bill built on the British governments announcement in October 2007 that as from 2009 air taxes would be levied on the flight, rather than directly on the passenger, a measure that is regarded as better targeted at cutting carbon emissions (Wintour & Elliott, 2007), and which was also a policy of the conservative opposition. The policy is supported by budget airline easyJet who argued that air passenger duty should be dropped in favour of a scheme that grades aircraft according to their emissions and length of journey (Milmo, 2007). However, given the potential lifespan of aircraft this may also have the effect of moving older aircraft with higher emissions to regions with weaker emissions rules such as Africa, Asia or Latin America.

Also of great significance in policy terms has been the awarding of the Nobel Peace Prize to former US Vice-President Al Gore and to the IPCC in October 2007. In awarding the prize the Nobel committee signalled its view that climate change is now a global security issue which therefore means its policy importance is reinforced beyond an economic or environmental interpretation (Black, 2007). However, it is important to recognize the extent to which such an award lends credibility to the science of climate change despite attempts by various interests to discredit scientific findings and recommendations (Monbiot, 2006). Nevertheless, it is often the coincidence of weather related natural disasters combined with awareness of climate change that arguably leads to the most volatile policy environments such as in the case of the impacts of Hurricane Katrina in the United States; wildfires in Greece, the Iberian Peninsula and the United States; drought in Australia and the United States; and flooding in Africa and Mexico. In such cases, which involve major temporary and, in the case of Hurricane Katrina and the Australian drought, potentially permanent displacement of people, it is perhaps not surprising that the spectre of environmental refugees (Myers, 2002) wrought by climate change is regarded as occurring in the present rather than just being in the future.

Tourism Within Climate Change Reports

The global climate change political and public debate has been strongly influenced by the contents and media reporting of a number of climate change reports in which tourism is noted. In policy terms the 2006 review on the economics of climate change commissioned by the UK government from Sir Nicholas Stern, a former chief economist with the World Bank, was extremely important. In media terms, it generated substantial interest (e.g. Hinsliff, 2006b, 2006b; Randerson, 2006). For example, the UK's *Observer* newspaper ran a headline in covering the review: 'Ten years to save the planet from mankind' (Hinsliff, 2006b).

The Stern review arguably did not bring anything new to the understanding of climate change in a scientific or environmental sense. But it did have the important function of 'independently' affirming scientific argument from the respectability of a conservative economic position rather than one that may be perceived as solely 'academic'. Just as significantly for economic thinking it placed a price on the cost of failing to act on climate change unless it is tackled immediately – a figure of £3.68 trillion. Stern (2006) forecast that the world needed to spend 1% of global GDP – equivalent to about £184 billion per year – in dealing with climate change, or face a bill between five and 20 times higher for damage caused by letting it continue. According to Hinsliff (2006a) unchecked climate change could therefore cost as much as £566 for every man, woman and child on the planet – roughly 6.5 billion people – in 2006 terms.

The debate surrounding the review's release focused on a number of areas of interest to tourism, particularly with respect to the potential imposition of green taxes on aviation (e.g. Hinsliff, 2006a). Yet the review itself gave only minor consideration to tourism. Table 13.1 notes that there were only 19 references to tourism, tourist and recreation and cognate terms in the review with most of these being brief mentions. The most substantive discussion of tourism was in relation to some of the positive effects that climate change will bring for a few developed countries for moderate amounts of warming, although it is also noted that they will become very damaging at the higher temperatures expected in the second half of this century. On page 126 it is stated that 'tourism may shift northwards, as cooler regions enjoy warmer summers, while warmer regions like southern Europe suffer increased heat wave frequency and reduce water availability. One study projected that Canada and Russia would both see a 30% increase in tourists with only 1°C of warming' (Stern, 2006: 126; referring to Hamilton *et al.,* 2005). Although the discussion then goes on to note the potentially harmful effects of climate change to alpine and winter tourism and to tourism resources such as the Great Barrier Reef. Tourism is thus only superficially considered in the Stern Review, and represented by a selective, non-comprehensive choice of references.

Table 13.1 References to tourism, tourist and recreation and cognate terms in Stern Review

Part	*chapters*	*Focus*	*citations in text**
1	1–2, technical annex to 2	The nature of the scientific evidence for climate change, and the nature of the economic analysis required by the structure of the problem which follows from the science	0
2	3–6	How climate change will affect people's lives, the environment and the prospects for growth and development in different parts of the world.	10
3	7–13	The economic challenges of achieving stabilisation of greenhouse gases in the atmosphere	4
4	14–17	Policy responses to mitigation	0
5	18–20	Policy responses for adaptation	5
6	21–27	International collective action	0

In contrast tourism is given much greater emphasis in the IPCC climate change reports. The IPCC's role is to 'assess on a comprehensive, objective, open and transparent basis the best available scientific, technical and socio-economic information on climate change from around the world. The assessments are based on information contained in peer-reviewed literature and, where appropriately documented, in industry literature and traditional practices' (IPCC, 2004: 1). The First Assessment Report was released in 1990 and outlined the scientific basis for concern about climate change. The Second Assessment Report was released in 1995. The Third Assessment Report, released in 2001, consisted of three working group reports on The Scientific Basis; Impacts, Adaptation and Vulnerability; and Mitigation, and a Synthesis Report primarily aimed at policy makers. The Fourth Assessment Report in 2007 was released with a similar structure with three working group reports on The Physical Science Basis; Impacts, Adaptation and Vulnerability; Mitigation of Climate Change and a Synthesis Report.

Although tourism has not been cited in reports on the scientific basis for climate change and only marginally with respect to mitigation, it has become increasingly recognized in the reports of the IPCC Working Group on Impacts, Adaptation and Vulnerability (Wall, 1998; see also Amelung *et al.,* 2008). Table 13.2 compares reference to tourism and cognate terms

Table 13.2 References to tourism, tourist and recreation and cognate terms in regional chapters of IPCC Working Group II 2001 and 2007 reports on impacts, adaptation and vulnerability

Part	*chapters*	*Focus*	*citations in text**
1	1–2, technical annex to 2	The nature of the scientific evidence for climate change, and the nature of the economic analysis required by the structure of the problem which follows from the science	0
2	3–6	How climate change will affect people's lives, the environment and the prospects for growth and development in different parts of the world.	10
3	7–13	The economic challenges of achieving stabilisation of greenhouse gases in the atmosphere	4
4	14–17	Policy responses to mitigation	0
5	18–20	Policy responses for adaptation	5
6	21–27	International collective action	0

in comparable chapters of the 2001 (McCarthy *et al.,* 2001) and 2007 (Parry *et al.,* 2007) reports of the Working Group. It is noticeable that although there are similar total numbers of references to tourism, tourist, recreation and cognate terms in the regional chapters of the two reports, there is a substantial variation between chapters. In particular there are notable increases (greater than 10%) in citations in the chapters on Africa, Australia and New Zealand, Europe and Small Island states. There are decreases in citations with respect to Asia and the polar regions although these only have a small overall number of citations. The most significant decrease in citations of tourism related words occurring in the chapter on North America, down from 86 in 2001 to 27 in 2007.

Changes in word counts are important as it can be argued that they reflect aspects of the issue ecology of content in IPCC reports and are therefore a crude surrogate measure of importance and levels of knowledge on particular subjects. Of course, this is also partly a reflection of the authors of the various chapters as well with respect to their levels of knowledge of tourism and the selection of references they cite with the range of academic publications and approaches to tourism and climate change (Scott, Wall *et al.,* 2005; Gössling & Hall, 2006c; Scott, Amelung *et al.,* 2007) being far greater than those actually utilized. Nevertheless, the relative emphasis of subjects covered by the report may broadly correspond to the relative research emphasis on tourism and climate change in general. Table 13.3 outlines the substantive comments on tourism issues (a sentence or more) in Parry *et al.* (2007). Although not the sole source, the identification of particular topics is used in the next section to provide an overview of the major current research foci on tourism and climate change.

Research Issues and Topics

The aim of this section is not to restate the IPCC's conclusions and forecasts with respect to the effects of climate change on the global environment and the role of anthropogenic warming (see Solomon *et al.,* 2007). Instead, it seeks to identify some of the key topics and issues addressed by research on tourism and climate change.

The relationship between GHG emissions and the nature of economic development is regarded as extremely important to climate change mitigation in the longer term with a number of factors influencing the amount of emissions:

- Structural changes in the production system in relation to the role of high or low energy-intensive industries and services.
- Technological patterns in sectors such as energy, transportation, building, waste, agriculture and forestry.

Table 13.3 Dimensions of the tourism and climate change relationships covered in the IPCC Working Group II 2007 reports on impacts, adaptation and vulnerability

Dimension	*Specified locations*	*Chapter source*
Sensitivity to climate change	Africa, Asia, Australia and New Zealand, Europe, North America, Tropical destinations, Small islands	Alcamo *et al.* 2007: 543–4; Boko *et al.* 2007: 450, 459; Cruz *et al.* 2007: 489; Hennessy *et al.* 2007: 523; Mimura *et al.* 2007: 689, 697; Schneider *et al.* 2007: 790; Wilbanks *et al.* 2007: 363, 368, 375, 380.
Issues in adaptation to climate change	Small islands	Alcamo *et al.* 2007: 561; Anisimov *et al.* 2007: 673, 676; Mimura *et al.* 2007: 705; Wilbanks *et al.* 2007: 380.
Effects on coastal tourism	Africa, Americas, Caribbean, Mediterranean, Florida, Thailand, Maldives, small islands	Alcamo *et al.* 2007: 543–4; Boko *et al.* 2007: 440, 449; Field *et al.* 2007: 634; Magrin *et al.* 2007; 584, 599, 600; Mimura *et al.* 2007: 689, 696, 698, 701, 703; Nicholls et al 2007: 335–6, 337.
Degradation of coral reef, coral bleaching	Africa, Australia, small islands	Boko *et al.* 2007: 439; Fischlin *et al.* 2007: 235; Hennessy *et al.* 2007: 523; IPCC 2007: 13, 15; Mimura *et al.* 2007: 689, 696; Nicholls *et al.* 2007: 320.
Effects on summer tourism	Europe, especially Mediterranean	Alcamo *et al.* 2007: 565.
Effects on winter tourism	Asia, Europe, Bolivia	Adger *et al.* 2007: 721, 722, 734; Alcamo *et al.* 2007: 543–4, 561, 565; Cruz *et al.* 2007: 489; Field *et al.* 2007: 634; Hennessy *et al.* 2007: 523; IPCC 2007: 14, 18; Magrin *et al.* 2007; 589; Rosenzweig *et al.* 2007: 117; Wilbanks *et al.* 2007: 363.
Skiing	Asia, Australia and New Zealand, Europe	Adger *et al.* 2007: 721, 722, 734; Alcamo *et al.* 2007: 557, 561; Field *et al.* 2007: 634; Hennessy *et al.* 2007: 523; Magrin *et al.* 2007; 589; Rosenzweig *et al.* 2007: 89, 111.
Effects on mountain tourism	Europe	Fischlin *et al.* 2007: 223; Rosenzweig *et al.* 2007: 88, 89.

Table 13.3 Continued

Dimension	*Specified locations*	*Chapter source*
Effect on wild faunal diversity/ nature-based tourism	Marine ecosystems, Mediterranean-type ecosystems, Small islands, Southern Africa, Tropical savanna systems	Boko *et al.* 2007: 435, 459; Field *et al.* 2007: 634; Fischlin *et al.* 2007: 225, 226, 234; Mimura *et al.* 2007: 696.
Reservation economies	North America	Field *et al.* 2007: 625.
Effect of extreme events	Mexico, North America, small islands	Field *et al.* 2007: 626; Magrin *et al.* 2007: 585; Mimura *et al.* 2007: 693, 702.

- Geographical distribution of activities encompassing both human settlements and urban structures in a given territory, and its two-fold impact on the evolution of land use, and on mobility needs and transportation requirements.
- Consumption patterns – for a given income per person, parameters such as housing patterns, leisure styles or the durability and rate of obsolescence of consumption goods will have a critical influence on long-run emission profiles.
- Trade patterns – which may influence access to the best available technologies and to finance may limit the building of infrastructure (Fisher *et al.,* 2007: 177).

Surprisingly, even though mobility, transport and leisure consumption were noted as long-term determinants of emissions there was very little discussion in the IPCC reports on the role of tourism in mitigation, even with respect to tourism transport. In contrast several studies (Gössling, 2002; Gössling & Hall, 2006d; Peeters *et al.,* 2006) have noted the rapid growth of tourism as a contributor to emissions. Scott, Amelung *et al.* (2007) estimated that 5% of the global share of CO_2 emissions are attributable to tourism within a range of 4–6%. Measured in radiative forcing, a measure of the influence a factor has in altering the balance of incoming and outgoing energy in the Earth–atmosphere system that is an index of the importance of the factor as a potential climate change mechanism, the contribution of tourism to global warming may be up to 14%, even though there is great uncertainty with regard to the contribution of cirrus-related effects. The majority of tourist trips create only small amounts of emissions. However, air travel, and especially long-haul travel, are major contributors. For example, long-haul travel between the five world regions accounts for only 2.7% of all tourist trips, but contributes 23% to global tourist transport emissions. Furthermore, given current predictions for travel growth under a 'business as usual' scenario it is estimated that tourist attributable CO_2 emissions will grow by 152% and radiative forcing by 171% by 2035 (Scott, Amelung *et al.,* 2007). Therefore, given such forecasts it is perhaps not surprising that reducing the effects of air travel through technological, regulatory and/or behavioural mechanisms has become a significant focus for tourism and climate change research (e.g. Peeters *et al.,* 2006, 2007; Peeters, 2007; Gössling & Broderick *et al.,* 2007; Gössling & Lindén *et al.*, 2007), although other parts of the tourism system are also of importance (e.g. Chan & Lam, 2003; Gössling *et al.,* 2005).

Wilbanks *et al.*'s (2007) chapter on industry, settlement and society probably provided the most substantial assessment of the impact of climate change on tourism within the IPCC context. Tourism was identified as a 'climate-sensitive human activity' with the chapter concluding that vulnerabilities of industries to climate change are 'generally greater in certain high-risk locations, particularly coastal and riverine areas, and areas whose economies are closely linked with climate sensitive resources, such as ... tourism; these vulnerabilities tend to be localized but are often large and growing' (Wilbanks *et al.,* 2007: 359). However, one of the greatest problems with assessing the impacts of climate change on tourism is that both direct and indirect effects will vary greatly with location (Gössling & Hall, 2006d). Direct effects include the role climate variables, such as temperature, sunshine hours, precipitation, humidity and storm frequency and intensity, play with respect to tourist decision-making, including activity and destination choice (Scott, Gössling *et al.,* 2008). Another effect is the extent to which particular environments, such as tropical or alpine resorts, gain some of their appeal from climatic variables. Finally, indirect effects of climate change such as heatwaves, fires, disease outbreaks, landscape change and biodiversity change can also have substantial effects on tourism activities, perceptions of a location and the capacity of firms to do business. However, as Rosenzweig *et al.* (2007: 111) identified, 'as a result of the complex nature of the interactions that exist between tourism, the climate system, the environment and society, it is difficult to isolate the direct observed impacts of climate change upon tourism activity. There is sparse literature about this relationship at any scale'.

Climate change is likely to have a long-term effect on domestic and international tourist flows. Higher temperatures are potentially likely to change summer and winter destination preferences in the longer term, either through direct effects on a tourism resource such as snow availability, or in terms of making competing destinations more or less attractive climatically. However, the capacity of potential tourists to accurately judge the implications of changes in average temperatures or climate conditions for a destination is quite debatable (Gössling *et al.*, 2006) with the exception of tourism resource loss, such as snow. Deterministic models such as the Tourism Comfort Index (Amelung & Viner, 2006) indicated improved conditions for tourism in northern Europe, while econometric modelling by Hamilton *et al.* (2005) indicated that an arbitrary climate change scenario of 1°C would also lead to a gradual shift of tourist destinations further north and up mountains thereby affecting the preferences of European summer tourists (Alcamo *et al.*, 2007). Nevertheless, Gössling and Hall (2006a, 2006b, 2006d) have noted that there are a number of major weaknesses with respect to current models in predicting travel flows under conditions of climate change (Table 13.4) and have urged substantial caution in utilizing the results of deterministic approaches to climate change adaptation given high levels of behavioural uncertainty.

In light of concern with the role of higher temperatures on tourist flows it is perhaps not surprising that the environments that have attracted the most research attention are those of coastal, including those of small islands, and alpine areas. Arguably, these are the same environments that have attracted tourism research on environmental impacts in general (Hall & Page, 2006). Coasts and small islands are projected to be exposed to increasing risks, including coastal erosion, due to climate change and

Table 13.4 Major weaknesses of current models in predicting travel flows

• Validity and structure of statistical databases
• Temperature assumed to be the most important weather parameter
• Importance of other weather parameters largely unknown (rain, storms, humidity, hours of sunshine, air pollution)
• Role of weather extremes unknown
• Role of information in decision -making unclear
• Role of non -climatic parameters unclear (e.g., social unrest, political instability, risk perceptions, destination perception)
• Existence of fuzzy -variables problematic (terrorism, war, epidemics, natural disasters)
• Assumed linearity of change in behaviour unrealistic
• Future costs of transport and availability of tourism infrastructure uncertain
• Future levels of personal disposable income (economic budget) and availabilit y of leisure time (time budget) that are allocated to travel uncertain

Source: Gössling and Hall (2006a, 2006b, 2006d).

sea-level rise (IPCC, 2007). In addition, the impact of climate change on tourism-dependent coastal economies may be exacerbated because of the increased human-induced pressures tourism brings to coastal areas (Hall, 2006c; Nicholls *et al.,* 2007). Nicholls *et al.*'s (2007: 331) summary of climate-related impacts in relation to recreation and tourism in coastal areas argued that temperature rise (air and seawater), extreme events (storms, waves), erosion (sea level, storms, waves) and biological effects will have strong impacts; floods (sea level, runoff) will have a weak impact; and rising water tables (sea level) and saltwater intrusion (sea level, runoff) will have a negligible impact or that an impact is not established. However, such relative impacts will not be the same in all locations (Uyarra *et al.,* 2005).

Extreme events, such as cyclones and hurricanes, for example, are regarded as significant because of both their physical impact on environment and infrastructure (Nurse & Moore, 2005) as well as the potential negative contribution to destination image. Under climatic change more frequent high-magnitude events may mean there is less time for physical and human systems to recover, meaning that recovery may never be complete and thereby resulting in long-term environmental deterioration. Such effects are already observed in coral reef ecosystems, even though the consequences for tourism remain unclear (Gössling & Lindén *et al.,* 2007).

With respect to small islands Mimura *et al.* (2007) concluded that there was a high degree of confidence that the effects of climate change on tourism are likely to be direct and indirect, and largely negative:

> Tourism is the major contributor to GDP and employment in many small islands. Sea-level rise and increased sea water temperature will cause accelerated beach erosion, degradation of coral reefs, and bleaching. In addition, a loss of cultural heritage from inundation and flooding reduces the amenity value for coastal users. Whereas a warmer climate could reduce the number of people visiting small islands in low latitudes, it could have the reverse effect in mid- and high-latitude islands. However, water shortages and increased incidence of vectorborne diseases may also deter tourists. (Mimura *et al.*, 2007: 689)

In the western Indian Ocean region, a 30% loss of corals resulted in reduced tourism in Mombasa and Zanzibar, and caused financial losses of about US$ 12–18 million (Payet & Obura, 2004). Australia's Great Barrier Reef has experienced eight mass bleaching events since 1979 (1980, 1982, 1987, 1992, 1994, 1998, 2002 and 2006). The most widespread and intense events occurred in 1998 and 2002, with about 42% and 54% of reefs affected, respectively (Berkelmans *et al.,* 2004). Climate change related effects such as rising sea temperatures and ocean acidification are being exacerbated by coral reefs exposure to local anthropogenic impacts, including sedimentation, pollution and reduction of fish stocks (Hoegh-Guldberg, 1999; 2004). However, in addition to marine environments climate change related biodiversity loss is also significant for nature-based tourism in terrestrial environments (Hall, 2006b; Scott, Amelung *et al.,* 2007).

The potential impact of climate change on alpine and winter tourism has been a major focus of research in Europe (e.g. Burki *et al.,* 2005; Harrison *et al.,* 2005) and North America (e.g. Scott *et al.,* 2003; Scott *et al.,* 2006; Scott, Amelung *et al.,* 2007, 2007b, 2007c). Declines in mountain snowpack in western North America and in the Swiss Alps are largest at lower, warmer elevations with corresponding impacts on skiing, ice climbing and scenic activities in areas affected (Rosenzweig *et al.,* 2007: 89). Over the past century snow cover has decreased in most regions, especially in spring. Northern Hemisphere snow cover over the 1966 to 2005 period decreased in every month except November and December (Solomon *et al.,* 2007).

The challenge for many alpine resorts is that mountain snow can be especially sensitive to only small changes in temperature, particularly in temperate regions where the transition from rain to snow is generally closely associated with the freezing level altitude. According to Agrawala (2007) under present climate conditions, 609 out of the 666 (or 91%) Alpine ski areas in Austria, France, Germany, Italy and Switzerland can be considered as naturally snow-reliable with the remaining 9% already operating under marginal conditions. Agrawala (2007) estimated that the number of naturally snow-reliable areas would drop to 500 under a 1°C increase, to 404 under 2°C, and to 202 under a 4°C warming of the climate. Responses to climate change in alpine and winter tourism destinations include artificial snow-making and associated structures such as high altitude water reservoirs; grooming of ski slopes; moving ski areas to higher altitudes and glaciers; use of white plastic sheets as protection against glacier melt; market, economic and regional diversification; and the use of market-based instruments such as weather derivatives and insurance (Scott, 2006; Scott, Amelung *et al.*, 2007; Scott, McBoyle *et al.*, 2007; Scott, Jones *et al.*, 2007; Agrawala, 2007).

Ecotourism has been seen as a potential substitute for the ski industry in Asia (Fukushima *et al.,* 2002; Cruz *et al.,* 2007). However, as Scott, Jones *et al.* (2007) point out climate and associated environmental change will also affect the viability of nature-based tourism in alpine areas. Similar concerns exist in the Arctic. For example, Anisimov *et al.* (2007) noted the potential significance of ecotourism as an opportunity for adaptation of indigenous peoples, even though the biodiversity and landscape of the region is experiencing some of the most rapid climate related environmental changes on the planet today, with average Arctic temperatures having increased at almost twice the rate of the rest of the world in the past 100 years (ACIA, 2005; Solomon *et al.,* 2007).

Even one of the most high-profile dimensions of climate change in 2007 – the opening of the Northwest Passage as the result of the loss of summer ice – is seen as potentially beneficial for tourism. Anisimov *et al.* (2007: 676) state 'the Northern Sea Route will create new opportunities for cruise shipping. Projections suggest that by 2050, the Northern Sea Route will have 125 days/yr with less than 75% sea-ice cover'. Similarly, Instanes *et al.* (2005) noted that increased possibilities for marine navigation and the extension of the warm-weather season will improve conditions for tourism.

The fact that there are potentially 'winners', at least in the short-term, as well as 'losers' from climate change also reflects on the adaptation capacities of different regions, sectors, actors or firms. 'Adaptive capacity is the ability or potential of a system to respond successfully to climate variability and change, and includes adjustments in both behaviour and in resources and technologies' (Adger *et al.,* 2007: 727). Because of their explicit focus on real-world behaviour, assessments of adaptation practices differ from the more theoretical assessments of potential response. However, at this stage an understanding of the adaptive capacities and practices of the various elements of tourism in relation to climate change are quite limited (Becken, 2005; Gössling & Hall, 2006c; Scott, Amelung *et al.,* 2007). In one sense this situation reflects the large knowledge gaps that surround a number of areas of tourism and climate change research.

Knowledge Gaps at a Regional Level

The level of knowledge with respect to tourism and climate change can be assessed regionally as well as thematically. Indeed, a number of knowledge gaps have been highlighted in the various IPCC reports. Table 13.5 provides a summary of the relative levels of knowledge for a region in relation to the estimated impact of climate change on tourism.

In the case of Africa, Boko *et al.* (2007: 450) stresses that 'very few assessments of projected impacts on tourism and climate change are available' and later notes:

> There is a need to enhance practical research regarding the vulnerability and impacts of climate change on tourism, as tourism is one of the most important and highly promising economic activities in Africa. Large gaps appear to exist in research on the impacts of climate variability and change on tourism and related matters, such as the impacts of climate change on coral reefs and how these impacts might affect ecotourism. (Boko *et al.,* 2007: 459)

Tourism is similarly recognized by the IPCC as one of the most important industries in Asia, although the lack of research is bemoaned. 'Nature-based

Table 13.5 Relative level of tourism specific climate change knowledge and estimated impact of climate change on tourism by region

Region	*Estimated impact of climate change on tourism*	*Relative level of tourism specific climate change knowledge*
Africa	Moderately-strongly negative	Extremely poor
Asia	Weakly-moderately negative	Extremely poor
Australia and New Zealand	Moderately-strongly negative	Poor-Moderate (high in Great Barrier Reef)
Europe	Weakly-moderately negative	Moderate (high in alpine areas)
Latin America	Weakly-moderately negative	Poor
North America	Weakly negative	Moderate (high in coastal and ski areas)
Polar regions	Weakly negative – weakly positive	Poor
Small islands	Strongly negative	Moderate

Sources: Derived from Gössling and Hall (2006c); Parry *et al.* (2007); see also Scott, Amelung *et al.* (2007).

tourism is one of the booming industries in Asia, especially ski resorts, beach resorts and ecotourist destinations which are likely vulnerable to climate change; yet only a few assessment studies are on hand for this review' (Cruz *et al.,* 2007: 489).

Even in North America, which is one of the better studied regions, substantial knowledge gaps exist. For example, Field *et al.* (2007: 634) note that, although 'coastal zones are among the most important recreation resources in North America, the vulnerability of key tourism areas to sea-level rise has not been comprehensively assessed'. Such assessment is extremely significant for policy makers and industry stakeholders as it can provide increased understanding of the relative vulnerabilities of destinations and attractions to climate change. For example, Scott, McBoyle *et al.* (2007) highlight that early studies of climate change impact on the ski industry did not account for snow-making capacity, which substantially lowers the vulnerability of some ski resorts to climate change.

In the case of Australia and New Zealand, Hennessy *et al.* (2007) note that few regional studies have assessed potential impacts of tourism, although it is still argued that 'Some tourist destinations may benefit from drier and warmer conditions, e.g., beach activities, viewing wildlife and geothermal activity, trekking, camping, climbing, wine tasting and fishing' (Hennessy *et al.,* 2007: 523) but that there is likely to be greater risks to tourism as a result of increased hazards. Similarly, positive change with respect to summer and winter tourism in parts of Europe (Alcamo *et al.*, 2007: 565), even though, as noted above, understanding of many of the behavioural and adaptive capacities with respect to tourism is currently relatively weak.

This review has sought to outline some of the key issues that have emerged in recent studies and debate on tourism and climate change. Although, it has noted some of the findings with respect to the potential impacts of climate change on tourism it has indicated that the lack of knowledge in many areas, including primary tourism processes and systems, is severely constraining capacity to better understand the relationship between tourism and climate change (Table 13.6). In particular, it reinforces the need to improve understanding 'as to how direct and indirect impacts of climate change affect human behaviour with respect to recreation patterns and holiday destination choice' (Hennessy *et al.*, 2007: 530). Undoubtedly, tourism will see changes in travel flows and patterns in the short-, e.g. reduction in ski season, and long-term, e.g. changed competitiveness of destinations because of changes in climate and high-magnitude events (Hamilton *et al.*, 2005; Schneider *et al.*, 2007; Scott, Amelung *et al.*, 2007). However, it is extremely important that the research-based advice available to policy makers and stakeholders in tourism moves beyond deterministic modelling and is based on a deeper understanding of tourism behaviours and flows.

It is also important that a better understanding of the adaptive capacities of destinations, environments and firms is developed and, where possible, that adaptations are charted over time so as to better transfer innovations from one location or firm to another as well as gain a more accurate account of such capacities. For example, research on small tourism firms by Hall (2006a) in New Zealand and Finland (Saarinen & Tervo, 2006) indicates that although firms may be seeking to both adapt and mitigate with respect to climate change, other immediate and more pressing business needs mean that climate change cannot be a primary focus of business activity – thereby potentially hindering firm and destination innovation with respect to climate change (Hall, 2007).

As well as being subject to climate change, tourism is also seen as a non-climatic stress. For example, with respect to Latin America the IPCC comments 'The rapidly expanding tourism industry is driving much of the transformation of natural coastal areas, paving the way for resorts, marinas and golf courses' (Magrin *et al.*, 2007: 587). Similarly, tourism is also regarded as a competitor in a number of locations for potentially scarce water resources (Gössling *et al.*, 2002; Mimura *et al.*, 2007). Increasingly, demands for biofuels for transport, including for leisure travel, may also be placing pressure on food production leading to competition between mobility for the world's rich and food for the world's poor (Vidal, 2007). Therefore, a

Table 13.6 Adequacy of tourism knowledge with respect to climate change adaptation, mitigation and impacts

Factor	*Very inadequate*	*Inadequate*	*Adequate*	*Very adequate*
Generic				
Understanding of tourism system at various scales		X		
Understanding of tourism within human, environmental and innovation systems at various scales		X		
Understanding of human behaviour with respect to destination and activity choice		X		
Impact of climate change on different environments in which tourism occurs				
Alpine environments			X	
Arid environments		X		
Coastal environments			X	
Coral reef environments				X
Polar environments	X			
Small islands (warm-water)			X	
Small islands (cold-water)		X		
Temperate forests		X		
Temperate grasslands		X		
Tropical forests	X			
Tundra	X			
Urban	X			
Impact of climate change on different tourism activities				
Adventure	X			
Cycling	X			
Eating out/restaurants	X			
Farm tourism	X			
Food and wine	X			
Fishing	X			
Garden tourism	X			
Health and spa	X			

Table 13.6 Continued

Factor	*Very inadequate*	*Inadequate*	*Adequate*	*Very adequate*
Historic buildings	X			
Indigenous	X			
Museums and art galleries	X			
National parks		X		
Nightlife	X			
Scenic drives	X			
Shopping	X			
Sightseeing (urban and non-urban)	X			
Snow-based activities (e.g. skiing)				X
Sports tourism (e.g. golf)	X			
Theme Parks	X			
Water-based (lakes and rivers)	X			
Water-based (ocean)		X		
Wildlife viewing		X		
Adaptive capacities				
Airlines		X		
Attractions	X			
Coach and Bus		X		
Cruise ships	X			
Destination communities	X			
Hotels/resorts		X		
Industry associations	X			
Intermediaries	X			
Railways			X	
Small operators	X			
Tourists	X			
Transnationals		X		
Mitigation capacities				
Airlines			X	
Attractions		X		
Coach and Bus				X

Table 13.6 Continued

Factor	*Very inadequate*	*Inadequate*	*Adequate*	*Very adequate*
Cruise ships				X
Destination communities	X			
Hotels/resorts				X
Industry associations	X			
Intermediaries		X		
Railways				X
Small operators		X		
Tourists		X		
Transnationals			X	
Adaptation and Mitigation innovations and strategies				
Adoption of innovations and strategies	X			
Behavioural change	X			
Education and information		X		
Energy conservation (designed)				X
Energy conservation (behavioural)		X		
Energy-efficient building construction				X
Extreme event risk insurance				X
Demarketing	X			
Diversification of market (destination)	X			
Diversification of market (firm)	X			
Diversification of product (destination)		X		
Diversification of product (firm)	X			
Finance and insurance			X	
GHG emission offset programmes		X		
Government assistance programmes		X		
Health risk			X	
Impact management planning		X		
Monitoring and evaluation			X	

Table 13.6 Continued

Factor	*Very inadequate*	*Inadequate*	*Adequate*	*Very adequate*
Policy formulation		X		
Pricing/'Green' taxes		X		
Recycling				X
Site location			X	
Snowmaking				X
Transport technology improvements			X	
Water conservation and recycling (designed)				X
Water conservation and recycling (behavioural)		X		

better understanding of tourism and its role in sustainability also needs to be generated with respect to its place in wider human and physical systems (Hall, 2008b).

This review has identified a number of significant knowledge gaps with respect to understanding tourism and climate change relationships. These gaps are important not only in terms of the sustainability of the tourism industry but also with respect to destination communities and the physical environment. Stern (2006) noted that climate change adaptation policies and measures, if implemented in a timely and efficient manner, can generate valuable co-benefits such as enhanced energy security and environmental protection. It is to be hoped that if the present review was to be repeated following the next IPCC assessment report then not only would a greater range of research have been conducted and acknowledged in scientific and policy terms, but that concrete steps will have been taken to address climate change and security concerns.

Acknowledgement

The author would like to acknowledge the comments of Associate Professors Stefan Gössling and David Duval on an earlier version of this paper.

13.2

Climate Change – Beyond the Hype

Susanne Becken

In the last 12 months we have been literally bombarded with media reports on climate change. As a result, the awareness of climate change by the general public – including tourists – has increased dramatically. This is desirable, of course, as climate change is a serious issue and we are likely to already lag behind with what should be achieved in terms of greenhouse gas (GHG) reductions. Similarly, there is a great urgency to develop and implement cost-effective adaptation measures to assist those destinations that are mostly affected by climate change impacts, for example small island states. In the following, however, I will largely focus on GHG emissions and mitigation.

C.M. Hall in this probe rightly mentioned the Stern Report and IPCC's Fourth Assessment Report as key drivers in the media frenzy. However, he also noted that tourism is only marginally covered or at best treated as part of more generic sections, such as mitigation for transport (e.g. aviation) in the IPCC's Working Group III Report.

There is probably a reason for tourism being slightly 'underreported' in these major publications. First, one could say that tourism is only one particular application or manifestation of a bigger problem. For example, aviation relates to freight, non-tourist travel and tourism travel. In the end, all of them use the same kinds of aircraft and rely on the same technological improvements for GHG reductions. Similar arguments could be made for road transport technologies and building design. Moreover, tourism researchers very rarely hold expertise in energy efficiency technology and are more concerned with the more geographic or humanistic discussions of travel as a modern world phenomenon.

However, tourism as a field of study and policy cannot be so easily dismissed, especially not in relation to such an important issue as climate change, which has the potential to substantially change the way people will travel in the future (Becken & Hay, 2007). After all, tourists are not the same as a few boxes of fresh food transported by plane, hotels are not the same as office blocks and road transport is part of the experience rather than a means to move from A to B. Thus, not only do technological innovations have to take into account some of these tourism-specific features, but policies that aim at changing behaviour need to delve into the tourist's psyche to understand what is possible and what is desirable.

In the last few years, tourism research has made some progress in finding ways to reduce tourism's contribution to climate change. First of all, there has been an increasing volume of 'measure to manage' research. Such research seeks to assess tourism's contribution to climate change by measuring energy use and (largely) Co_2 emissions. In the latest synthesis of knowledge in this area, Scott, Amelung *et al.* (2007) report that tourism contributes about 5% to global GHG emissions. Figures can be manipulated, of course, and interpreted according to the user's perspective, as has happened with the well-quoted 3% contribution of aviation reported by the IPCC in 1999 (Penner *et al.,* 1999). Measuring, however, does help to focus on reduction efforts in those areas that provide the largest potential for gains. Typically, this is in the tourism transport sector. The research strand of quantitative emission analyses is often extended to modelling of future emissions (e.g. Peeters *et al.,* 2004). This research is highly relevant for developing policies such as those relating to emission trading schemes or carbon taxes. If aviation is not included in any climate policies, for example, it is likely to dominate a country's carbon budget by the middle of this century (Bows *et al.,* 2005).

Most recently, research has investigated carbon offsetting as a means of 'neutralizing' tourism's emissions, with a focus on those from flying. Researchers have studied tourists' perceptions of offsetting (Becken, 2007), the awareness of travel agents (Dodds, 2007), and the credibility of offsetting schemes (Gössling & Broderick *et al.,* 2007). Such research follows the line of earlier work undertaken in the areas of voluntary initiatives, ecolabelling and certification.

In addition to tourism-specific research – largely published in tourism journals – there is a substantial number of highly relevant publications from research fields relating to energy, transport and buildings. For example, the *Energy Policy Journal* published a study by Meyer *et al.* (2007) about growth rates and climate change impact of international passenger transport between 2000 and 2050. *Renewable Energy Reports* on a feasibility study of stand-alone renewable energy supply options for a large hotel (Dalton *et al.*, 2008). Often, tourism professionals remain unaware of papers published in such journals.

The heightened research interest and action by governments and other organizations resembles the awakening of the sleeping beauty: task forces and working groups are established, workshops organized, high level meetings held and strategies developed. Organizations such as the United Nations World Tourism Organization, the Pacific Asia Tourism Association and Southpacific Travel (former SPTO) made climate change one of their priority topics. While this 'burst of activity' is likely to be helpful to the cause, it may raise some suspicions about climate change and tourism becoming a new academic and political battlefield with possibly more opportunism than altruism. In some way this reminds one of the hype around ecotourism in the early 1990s, where ecotourism was seen as the panacea for almost everything, and mostly of course to sustainable development in local communities.

Interestingly, there is also an important North-South element in this present climate debate. Not only are developing countries most likely to be affected by the negative impacts of climate change, but they are also likely to suffer the most from climate policies imposed on aviation. So the argument goes…In my view this is one of the most serious research gaps needing to be addressed. Key questions that come to mind are (a) what benefits do developing countries really receive from tourism (given that tourism often adds to the destinations' vulnerability to climate change, e.g. pressure on water resources and coastal zones), (b) what forms of tourism would provide the greatest net benefit, (c) how price sensitive are the types of tourists that visit developing destinations, and (d) are there substitution possibilities between markets so that the overall numbers of arrivals does not change but market composition does (with tourists coming from closer countries of origin), or at least spending power?

These questions take me to another major research gap. Tourism is very rarely seen in the context of the wider economy. I recall interesting debates about a decade ago around the difference between 'sustainable tourism development' and 'sustainable development'. The latter does not necessarily rely on tourism. The former focuses on developing tourism (in a sustainable way, of course); however, isolating tourism from wider economic dynamics runs the risk of misjudging its significance and overall impacts, and also devising policies that may be sub-optimal in the wider scheme of things. Policy analyses specific to tourism and climate change are notably absent from the current research effort, yet it is here where most research is required to achieve significant GHG reductions. Such policy analysis would, for example, examine the wider implications of measures such as carbon taxes, business incentives for energy efficiency, and biofuel targets. Detailed-yet-comprehensive and place-specific analyses in these areas would enhance the discussion, inform decision making and counteract simplistic and scare-mongering statements such as those made by the author of the discussion paper in relation to biofuel and food crises.

Clearly, research on climate change and tourism requires a wide set of expertise rather than just relying on 'tourism researchers' who pick up the 'flavour of the month'. Future research programmes need to integrate tourism with established disciplines such as engineering, planning, sociology and psychology. Tourism researchers are good at identifying high-level messages and communicating them to the tourism sector (e.g. the need for more energy efficient transport). However, very few are actually trained to understand how a combustion engine works and why certain biofuels are not suitable at high altitudes. Also, not all tourism researchers will be equipped to understand the major changes in Western lifestyles that might be required in the future to address climate change. Analyses of societal trends, tipping points and behavioural changes are more likely to be the domain of sociologists and psychologists (some of whom may well work in the field of tourism).

Having said all this, I believe that tourism studies have come a long way in their endeavour to delineate the relevance of climate change to tourism. What is needed now is a quantum leap towards research that provides solutions rather than descriptions, that integrates across disciplines and that works in partnership with tourism organizations.

The challenge of climate change is also an opportunity for tourism to become more systematic, smart, strategic and sustainable. At a research level there is a risk, however, that climate change research becomes fragmented with a myriad of case studies emerging, just like we experienced for ecotourism (Becken & Schellhorn, 2007). Hall (2008a) unfortunately steers us in this direction by providing a (rather arbitrary) assessment of climate change knowledge in relation to the many facets of tourism, such as cycling, fishing, theme parks etc., rather than providing a systematic framework for analysis. At a policy level, there is a risk that climate change overpowers other issues of importance to tourism, such as growing concerns about global oil production, water resources and biodiversity. While some of these are clearly related to climate change, it would be foolish to view them in isolation, through the climate change lens. Rather than adding climate change as a separate task for a decision-maker to consider, it should be 'mainstreamed' into every-day management and planning , for example in the form of risk management, resource efficiency and sustainability (see also the discussion paper author's comment about the lack of interest by small businesses).

I strongly believe that there is an important place for research and policy in the area of tourism and climate change. High-quality research and sound policies will help to increase the 'voice of tourism' and, as a result, lead to greater support for the tourism sector. Clearly, there are some areas where tourism has the potential to show leadership and excellence (e.g. research on human-environment coupled systems), and there are other areas where tourism is likely to be a taker (rather than provider), for example, in specific technology. For all these activities, partnerships and collaborations will be crucial, both for the goal of reducing GHG emissions and for the future of tourism research.

13.3

Climate Change: Tourism Destination Dynamics

Ralf Buckley

The increasing attention given to the tourism sector within international negotiations on climate change is documented by Hall (2008a); and the industry's potential role in mitigation is explored by Becken in her probe on climate change. Here, therefore, I shall outline the ways in which tourist destinations of various types are likely to be affected by climate change, the ways in which they may respond and the research they will need to inform those responses. In particular, Table 13.3 of Hall's probe lists much of the relevant literature currently available, and Table 13.6 in that review compares the relative frequencies with which more detailed subtopics have been referred to. Further information is available in Hall and Higham (2005); Becken and Hay (2007). As noted by Becken (2008) in her response, however, this does not in itself provide a framework for further analysis; nor does that commentary itself adopt such an aim. Here, therefore, I attempt to provide such a framework, make some predictions, and identify research priorities.

Analytical Framework

There are four main links between tourism and climate change:

- contributions of tourism to climate change, and ways to mitigate or offset them;
- increased travel costs because of mitigation measures, and consequences for travel patterns;
- changing climates in countries of origin, and effects on outbound and domestic tourism;
- changing climates at destinations, and effects on their attractiveness, safety and comfort.

My focus here is on the last of these. It can be considered in three sub-categories: artificial climates, climatic variability, and climatic means.

Artificial climates provide a buffer against climate change, particularly in urban destinations, simply through changing heating and air-conditioning. If the natural attractions at a tourism destination are affected by climate change, the simplest response is to substitute artificial attractions: swimming pools instead of beaches, shops instead of ski slopes. Changes to climatic variability include, e.g. more storms and floods, and more droughts and heatwaves leading to wildfires and coral bleaching. Most of these effects are temporary, and various protective measures are available, either through engineering or insurance approaches (Buckley, 2007). The tourism industry too can actively adopt such protective measures, either independently or in conjunction with government and other landholders. The main approaches include upgrading buildings and infrastructure, and improving emergency provisions. Changes to climatic means may affect, e.g. skiable snow, beach configurations, river flows and lake levels, icon wildlife species, and so on. The principal response is to change the tourism products on offer, re-position the destination accordingly, and target new markets.

Regional Patterns

The ski industry is particularly vulnerable to warming and to reduced precipitation, since these changes reduce both natural snow cover and opportunities for snow-making. Ski resorts in North America and Europe, Australia and New Zealand are much more vulnerable than those of Hokkaido in Japan, or those of northern Canada and Alaska. The industry's response to date has been to increase snow-making, snow grooming and terrain modification; to re-position resorts from winter to four-season; and to increase financial reliance on property and retail rather than lift tickets.

The coastal resort industry has responded to climate change much more slowly than ski resorts (Buckley, 2008b). Possible reasons include: lags in oceanic as compared to atmospheric changes; more distributed ownership of land and infrastructure in coastal settlements; controversy over liabilities and responsibilities; and misperceptions of likely climatic impacts (Buckley, 2008b). Tropical coral reefs are susceptible to bleaching from warm ocean temperatures, but to date this seems to have had relatively little effect on tourism.

Most of the world's reef tourism destinations have experienced bleaching events, but the only serious effect on visitation seems to have been a 9% drop at Palau (Glantz, 2000, cited in Becken & Hay, 2007). Experienced divers descend below the impacts of bleaching, and snorkellers seem unconcerned as long as they can still see fish. Future effects may be more severe if the ocean becomes increasingly acidic (Hoegh-Guldberg *et al.*, 2007).

In forests, the main risks of climate change are from increased wildfires, which damage infrastructure and affect both access and attractiveness for tourism. In grasslands and deserts, droughts will affect populations and migrations of icon watchable wildlife species, as well as grazing pressure from domestic livestock, cultural landscapes and the availability of water.

Research Priorities

The framework outlined above, as well as Hall's probe, suggest three major research themes at the destination scale. In urban tourist destinations, the key is the degree to which market share can be maintained by modifying the mix of activities and attractions. This is essentially a question of tourist behaviour, and there is already a considerable body of academic literature on tourism economics and geography, marketing and market research, and destination life-cycles which could be adapted to examine this question.

The second major theme is that of extreme weather events: their immediate impacts on tourist safety, attractions and infrastructure; and their longer term effects on competition between destinations. There is already a significant academic literature on disaster response and crisis management in tourism, which can be applied here.

The third and perhaps the most complex research theme is in predicting how natural attractions for tourism may be modified by climate change, and how tourists may change their travel patterns accordingly. This theme covers not only direct impacts of climate change on parks and beaches, water and wildlife, but also the indirect impacts on tourism as the social frameworks for access to these attractions change in response to changing climates (Buckley, 2007).

Climate change is indeed a highly significant issue for the global tourism industry: less acute than wars and terrorism, but larger scale and longer lasting. The subsectors of the tourism industry most affected by climate change will be smaller fixed-site operators or destinations which rely heavily on a single natural attraction – such as snow, reefs or wildlife – that happens to be particularly susceptible to climate change. Even in these cases, however, there are opportunities to re-position either the product or the target market segment so as to maintain revenue. The key in these cases will be forethought and innovation, to maintain a competitive position as corresponding adjustments occur worldwide.

As suggested by Becken in her further probe, climate change is likely to become so pervasive that it should be mainstreamed into all aspects of tourism research. Already, many analyses in tourism economics refer routinely to various climate change scenarios in much the same way as they routinely specify discount rates. Future research in tourism planning, policy and geography will no doubt refer to regional climate change patterns and predictions in a similar way.

In the short-term, however, it would seem that if we were to pick a single top priority for research, it would be the social and environmental consequences of extreme weather events at destinations, and the flow-on effects for tourism.

13.4

Climate Change and Tourism: Time for Critical Reflection

Daniel Scott

The New Realities of Tourism in an Era of Global Climate Change

In 2007, the Intergovernmental Panel on Climate Change (IPCC) issued its Fourth Assessment Report (4AR), which made the most definitive statement on the anthropogenic causes of observed global climate change, stating that human interference with the global climate system was 'unequivocal'. In the words of Achim Steiner (Head of the United Nations Environment Programme), 'this year may be remembered as the day the question mark was removed from whether human activity has anything to do with climate change' (Giles, 2007: 579). The environmental and economic risks associated with climate change in the 21st century featured prominently in international policy debates that culminated at the 13th Conference of the Parties (COP) in Bali, Indonesia. While the 'Bali Road Map' did not produce new greenhouse gas emission targets to succeed the Kyoto Protocol, many are now convinced that the climate change science has progressed sufficiently that it will engender a serious response from global policymakers.

The year 2007 was also a witness to a flourish of high-level activities related to tourism and climate change. The United Nations World Tourism Organization (UNWTO) organized the Second International Conference on Climate Change and Tourism (Davos, Switzerland). The Davos Declaration that emerged was strongly endorsed by the Ministers' Summit which met during the World Travel Market (London, UK) and later at the UNWTO 2007 General Assembly (Cartagena, Colombia), and became a core component of the UNWTO submission to COP-13 in Bali. It is also not a coincidence that the title of one of the main sessions at the CEO Challenge: Confronting Climate Change conference, organized by Pacific Asia Travel Association in April 2008, was titled 'Environmental Regulations: Preparing for the Inevitable.' This and other sentiments expressed as part of the CEO Challenge (e.g. 'unless we take positive action to reduce our carbon footprint – and [are] seen to be doing it – others will act for us'), are highly revealing of industry concerns about how governments are preparing new climate policies and regulations that may have a considerable effect on the tourism sector.

What substance will emerge from these high-profile events remains to be seen. Regardless of whether rhetorical discussion is replaced by meaningful (and binding) emission reduction targets or adaptation plans with necessary financial and human resources, the nature of the climate change dialogue within the tourism sector in 2007 represents a significant advance in only a few short years. Not very long ago, Butler and Jones (2001: 300), in their concluding summary of the International Tourism and Hospitality in the 21st Century conference, forthrightly stated, '(Climate change) could have greater effect on tomorrow's world and tourism and hospitality in particular than anything else we've discussed (at this conference)...The most worrying aspect is that...to all intents and purposes the tourism and hospitality industries...seem intent on ignoring what could be the major problem of the century' (*original emphasis*).

The research probe led by Hall (2008a) has offered a discussion of the place of tourism in the IPCC Fourth Assessment Report (4AR) and the Stern Review, which have featured prominently in recent international climate policy discussions and outlined major knowledge gaps and research priorities to guide this rapidly developing field of tourism studies. Each of these topics is further considered below; however I shall restrict my remarks to the themes of impacts and adaptation.

The Place of Tourism in the IPCC 4AR

As indicated in Hall's research probe, word counts serve only as a crude metric of importance or level of knowledge and we must be careful not to place too much concern or optimism in relative changes from IPCC assessment to assessment.

While the tourism sector has been coming to terms with the new realities posed by an increasingly carbon-constrained global economy, environmentally conscious and carbon-wise consumers, and observable destination level impacts, climate change is also a rapidly developing area of tourism research. Contributions from multiple disciplines have doubled the number of scientific publications that examine the interactions of tourism and climate change between 1996–2000 and 2001–2005 (Scott, Jones *et al.,* 2005). With more information available, might we not have expected more tourism content than in the IPCC 4AR? Similar growth patterns in climate change research in other sectors have occurred as well, which has put coordinating lead authors in the unenviable position of setting extremely tight word limits for all sectors and cross-cutting issues in order to achieve the hard page limits set by the IPCC. As the author of the tourism section in the North American chapter of both the Third Assessment Report (TAR) and 4AR, this is the explanation for why the length of the tourism section was cut in half. Conversely, while the tourism content is noted to have increased substantially in the African chapter (Boko *et al.*, 2007), a review of the tourism section and other references to tourism throughout the chapter reveals a tremendous knowledge gap, as there is not a single tourism specific study cited from the continent and the entire discussion of tourism related impacts is based on conjecture and extrapolation of studies from other continents.

The place of tourism in IPCC assessments has also had much to do with people involved in past assessments. In the TAR and 4AR no coordinating lead author has had a tourism background and it was not until the 4AR that a chapter lead author (in the Industry, Settlement and Society chapter) had the requisite knowledge of the tourism sector and related literature to champion further discussion of the tourism sector. Jean-Paul Ceron (France) deserves credit for advancing the place of tourism in the 4AR, as does UNWTO for recommending a panel of experts to the IPCC for consideration as contributing authors or expert reviewers.

An equally important question from my perspective is how far has climate change discourse penetrated the ranks of the tourism research community? While there are encouraging signs of progress here as well, including three recent books (Gössling & Hall, 2006c; Becken & Hay, 2007), a science review for UNWTO (Scott, Amelung *et al.*, 2008), conference proceedings (Peeters, 2007) and many journal articles, my impression from discussions with colleagues from the tourism community is that much remains to be done to integrate climate change considerations into mainstream tourism literature and professional practice. At the same time, those researching the interface of climate change and tourism have much to gain from collaborating with the tourism research community, particularly to deepen our understanding of the response of tourists to climate change impacts and climate policies.

'Mind the Gaps'

In the same way we are advised not to ignore the 'gap' on the subway, the tourism sector should not ignore the important gaps in knowledge about the implications of climate change, for future regulatory changes or climate-induced impacts may have equally unpleasant consequences. If the climate change issue is not adequately considered, many of the other sustainability initiatives in tourism may become wasted efforts. I concur with the assessment of regional knowledge on climate change in Table 13.5 of Hall's (2008a) research probe, and believe that addressing the major regional gaps in our knowledge of how climate change will affect the natural and cultural resources critical for tourism in Africa, the Caribbean, South America and large parts of East Asia must be a priority considering the relative importance of tourism to the economies of some nations in these regions.

While I also applaud the effort that has gone into the evaluation of tourism knowledge in Table 13.6 of Hall's research probe, it must be considered a starting point for such a dialogue, which is surely the spirit in which it was intended. Given the contested nature of anthropogenic climate change itself, it should be no surprise that experts will hold differing views about what climate change might mean for the tourism sector and that ratings of the adequacy of knowledge by other leading experts would probably yield differing results. With this in mind, there are two important terms in Table 13.5 ('relative') and 13.6 ('adequacy') that deserve elaboration in my opinion. While alpine areas and snow-based/ski tourism are identified as having high or very adequate levels of knowledge, I would contend that our knowledge of the implications of climate change on these tourism markets remains very limited in many respects and is not yet 'adequate' for use in decision-making regarding climate change adaptation. For example, while the ski industry in the European Alps is probably the most frequently studied tourist activity, large uncertainties remain as to what the vulnerability of this industry is and what the response of national governments and destination communities should be. The most comprehensive analysis conducted for the Organization for Economic Cooperation and Development (OECD) pronounced that the number of ski areas that were considered 'snow reliable' dropped from 609 to 404 under a +2°C warming scenario and further declined to 202 under a +4°C warming scenario (Agrawala, 2007).

Yet, in the same report it is noted that over 50% of skiable terrain in Austria, and somewhat lesser proportions in other nations, have snow-making capacity that is not incorporated in the impact assessment. Thus, these widely cited impacts do not represent the operating realities of many ski areas today let alone 25 years from now, when Wolfsegger *et al.* (2008) showed most ski area managers plan to have enhanced snow-making capabilities. If snow-making capacity were factored into the OECD assessment, how would vulnerability change and for which locations? If available water supply was also considered, the relative vulnerabilities of communities with ski tourism would be further refined, perhaps in favour of ski areas further away from cities/towns that are major competitors for winter water supply. Even if we overlook the omission of snow-making as a key adaptation to climate change, the OECD report indicates that approximately one-third of existing ski areas would remain snow reliable with natural snowfall even under a +4°C scenario. The implication is that the ski tourism industry would contract to a smaller number of destinations.

Communities that are at risk of losing ski operators and those where ski operations are likely to persist will need to adapt to climate change, though for very different reasons. The former will need to adjust to reduced winter tourism spending, lost employment, and potentially declining real estate prices, while the latter will need to plan for increased visitation, congestion and perhaps greater development pressures. How will consumers respond to this supply-side contraction? How will factors like proximity to major markets or destination loyalty influence the vulnerability of destinations? Tourism studies have much to say on these issues. A comprehensive assessment of supply- and demand-side responses to climate change and how they will interact to affect the competitiveness of destinations and the restructuring of tourism markets is what is required by decision-makers. We are not there yet.

This single example highlights three other key knowledge gaps that I deem critical and were also identified by Hall (2008a) in his probe. The IPCC (2007) has indicated that the need for societies and economic sectors like tourism to adapt to further climate change in the decades ahead is inevitable. Adaptation has figured less prominently in climate change research on tourism than in some other economic sectors (e.g. agriculture) and remains an important knowledge gap, particularly with respect to destinations (Scott *et al.,* 2009). The dynamic nature of the tourism industry and its ability to cope with a range of recent shocks, including SARS, terrorism attacks or the Asian tsunami, suggests a relatively high climate change adaptive capacity within the tourism industry overall (Scott, Amelung *et al.,* 2008), however knowledge of the capacity of current climate adaptations to cope successfully with future climate regimes remains rudimentary.

The cavalier representation of adaptation by Buckley (2008a) is therefore to some extent disconcerting. First, we must be careful not to confuse adaptation to past-current climate variability with anticipatory adaptation to future climate change, as studies that have examined climate change risk appraisal among tourism operators have consistently found low awareness of climate change and little evidence of strategic planning in anticipation of future changes in climate (see Scott, Amelung *et al.,* 2008 for a summary). There is also some evidence that tourism operators are overestimating their capacity to adapt to future climate change, especially if high emission scenarios are realized (Wolfsegger *et al*., 2008).

Second, we should not assume that climate change adaptation will be simple or even successful. Adaptation will be a long-term process that must be anticipated and carefully planned. The information requirements, policy changes and investments required for effective adaptation by tourism destinations will require decades in some cases, and therefore the process of adaptation needs to commence in the very near future for destinations anticipated to be among those impacted by mid-century (Scott, Amelung *et al.,* 2008). In the words of the UNWTO Secretary-General (2007):

> ...it is not easy to see (adaptation) through successfully, because it entails, all at the same time, modifying economic circuits, introducing new technologies, carrying out intensive training, investing in the creation of new products, ... changing the minds of public authorities, entrepreneurs, host communities and tourists.

Using the ski tourism example above to illustrate this point, it is ill-considered to simply suggest that ski destinations can be rebranded through market diversification. There is only so much room for more conference centres and spas and if a market existed for these attractions, resorts would fill them now and not wait for climate change. Furthermore, from a revenue replacement perspective, golf courses and spas which can only put hundreds of people through in a day, are not substitutes for ski slopes that can accommodate thousands of visitors a day.

I concur with probe authors (Hall, 2008a; Becken, 2008), that with their capacity to adapt to the effects of climate change by substituting the place, timing and type of holidays in their travel decision, tourists will play a pivotal role in the eventual impacts of climate change on the tourism industry and destinations. Information on tourist climate preferences and tourist perceptions of the environmental impacts of global climate change at destinations (i.e. perceptions of coral bleaching, diminished or lost glaciers, degraded coastlines, reduced biodiversity or wildlife prevalence, insect harassment) remain important knowledge gaps that must be addressed if potential long-range shifts in tourism demand are to be more accurately projected. All models of what happens to tourism flows from regional to international scales ultimately depend on understanding what individuals will do – these causal mechanisms remain poorly understood. There is also limited understanding of how climate change impacts will interact with other long-term social and market trends influencing tourism demand, including: ageing populations in industrialized countries, increasing travel safety and health concerns, increased environmental and cultural awareness, advances in information and transportation technology, and shifts toward shorter and more frequent holidays. Here a vexing conceptual challenge exists that has been largely overlooked in climate change impact assessments. Can the potential behavioural response of tourists a generation or two into the future be inferred from responses from contemporary visitors? Can behavioural transference between adjoining generational cohorts be considered reliable?

The call for the development of a more comprehensive framework for understanding the potential implications of climate change for the tourism sector are echoed here. Analyses of potential impacts at destinations have focused on a narrow range of impacts (e.g. natural snow cover for skiing), but to understand the implications of climate change for a destination requires the development of a systems approach that holistically considers four-season supply and demand-side impacts and adaptation options, as well as the fate of key marketplace competitors and interactions with other major influencing variables in the tourism sector (e.g. fuel prices, ageing populations, increasing travel safety and health concerns (Gössling & Scott, 2008)). Multi-disciplinary collaborations will be increasingly needed to develop such a comprehensive framework and closer coordination with governments and the private sector will be needed to ensure that possible effects of climate change are effectively factored into relevant tourism policies and development and management plans (i.e. 'mainstreaming climate change in decision making').

Finally, the climate change and tourism research community must not overstate the case of climate change impacts, since the real problems of climate change will be serious enough if we remain on the current emissions trajectory for much longer. A number of media stories have foretold the major threat that increased future summer temperatures poses for tourism in the Mediterranean. Indeed some stated that 'The likelihood [is] that Mediterranean summers may be too hot for tourists after 2020' (*Guardian*, 2006) and that 'by 2030, the traditional British package holiday to a Mediterranean beach resort may be consigned to the "scrap-heap of history"' (Halifax Travel Insurance, 2006). As Scott, Amelung *et al.* (2008) demonstrate empirically, such pronouncements are scientifically unfounded and diminish the credibility of all other climate change studies.

Time for Critical Reflection

As with all maturing fields of scholarly work, there comes a time when a sufficient critical mass of research exists to warrant critical reflection on what has been done, its limitations, major knowledge gaps and new pathways forward. With the earliest scholarly publications on climate change and tourism now 20 years old and rapid increase in the volume of research over the past 10 years, this time has come. I believe there is a strong need for critical reflection on this collective body of work to enable the emergence of a new generation of climate change and tourism assessments that deliver far more rigorous insight than some of the past pedestrian pronouncements about the possible impacts of climate change that lack information on the magnitude and timing of impacts and under what climate change scenarios or adaptation initiatives that destination communities or tourism operators are vulnerable.

This research probe, particularly the constructive exchange on the limitations of models used to predict travel demand patterns under climate change (summarized in Table 13.4, but see also – Gössling & Hall, 2006d), and other recent synthesis works (Gössling & Hall, 2006c; Becken & Hay, 2007; Scott, Amelung *et al.*, 2008) represent a starting point for such a critical dialogue, but more is required. I would hope that a future conference or series of review papers would take on this much needed task.

Concluding Remarks

The intention of the original article by Hall was to provide a snapshot of tourism-related climate change knowledge in order to illustrate its unevenness not only with respect to place but also various topics. Such an approach is of interest to both researchers as well as industry and destinations, who are often seeking guidance as to what their response to climate change should be. As noted by Becken in her response this does not in itself provide a framework for further analysis. However, as Buckley observed, the initial commentary itself never adopted such an aim. Becken argued that 'research on climate change and tourism requires a wide set of expertise rather than just relying on "tourism researchers" who pick up the "flavour of the month". Future research programmes need to integrate tourism with established disciplines such as engineering, planning, sociology, and psychology'. Of course, as suggested in the introduction to this topic, the expertise required to respond to the problem of climate change also depends very much on how the problem is defined in the first place. Unfortunately, as also suggested in the introduction, in many cases some 'experts' may well define the problem in narrow reductionist and instrumental terms that are usually expressed in terms of formal modelling and neoclassical economics which may fail to embrace broader social scientific perspectives. Yet the need to integrate social and biophysical perspectives in climate and environmental change research is becoming widely recognized (Füssel & Klein, 2006; Conrad, 2009; Demeritt, 2009), including with respect to the need for ontological shifts (Turnpenny *et al.*, 2010), and the production of local knowledge(s) (Slocum, 2010). Indeed, such a perspective clearly reinforces Scott's observation of the need for more critical dialogue with respect to tourism and climate change.

The need for critical analysis can also be extended to the institutional arrangements that surround tourism and climate change policy which remain wedded to assumptions based on the compatibility between the environment and economic growth and acceptance of market forces. Indeed, far too much attention is given to the assumption that a well-designed institution is 'good' because it facilitates cooperation and partnership, rather than a focus on norms and institutionalization as first and necessary steps in the assessment of what kind of changes certain research, policies and institutional arrangements are promoting and their potential outcomes. Instrumental institutional approaches only reinforce limited and incremental change in tourism policy and research rather than encourage conceptual policy learning and paradigm change via the examination of policy alternatives (Hall, 2011). Rather than seeking to increase the 'voice of tourism' good tourism research needs to understand the situated and social constructed nature of knowledge as a means of being able to work through the various policy demands by different interests as well as better communicate the results of research in a manner so that emissions and environmental impacts are actually reduced. Most importantly, a more critical perspective is required not only with respect to policy and the framing of climate change research but also as to whether tourism is the best development alternative.

Discussion Questions

(1) To what extent could reduced long-distance travel contribute to a lowering of emissions from tourism?
(2) The UNWTO forecasts that there will be 1.8 billion international visitor arrivals by 2030, a growth rate of 3.3% (2010–2030). What will be the implications for climate change?
(3) Some regions may be more affected than others by climate change. Which do you believe will be the main regions that will be affected?
(4) Are people willing to travel less and/or shorter distances in order to reduce their emissions?
(5) Will climate change become so pervasive that it will be mainstreamed into all aspects of tourism research?
(6) Will increased partnership and collaboration reduce GHG emissions or will it just slow down the rate at which decisions are made or even mean that hard decisions will not be made at all?
(7) How can the impacts of climate change avoid being overstated in the media?
(8) What are the limitations of the IPCC models with respect to understanding the effects of climate change on tourism?

Note

(1) © C. M. Hall

References

Adger, W.N., Agrawala, S., Mirza, M.M.Q., Conde, C., O'Brien, K., Pulhin, J., Pulwarty, R., Smit, B. and Takahashi, K. (2007). Assessment of adaptation practices, options, constraints and capacity. In M.L. Parry, O.F. Canziani, J.P. Palutikof, P.J. van der Linden and C.E. Hanson (eds) *Climate Change 2007: Impacts, Adaptation and Vulnerability* (pp. 717–743). Cambridge: Cambridge University Press.

Agrawala, S. (ed) (2007). *Climate Change in the European Alps: Adapting Winter Tourism and Natural Hazards Management*. Paris: OECD.

Alcamo, J., Moreno, J.M., Nováky, B., Bindi, M., Corobov, R., Devoy, R.J.N., Giannakopoulos, C., Martin, E., Olesen, J.E. and Shvidenko, A. (2007). Europe. In M.L. Parry, O.F. Canziani, J.P. Palutikof, P.J. van der Linden and C.E. Hanson (eds) *Climate Change 2007: Impacts, Adaptation and Vulnerability* (pp. 541–580). Cambridge: Cambridge University Press.

Amelung, B. and Viner, D. (2006). Mediterranean tourism: Exploring the future with the tourism climatic index. *Journal of Sustainable Tourism* 14: 349–366.

Amelung, B., Moreno, A. and Scott, D. (2008). The place of tourism in the IPCC Fourth Assessment Report: A review. *Tourism Review International* 12(1): 5–12.

Anisimov, O.A., Vaughan, D.G., Callaghan, T.V., Furgal, C., Marchant, H., Prowse, T.D., Vilhjálmsson, H. and Walsh, J.E. (2007). Polar regions (Arctic and Antarctic). In M.L. Parry, O.F. Canziani, J.P. Palutikof, P.J. van der Linden and C.E. Hanson (eds) *Climate Change 2007: Impacts, Adaptation and Vulnerability* (pp. 653–685). Cambridge: Cambridge University Press.

Arctic Climate Impacts Assessment (ACIA) (2005). *Impacts of a Warming Arctic: Arctic Climate Impacts Assessment*. Cambridge: Cambridge University Press.

BBC News (2007). Rudd unveils Barrier Reef plan. *BBC News,* October 29. Online at: http://news.bbc.co.uk/2/hi/asia-pacific/7067158.stm. Accessed 30 October 2007.

Becken, S. (2005). Harmonizing climate change adaptation and mitigation: The case of tourist resorts in Fiji. *Global Environmental Change* – Part A 15(4): 381–393.

Becken, S. (2007). Tourists' perception of international air travel's impact on the global climate and potential climate change policies. *Journal of Sustainable Tourism* 15(4): 351–368.

Becken, S. (2008). Climate change – Beyond the hype. *Tourism Recreation Research* 33(3): 351–353.

Becken, S. and Hay, J. (2007). *Tourism and Climate Change: Risks and Opportunities*. Clevedon: Channel View Publications.

Becken, S. and Schellhorn, M. (2007). Ecotourism, energy use and the global climate – Widening the local perspective. In J. Higham (ed) *Critical Issues in Ecotourism – Paradoxes, Problems and Pathways for the Future* (pp. 85–101). Oxford: Butterworth-Heinemann.

Berkelmans, R., De'ath, G., Kininmonth, S. and Skirving, W.J. (2004). A comparison of the 1998 and 2002 coral bleaching events of the Great Barrier Reef: Spatial correlation, patterns and predictions. *Coral Reefs* 23: 74–83.

Black, R. (2007). Nobel prize recognises climate crisis. *BBC News* (International Version), October 12.

Boko, M., Niang, I., Nyong, A., Vogel, C., Githeko, A., Medany, M., Osman-Elasha, B., Tabo, R. and Yanda, P. (2007). Africa. In M.L. Parry, O.F. Canziani, J.P. Palutikof, P.J. van der Linden and C.E. Hanson (eds) *Climate Change 2007: Impacts, Adaptation and Vulnerability* (pp. 433–467). Cambridge: Cambridge University Press.

Bows, A., Upham, P. and Anderson, K. (2005). *Growth Scenarios for EU and UK Aviation: Contradictions with Climate Policy.* Report for Friends of the Earth Trust Ltd. Tyndall Centre for Climate Change (North). UK. The University of Manchester.

Buckley, R. (ed) (2007). *Climate Response*. Gold Coast and Brisbane: Griffith University.

Buckley, R. (2008a). Climate change: Tourism destination dynamics. *Tourism Recreation Research* 33(3): 354–355.

Buckley, R. (2008b). Misperceptions of climate change damage coastal tourism: Case study of Byron Bay, Australia. *Tourism Review International* 12(1): 71–78.

Burki, R., Elsasser, H., Abegg, B. and Koenig, U. (2005). Climate change and tourism in the Swiss Alps. In C.M. Hall and J. Higham (eds) *Tourism, Recreation and Climate Change* (pp. 155–163). Clevedon: Channel View Publications.

Butler, R. and Jones, P. (2001). Conclusions – Problems, challenges and solutions. In A. Lockwood and S. Medlik (eds) *Tourism and Hospitality in the 21st Century* (pp. 296–309). Oxford, UK: Butterworth-Heinemann.

Chan, W.W. and Lam, J.C. (2003). Energy-saving supporting tourism: A case study of hotel swimming pool heat pump. *Journal of Sustainable Tourism* 11(1): 74–83.

Conrad, J. (2009) Climate research and climate change: Reconsidering social science perspectives. *Nature and Culture* 4(2): 113–122.

Cruz, R.V., Harasawa, H., Lal, M., Wu, S., Anokhin, Y., Punsalmaa, B., Honda, Y., Jafari, M., Li, C. and Huu Ninh, N. (2007). Asia. In M.L. Parry, O.F. Canziani, J.P. Palutikof, P.J. van der Linden and C.E. Hanson (eds) *Climate Change 2007: Impacts, Adaptation and Vulnerability* (pp. 469–506). Cambridge: Cambridge University Press.

Dalton, G.J., Lockington, D.A. and Baldock T.E. (2008). Feasibility analysis of standalone renewable energy supply options for a large hotel. *Renewable Energy* 33(7): 1475–1490.

Demeritt, D. (2001) The construction of global warming and the politics of science. *Annals of the Association of American Geographers,* 91: 307–37.

Demeritt, D. (2009) Geography and the promise of integrative environmental research. *Geoforum,* 40: 127–29.

Demeritt, D. and Rothman, D. (1999) Figuring the costs of climate change: An assessment and critique. *Environment and Planning A,* 31: 389–408.

Dodds, R. (2007). *Climate Change and Carbon Offsetting Awareness by Travel Agents and Travellers.* Paper presented at 5th Bi-Annual Symposium of the International Society of Culture, Tourism, and Hospitality Research, Charleston, South Carolina, 4–6 June, 2007.

Field, C.B., Mortsch, L.D., Brklacich, M., Forbes, D.L., Kovacs, P., Patz, J.A., Running, S.W. and Scott, M.J. (2007). North America. In M.L. Parry, O.F. Canziani, J.P. Palutikof, P.J. van der Linden and C.E. Hanson (eds) *Climate Change 2007: Impacts, Adaptation and Vulnerability* (pp. 617–652). Cambridge: Cambridge University Press.

Fischlin, A., Midgley, G.F., Price, J.T., Leemans, R., Gopal, B., Turley, C., Rounsevell, M.D.A., Dube, O. P., Tarazona, J. and Velichko, A.A. (2007). Ecosystems, their properties, goods, and services. In M.L. Parry, O.F. Canziani, J.P. Palutikof, P.J. van der Linden and C.E. Hanson (eds) *Climate Change 2007: Impacts, Adaptation and Vulnerability* (pp. 211–272). Cambridge: Cambridge University Press.

Fisher, B.S., Nakicenovic, N., Alfsen, K., Corfee Morlot, J., De La Chesnaye, F., Hourcade, J., Jiang, K., Kainuma, M., La Rovere, E., Matysek, A., Rana, A., Riahi, K., Richels, R., Rose, S., Van Vuuren, D. and Warren, R. (2007). Issues related to mitigation in the long term context. In B. Metz, O.R. Davidson, P.R., Bosch, R. Dave and L.A. Meyer (eds) *Climate Change 2007: Mitigation* (pp. 169–250). Contribution of Working Group III to the Fourth Assessment Report of the Inter-governmental Panel on Climate Change. Cambridge: Cambridge University Press.

Fukushima, T., Kureha, M., Ozaki, N., Fujimori, Y. and Harasawa, H. (2002). Influences of air temperature change on leisure industries: Case study on ski activities. *Mitigation and Adaptation Strategies for Global Change* 7: 173–189.

Füssel, H-M. and Klein, R.J.T. (2006) Climate change vulnerability assessments: An evolution of conceptual thinking. *Climatic Change,* 75: 301–329.

Giles, J. (2007). From words to action. *Nature* 445: 578–579.

Gössling, S. (2002). Global environmental consequences of tourism. *Global Environmental Change* 12(4): 283–302.

Gössling, S. and Hall, C.M. (2006a). An introduction to tourism and global environmental change. In S. Gössling and C.M. Hall (eds) *Tourism and Global*

Environmental Change: Ecological, Social, Economic and Political Interrelationships (pp. 1–34). London: Routledge.
Gössling, S. and Hall, C.M. (2006b). Conclusion: Wake up... This is serious. In S. Gössling and C.M. Hall (eds) *Tourism and Global Environmental Change: Ecological, Social, Economic and Political Interrelationships* (pp. 305–320). London: Routledge.
Gössling, S. and Hall, C.M. (eds) (2006c). *Tourism and Global Environmental Change: Ecological, Social, Economic and Political Interrelationships*. London: Routledge.
Gössling, S. and Hall, C.M. (2006d). Uncertainties in predicting tourist flows under scenarios of climate change. *Climatic Change* 79(3–4): 163–173.
Gössling, S. and Scott, D. (2008). Editorial – Special issue on climate change and tourism: Exploring destination vulnerability. *Tourism Review International* 12(1): 88.
Gössling, S., Borgström-Hansson, C., Hörstmeier, O. and Saggel, S. (2002). Ecological footprint analysis as a tool to assess tourism sustainability. *Ecological Economics* 43(2–3): 199–211.
Gössling, S., Bredberg, M., Randow, A., Svensson, P. and Swedlin, E. (2006). Tourist perceptions of climate change: A study of international tourists in Zanzibar. *Current Issues in Tourism* 9(4–5): 419–435.
Gössling, S., Lindén, O., Helmersson, J., Liljenberg, J. and Quarm, S. (2007). Diving and global environmental change: A Mauritius case study. In B. Garrod and S. Gössling (eds) *New Frontiers in Marine Tourism: Diving Experiences, Management and Sustainability* (pp. 67–92). Amsterdam: Elsevier.
Gössling, S., Peeters, P., Ceron, J.-P., Dubois, G., Pattersson, T. and Richardson, R. (2005). The eco-efficiency of tourism. *Ecological Economics* 54(4): 417–434.
Gössling, S., Broderick, J., Upham, P., Peeters, P., Strasdas, W., Ceron, J.-P. and Dubois, G. (2007). Voluntary carbon offsetting schemes for aviation: Efficiency and credibility. *Journal of Sustainable Tourism* 15(3): 223–248.
Guardian (2006). Climate change could bring tourists to UK – report. *The Guardian,* 28 July, 2006. Online at: http://www.guardian.co.uk/travel/2006/jul/28/travelnews.uknews. Accessed 10 March 2008.
Halifax Travel Insurance (2006). *Holiday 2030*. Online at: http://www.hbosplc.com/media/pressreleases/articles/halifax/2006-09-01-05.asp?section=Halifax. Accessed 10 March 2008.
Hall, C.M. (2006a). New Zealand tourism entrepreneur attitudes and behaviours with respect to climate change adaption and mitigation. *International Journal of Innovation and Sustainable Development* 1(3): 229–237.
Hall, C.M. (2006b). Tourism, biodiversity and global environmental change. In S. Gössling and C.M. Hall (eds) *Tourism and Global Environmental Change: Ecological, Social, Economic and Political Interrelationships* (pp. 211–225). London: Routledge.
Hall, C.M. (2006c). Tourism urbanization and global environmental change. In S. Gössling and C.M. Hall (eds) *Tourism and Global Environmental Change: Ecological, Social, Economic and Political Interrelationships* (pp. 142–156). London: Routledge.
Hall, C.M. (2007). An innovation systems approach to tourism and climate change and sustainable tourism. Presentation at Helsingborg Expert Meeting on Sustainable Mass Tourism, September. Helsingborg. Lund University.
Hall, C.M. (2008a). Tourism and climate change: Knowledge gaps and issues. *Tourism Recreation Research* 33(3): 339–350.
Hall, C.M. (2008b). *Tourism Planning* (2nd edn). Harlow: Pearson.
Hall, C.M. (2011) Policy learning and policy failure in sustainable tourism governance: From first and second to third order change? *Journal of Sustainable Tourism*, 19(4–5), 649–671.

Hall, C.M. (2012) The natural science ontology of environment. In A. Holden and D. Fennell (eds) *A Handbook of Tourism and the Environment*, London: Routledge.

Hall, C.M. and Higham, J. (eds) (2005). *Tourism, Recreation and Climate Change*. Clevedon: Channel View Publications.

Hall, C.M. and Page, S.J. (2006). *The Geography of Tourism and Recreation* (3rd edn). London: Routledge.

Hamilton, J.M., Maddison, D.J. and Tol, R.S.J. (2005). Climate change and international tourism: A simulation study. *Global Environmental Change* 15: 253–266.

Harrison, S., Winterbottom, S. and Johnson, R.C. (2005). Changing snow cover and winter tourism and recreation in the Scottish Highlands. In C.M. Hall and J. Higham (eds) *Tourism, Recreation and Climate Change* (pp. 143–154). Clevedon: Channel View Publications.

Hennessy, K., Fitzharris, B., Bates, B.C., Harvey, N., Howden, S.M., Hughes, L., Salinger, J. and Warrick, R. (2007). Australia and New Zealand. In M.L. Parry, O.F. Canziani, J.P. Palutikof, P.J. van der Linden and C.E. Hanson (eds) *Climate Change 2007: Impacts, Adaptation and Vulnerability* (pp. 507–540). Cambridge: Cambridge University Press.

Hinsliff, G. (2006a). £3.68 Trillion: The price of failing to act on climate change. *The Observer,* October 29.

Hinsliff, G. (2006b). Ten years to save the planet from mankind. *The Observer,* October 29.

Hoegh-Guldberg, O. (1999). Climate change, coral bleaching and the future of the world's coral reefs. *Marine and Freshwater Research* 50: 839–866.

Hoegh-Guldberg, O. (2004). Coral reefs in a century of rapid environmental change. *Symbiosis* 37: 1–31.

Hoegh-Guldberg, O., Mumby, P.J., Hooten, A.J., Steneck, R.S., Greenfield, P., Gomez, E., Harvell, C.D., Sale, P.F., Edwards, A.J., Caldeira, K., Knowlton, N., Eakin, C.M., Iglesias-Prieto, R., Muthiga, N., Bradbury, R.H., Dubi, A. and Hatziolos, M.E. (2007). Coral reefs under rapid climate change and ocean acidification. *Science* 318 (5857): 1737–1742.

Instanes, A., Anisimov, O., Brigham, L., Goering, D., Ladanyi, B., Larsen, J.O. and Khrustalev, L.N. (2005). Infrastructure: Buildings, support systems, and industrial facilities. In C. Symon, L. Arris and B. Heal (eds) *Impacts of a Warming Arctic: Arctic Climate Impacts Assessment* (pp. 907–944). Cambridge: Cambridge University Press.

Intergovernmental Panel on Climate Change (IPCC) (2004). *Introduction: The Intergovernmental Panel on Climate Change (IPCC)*. Geneva: IPCC Secretariat.

Intergovernmental Panel on Climate Change (IPCC) (2007). Summary for policymakers. In M.L. Parry, O.F. Canziani, J.P. Palutikof, P.J. van der Linden and C.E. Hanson (eds) *Climate Change 2007: Impacts, Adaptation and Vulnerability* (pp. 7–22). Cambridge: Cambridge University Press.

Magrin, G., Gay García, C., Cruz Choque, D., Giménez, J.C., Moreno, A.R., Nagy, G. J., Nobre, C. and Villamizar, A. (2007). Latin America. In M.L. Parry, O.F. Canziani, J.P. Palutikof, P.J. van der Linden and C.E. Hanson (eds) *Climate Change 2007: Impacts, Adaptation and Vulnerability* (pp. 581–615). Cambridge: Cambridge University Press.

McCarthy, J.J., Canziani, O.F., Leary, N.A., Dokken, D.J. and White, K.S. (eds) (2001). *Climate Change 2001: Impacts, Adaptation, and Vulnerability*. Contribution of Working Group II to the Third Assessment Report of the Intergovernmental Panel on Climate Change. Cambridge: Cambridge University Press.

Meyer, I., Leimbach, M. and Jaeger, C.C. (2007). International passenger transport and climate change: A sector analysis in car demand and associated CO_2 emissions from 2000 to 2050. *Energy Policy* 35: 6332–6345.

Milmo, D. (2007). EasyJet calls for tax on planes, not passengers. *The Guardian,* September 19.

Mimura, N., Nurse, L., McLean, R. F., Agard, J., Briguglio, L., Lefale, P., Payet, R. and G. Sem, G. (2007). Small islands. In M.L. Parry, O.F. Canziani, J.P. Palutikof, P.J. van der Linden and C.E. Hanson (eds) *Climate Change 2007: Impacts, Adaptation and Vulnerability* (pp. 687–716). Cambridge: Cambridge University Press.

Monbiot, G. (2006). *Heat: How to Stop the Planet Burning*. London: Allen Lane.

Myers, N. (2002). Environmental refugees: A growing phenomenon of the 21st century. *Philosophical Transactions of the Royal Society B* 357: 609–613.

Nicholls, R.J., Wong, P.P., Burkett, V.R., Codignotto, J.O., Hay, J.E., McLean, R.F., Ragoonaden, S. and Woodroffe, C.D. (2007). Coastal systems and low-lying areas. In M.L. Parry, O.F. Canziani, J.P. Palutikof, P.J. van der Linden and C.E. Hanson (eds) *Climate Change 2007: Impacts, Adaptation and Vulnerability* (pp. 315–356). Cambridge: Cambridge University Press.

Nurse, L. and Moore, R. (2005). Adaptation to global climate change: An urgent requirement for small island developing states. *Review of European Community and International Environmental Law* 14: 100–107.

Parry, M.L., Canziani, O.F., Palutikof, J.P., Van Der Linden, P.J. and Hanson, C.E. (eds) (2007). *Climate Change 2007: Impacts, Adaptation and Vulnerability.* Contribution of Working Group II to the Fourth Assessment Report of the Intergovernmental Panel on Climate Change, Cambridge: Cambridge University Press.

Payet, R. and Obura, D. (2004). The negative impacts of human activities in the Eastern African Region: An international waters perspective. *Ambio* 33: 24–33.

Peeters, P. (ed) (2007). *Tourism and Climate Change Mitigation – Methods, Greenhouse Gas Reductions and Policies*. Breda: Breda University.

Peeters, P., Gössling, S. and Becken, S. (2006). Innovation towards tourism sustainability: Climate change and aviation. *International Journal of Innovation and Sustainable Development* 1(3): 184–200.

Peeters, P., Szimba, E. and Duijnisveld, M. (2007). Major environmental impacts of European tourist transport. *Journal of Transport Geography* 15: 83–93.

Peeters, P.M., Van Egmond, T. and Visser, N. (2004). *European Tourism, Transport and Environment.* Breda: NHTV CSTT.

Penner, J., Lister, D., Griggs, D., Dokken, D. and McFarland, M. (eds) (1999). *Aviation and the Global Atmosphere* – A Special Report of IPCC Working Groups I and III. Published for the Intergovernmental Panel on Climate Change. Cambridge: Cambridge University Press.

Randerson, J. (2006). Tackle climate change or face deep recession, world's leaders warned. *The Guardian,* October 26.

Rosenzweig, C., Casassa, G., Karoly, D.J., Imeson, A., Liu, C., Menzel, A., Rawlins, S., Root, T.L., Seguin, B. and Tryjanowski, P. (2007). Assessment of observed changes and responses in natural and managed systems. In M.L. Parry, O.F. Canziani, J.P. Palutikof, P.J. van der Linden and C.E. Hanson (eds) *Climate Change 2007: Impacts, Adaptation and Vulnerability* (pp. 79–131). Cambridge: Cambridge University Press.

Ryan, R. and Stewart, E. (2007). Benn announces 'stronger' climate change bill. *Guardian,* October 29.

Saarinen, J. and Tervo, K. (2006). Perceptions and adaptation strategies of the tourism industry to climate change: The case of Finnish nature-based tourism entrepreneurs. *International Journal of Innovation and Sustainable Development* 1(3): 214–228.

Schneider, S.H., Semenov, S., Patwardhan, A., Burton, I., Magadza, C.H.D., Oppenheimer, M., Pittock, A.B., Rahman, A., Smith, J.B., Suarez, A. and Yamin, F.

(2007). Assessing key vulnerabilities and the risk from climate change. In M.L. Parry, O.F. Canziani, J.P. Palutikof, P.J. van der Linden and C.E. Hanson (eds) *Climate Change 2007: Impacts, Adaptation and Vulnerability* (pp. 779–810). Cambridge: Cambridge University Press.

Scott, D. (2006). Impacts of global environmental change on tourism in mountain regions. In S. Gössling and C.M. Hall (eds) *Tourism and Global Environmental Change: Ecological, Social, Economic and Political Interrelationships* (pp. 54–75). London: Routledge.

Scott, D., De Freitas, C. and Matzarakis, A. (2009). Climate change adaptation in the recreation and tourism sector. K. Ebi, I. Burton and G. McGregor (eds) *Biometeorology for Adaptation to Climate Variability and Change* (pp. 171–194). New York: Springer.

Scott, D., Gössling, S. and De Freitas, C. (2008). Preferred climates for tourism: Case studies from Canada, New Zealand and Sweden. *Climate Research* 38(1): 61–73.

Scott, D., Hall, C.M. and Gössling, S. (2012) *Tourism and Climate Change: Impacts, Adaptation & Mitigation*. London: Routledge.

Scott, D., Jones, B. and Konopek, J. (2007) Implications of climate and environmental change for nature-based tourism in the Canadian Rocky Mountains: A case study of Waterton Lakes National Park. *Tourism Management* 28: 570–579.

Scott, D., Jones, B. and McBoyle, G. (2005). *Climate, Tourism and Recreation: A Bibliography – 1936 to 2005*. Waterloo, Canada: University of Waterloo.

Scott, D., McBoyle, G. and Mills, B. (2003). Climate change and the skiing industry in Southern Ontario (Canada): Exploring the importance of snowmaking as a technical adaptation. *Climate Research* 23: 171–181.

Scott, D., McBoyle, G. and Minogue, A. (2007). The implications of climate change for the Québec Ski Industry. *Global Environmental Change* 17: 181–90.

Scott, D., Wall, G. and McBoyle, G. (2005). The evolution of the climate change issue in the tourism sector. In C.M. Hall and J. Higham (eds) *Tourism, Recreation and Climate Change* (pp. 44–60). Clevedon: Channel View Publications.

Scott, D., McBoyle, G., Mills, B. and Minogue, A. (2006). Climate change and the sustainability of ski-based tourism in Eastern North America: A reassessment. *Journal of Sustainable Tourism* 14: 376–398.

Scott, D., Amelung, B., Becken, S., Ceron, J.P., Dubois, G., Gossling, S., Peeters, P. and Simpson, M. (2007). Climate change and tourism: Responding to global challenges [Draft of technical report]. Madrid/Paris: World Tourism Organization and United Nations Environment Programme.

Scott, D., Amelung, B., Becken, S., Ceron, J.P., Dubois, G., Gössling, S., Peeters, P. and Simpson, M. (2008). *Climate Change and Tourism: Responding to Global Challenges*. Madrid/Paris/Geneva: World Tourism Organization, United Nations Environment Programme, World Meteorological Organization.

Slocum, R. (2010) The sociology of climate change: Research priorities. In J. Hagel, T. Dietz and J. Broadbent (eds) *Workshop on Sociological Perspectives on Global Climate Change*. Arlington: National Science Foundation and American Sociological Association.

Solomon, S., Qin, D., Manning, M., Chen, Z., Marquis, M., Averyt, K.B., Tignor, M. and Miller, H.L. (eds) (2007). *Climate Change 2007: The Physical Science Basis*. Contribution of Working Group I to the Fourth Assessment Report of the Intergovernmental Panel on Climate Change Cambridge: Cambridge University Press.

Stern, N. (2006). *The Economics of Climate Change: The Stern Review*. Cambridge: Cambridge University Press.

Turnpenny, J., Jones, M. and Lorenzoni, I. (2010) Where now for post-normal science? A critical review of its development, definitions, and uses. *Science Technology Human Values* 36(3): 287–306.

Uyarra, M.C., Cote, I., Gill, J., Tinch, R., Viner, D. and Watkinson, A. (2005). Island-specific preferences of tourists for environmental features: Implications of climate change for tourism-dependent states. *Environmental Conservation* 32: 11–19.

Vidal, J. (2007). Global food crisis looms as climate change and fuel shortages bite: Soaring crop prices and demand for biofuels raise fears of political instability. *The Guardian,* November 3.

Wainwright, J. (2010) Climate change, capitalism, and the challenge of trans-disciplinarity. *Annals of the Association of American Geographers,* 100: 983–91.

Wall, G. (1998). Climate change, tourism and the IPCC. *Tourism Recreation Research* 23(2): 65–68.

Wilbanks, T.J., Romero Lankao, P., Bao, M., Berkhout, F., Cairncross, S., Ceron, J-P., Kapshe, M., Muir-Wood, R. and Zapata-Marti, R. (2007). Industry, settlement and society. In M.L. Parry, O.F. Canziani, J.P. Palutikof, P.J. van der Linden and C.E. Hanson (eds) *Climate Change 2007: Impacts, Adaptation and Vulnerability* (pp. 357–390). Cambridge: Cambridge University Press.

Winter, D.D. and Koger, S.M. (2004) *The Psychology of Environmental Problems*. Mahwah, NJ: Lawrence Erlbaum Associates.

Wintour, P. and Elliott, L. (2007). Smash and grab: How Labour stole the Tories' big ideas. *The Guardian,* October 10.

Wolfsegger, C., Gössling, S. and Scott, D. (2008). Climate change risk appraisal in the Austrian ski industry. *Tourism Review International* 12(1): 13–23.

Wynne, B. (1997) Methodology and institutions: Value as seen from the risk field. In J. Foster (ed) *Valuing Nature: Economics, Ethics and Environment* (pp. 135–150). London: Routledge.

Further Reading

Becken, S. and Hay, J. (2007). *Tourism and Climate Change: Risks and Opportunities*. Clevedon: Channel View Publications. (Provides an accessible introduction to tourism and climate change.)

Buckley, R. (2008). Misperceptions of climate change damage coastal tourism: Case study of Byron Bay, Australia. *Tourism Review International* 12(1): 71–78. (Discusses some of the ways in which climate change may be misunderstood in the context of destination planning and development)

Cohen, S. and Higham, J. (2010). Eyes wide shut? UK consumer perceptions on aviation climate impacts and travel decisions to New Zealand. *Current Issues in Tourism* 14: 323–335. (Details consumer perceptions and perceptions of travel in relation to climate change.)

Gössling, S. and Hall, C.M. (2006). Uncertainties in predicting tourist flows under scenarios of climate change. *Climatic Change* 79(3&4): 163–173. (Outlines some of the issues associated with forecasting and predicting tourist flows when considering the effects of climate change.)

Gössling, S., Hall, C.M., Peeters, P. and Scott, D. (2010). The future of tourism: A climate change mitigation perspective. *Tourism Recreation Research* 35(2): 119–130. (Discusses the difficulties in mitigating the effects of climate change, with particular reference to less developed countries.)

Hall, C.M. (2011). Policy learning and policy failure in sustainable tourism governance: From first and second to third order change? *Journal of Sustainable Tourism* 19(4&5):

649–671. (Outlines some of the potential reasons for failures by policy-makers to tackle issues of climate and environmental change.)

Scott, D., Hall, C.M. and Gössling, S. (2012). *Tourism and Climate Change: Impacts, Adaptation and Mitigation*. London: Routledge. (Provides an outline of some of the most recent research and issues in relation to climate change and tourism.)

Zeppel, H. (2012). Climate change and tourism in the Great Barrier Reef Marine Park. *Current Issues in Tourism*, DOI:10.1080/13683500.2011.556247. (Provides an overview of issues associated with tourism and climate change in one of the world's most well-known natural tourism destinations.)

Chapter 14

Slow Tourism: Back to Bullock Cart Days!

Dennis Conway and Benjamin F. Timms, Alison Caffyn and Rachel Dodds

Context

'Slow tourism' is another form of alternative tourism and was born in the Italian village of Bra. As the name (slow tourism) suggests, it is the antithesis of contemporary fast tourism and, along with 'slow travel', was born out of the concept of slow food. The Slow Food Movement was launched in 1989, it now has around 100,000 members from 153 countries (Slowfood, 2012). It was a movement which grew out of local people's strong commitment to their culture (as they rebelled against the mono-culture of over-riding industrialization), where food, drink, homemade wine, and so on, became popular. This gave rise to a new type of tourism – with the idea of 'slow' living at its core – which saw people working to preserve their gastronomic traditions and reacting against faster lifestyles and the McDonaldization of tourism. Slow culture, however, is not confined to travel or tourism alone; it relates to a slowing down in everyday life (such as taking part in de-stressing activities, avoiding multi-tasking, healthy eating, taking slow holidays, eating local food and avoiding becoming impatient).

Here, Dennis Conway and Benjamin Timms attempt to define the two terms 'slow travel' and 'slow tourism' because, for many scholars, these terms are synonymous with each other and have little semantic difference. The authors question whether slow tourism and slow travel are viable alternatives for a new tourism model. Certainly, slow tourism has a lower ecological footprint and, with its softer-side, its holistic approach should fit in with the idea of sustainable tourism.

Alison Caffyn advocates slow tourism as an antidote to our 'fast', modern lifestyles. Proponents of travel place more emphasis on the *journey* rather than a single *destination*. Distances and means of travel are the important issues. She gives a long list of points that are important to incorporate into slow tourism. She is optimistic about the future of these new models.

Rachel Dodds believes that sustainability should be at the core of any form of appropriate tourism – and with slow tourism this is not dependent on carbon intensive transportation. The fundamentals of slow travel are provided clearly and the 'slow' mantra is employed tactfully and sustainably.

14.1

Are Slow Travel and Slow Tourism Misfits, *Compadres* or Different Genres?

Dennis Conway and Benjamin F. Timms

Introduction

At first glance, differentiating slow travel from slow tourism might appear to be predominantly a question of semantics, or at best an opinionated debate about minor differences between the two qualitative conceptualizations. We beg to differ, and hope to provide more substantive points of note that will advance the theoretical discourse that surrounds these two alternative, sustainable tourism genres. Both ideas arise out of the Italy-born Slow Food Movement's emergence as an antithesis to the 'fast life' of modern society. Beyond this commonality, the two are offered as alternatives to mass tourism, though their rationales for doing so diverge theoretically and practically. There is also a parallel concern about how these alternative genres are ecologically sensitive and how they fit, or do not fit, the tenets of sustainable tourism, in which the best interests of human-environmental relations in the near and more distant future are not compromised.

The structure of our argument is as follows. First, we compare and contrast slow travel and slow tourism, pointing out their common conceptual roots, their similarities and points of departure, and most significantly their decidedly different contextual 'spaces'. Of particular salience are the ecological and developmental conundrums the two genres pose or do not address sufficiently, and in the case of slow travel appear to avoid. We then proceed to a discussion that brings the challenging objectives of sustainable tourism and ecologically sustainable development into focus, with the following questions helping to guide the critique. Are slow travel and slow tourism equally viable alternatives as new tourism models? How much should sustainable tourism and sustainable development goals enter into our assessments of their future promise? How much do geographical and developmental contexts matter when evaluating these slow movement genres? Finally, in a concluding summary we press the point that slow travel and slow tourism are decidedly different genres with different ecological rationales, despite their common conceptual roots.

Comparing Slow Travel and Slow Tourism

Slow travel

To begin with, slow travel is defined as a qualitative focus on the journey traveled in which the main emphasis and explanation is upon the traveling tourist's consumption-oriented enjoyments and experiences. This means that it is primarily focused on slow traveling tourists' demand-side issues, while giving scant consideration to the supply-side dimensions of the sector's growth, differentiation and dynamism. Slow travel is expected to bring about added environmental benefits, because in its purest, ideal form it avoids 'fast travel' via air carriers or cars due to both the detachment they create from the 'journey as destination' focus and the high global greenhouse gas emissions they emanate (Dickinson *et al.*, 2011).

Instead, alternative slow modes that enjoy a lower ecological footprint are favored – such as walking, cycling, animal traction or the possibility of mass-transit options including local buses and trains, depending on their greenhouse gas emission profiles. It is this explicit emphasis on modes of transport that is central to the definition of slow travel, where its proponents argue that '...in many respects, [modes of transport] are the destination' (Dickinson *et al.*, 2011: 293). Their claim, which we will take issue with later, is that:

> Slow travel is a holistic approach in that the outward journey, destination and return are integral; they make one travel experience. For most of the experts, there is no division between slow travel as the journey and slow tourism as a way of enjoying the destination. They are one and the same. (Lumsdon & McGrath, 2011: 274)

Stripped of its ecological trappings, slow travel is a more consumption-oriented notion of tourist mobility – traveling, visiting or journeying – to whit, a demand-side explanation. The traveler's optic is privileged, the traveling tourist's experiences are valued uppermost, and the journey's slow pace and subsequent aesthetic enjoyment of passing the time away while experiencing the sights and sounds of the immediate natural environment, underpin its rationale. Gardner (2009: 11) provided this philosophical view of slow travel:

> Slow travel is about making conscious choices. It is about deceleration rather than speed. The journey becomes the moment to relax, rather than a stressful interlude imposed between home and destination. Slow travel re-engineers time, transforming it into a commodity of abundance rather than scarcity. And slow travel also reshapes our relationship with places, encouraging and allowing us to engage more intimately with the communities through which we travel.

To advocates such as Dickinson and Lumsdon (2010), slow travel is the way the tourist approaches the journey as slower travel times characterize the journey by bus, train, cycle or on foot. Slow travelers thereby experience 'localness' while undertaking 'day-tripping' around a base, where they stay-a-while in a vacation-rental, youth hostel or bed and breakfast and enjoy local cuisine, regional foods and beverages consumed at the same leisurely pace that locals have come to appreciate.

Slow travel is highly selective of geographical context and degree of infrastructure sophistication. It is most suitable for journeys to destinations closer to the home of the traveler, thereby limiting it to affluent regions of the world (Lumsdon & McGrath, 2011). Since comparatively short distances are bound to be favored given the modes of travel envisaged, slow travel experiences will be best enjoyed in local environs, where the richness, fullness and diversity of proximate landscapes offers a myriad, or multitude, of pleasurable and rewarding experiences while slow travelers undertake the relatively short-distanced journeys.

European 'slow travelers' enjoying mainland European diversity is one geographical region eminently suited to this genre of alternative tourism. Walking parts of the national footpaths of the United Kingdom is another, with Alfred Wainwright's guides to the Lake District fells initiating this slow travel tradition for generations of 'fell-walkers' and 'ramblers' that continues to this day (Wainwright, 1955–1966; Wainwright, 1973). The British Ordnance Survey further provided a detailed map series and its associated leisure guide books of Regional Short Walks, Pathfinder Guides and such have certainly continued Wainwright's mission, assuring that today's generations of walkers have the flexibility to build their own slow travel itineraries. Such slow travel enjoyment – rambling, hiking or walking, even cycle touring (Lumsdon, 2011) – can also be designed, planned and carried out in scenic locales elsewhere in Europe with the help of similar 'slow travel-itineraries'. One of note is Tippett's (2008) Guidebook that details a multitude of short bus trips and self-guided scenic walks around Sorrento in Italy, including the spectacular *Sentiero degli Dei* 'Pathway of the Gods' transect-walk high above the Positano and Amalfi coast. Spain's pilgrimage trail, *El Camino do Santiago*, is another prime example whereby slow travelers traverse northern Spain by foot to Santiago de Compostela's cathedral.

Slow travel can therefore be characterized as a highly varied set of alternative tourism offerings throughout Europe, Britain, Japan and New Zealand where the more affluent live in close geographic proximity to destinations and there exists well-developed infrastructure for slow-journeying modes of transport. However, in other affluent locales such as the United States most tourists/travelers have temporal limitations in terms of leisure time, suffer from limited mass-transit infrastructure and face daunting geographical distances to traverse. These intervening obstacles are problematic to slow travel beyond an immediate home locale, and can scarcely be attempted in that large continent without recourse to an initial 'fast travel' segment by air[1] that gets you to your distant destination – be it a National Park, an Appalachian Trail segment, a coastal hideaway or mountainous retreat.

Similarly, slow travel offerings in internationally-distant, tourist destinations for North American and European tourists, such as the Caribbean, Latin America, Asia, Oceania and Africa, also must rely upon 'fast travel' access by airline to travel to locales where slow tourism might be on offer as alternative, sustainable tourism options (Conway & Timms, 2010). Travel via ocean liner to these destinations could possibly be included, but a lack of analysis of global greenhouse gas emissions by this mode of transport has left it out of slow travel's ecological accounting at present (Dickinson *et al.*, 2011). However, as will be discussed, this geographical limitation to affluent regions of the world raises serious questions about this slow genre's applicability to less prosperous regions where tourism is an important contributor to their social, cultural and economic development prospects (Conway & Timms, 2003).

Slow tourism

Slow tourism shares with slow travel the underlying slow food movement's call to combat the increasing speed of the modern world, particularly as it relates to mass-tourism; this latter being a 'proxy' for fast-food. However, slow tourism goes further by linking the slow traveler's qualitative experiences and enjoyments on the journey and at their destinations with the benefits they provide for local stakeholders. These benefits of slow tourism explicitly advance sustainable tourism objectives and promote grassroots, local 'development from below' initiatives, which promote positive environmental outcomes and address pro-poor tourism goals (Renard, 2001; Torres & Momsen, 2004). As such, slow tourism places a greater priority on supply-side issues and concerns of tourism destinations.

Slow tourism's definition broadens the scope and widens the purview of mass-tourism alternatives by melding Campbell's (1996) three 'E's' of sustainable development – economy, equity and the environment. Borrowing from the vigorous debate that has waged over a universally-accepted definition of ecotourism, Weaver touches on these three principles of sustainable development that we can readily use as an appropriate analog for this slow-tourism construct:

> Ecotourism is a form of tourism that fosters learning experiences and appreciation of the natural environment, or some component thereof, within its associated cultural context. It is managed in accordance with industry best practice to attain environmentally and socioculturally sustainable outcomes as well as financial viability. (2008: 17)

However, slow tourism further contributes to this definition of ecotourism, and the slow movement in general, by adopting Daly's (1990, 1996) concept of slow growth which allows it to promote socio-economic development and equity while facilitating environmental sustainability (Conway & Timms, 2010). From an economic sustainability perspective, the disparities between hard growth and slow growth mimic the difference between 'fast life' and the slow movement's call for a 'slow life'. Hard growth emphasizes an increase in size and scale with a focus on raising production and consumption levels. Mass tourism exhibits hard growth maxims through increasing the quantity of rooms and tourists that flow through the system. Alternately, soft growth promotes improvements in qualitative efficiency by promoting economic development, rather than merely econometrically-fostered growth and capital accumulation, by developing local resources and improving efficiency levels of production, consumption and qualitative returns. By extension, we envisage this soft growth conceptualization of sustainability 'beyond hard growth' as a theoretical basis for not only slow tourism but the slow movement in general.

For example, reducing capital leakage through local provisioning of agricultural products and beverages, handicrafts, furnishings and service activities captures a greater amount of tourist expenditures at the destination. Small, locally-owned tourism facilities purchase a much larger percentage of their inputs from local sources than larger foreign-owned enterprises due to economies of scale and well-developed local relationships (Momsen, 1998; Timms, 2006). Further, increasing local multipliers has positive ramifications for local development whereby productive activities associated with tourism development spill-over to the local market itself, creating more diversified, articulated and resilient local economies.

Concerning equity goals, local control of planning, managing and implementing slow tourism enterprises should not only combat capital leakage but also promote equitable socio-economic benefits to local communities (Renard, 2001). We argue that slow tourism shifts the balance of social power and decision-making to local stakeholders, including both those in the formal economy but also in the informal (complementary) economy. In doing so, it broadens the stakeholders who participate to include impoverished community members, addressing many of the poverty reduction goals of pro-poor tourism (Renard, 2001; Torres & Momsen, 2004). Furthermore, slow tourism promotes local distinctiveness through understanding others' cultures and developing common interests between hosts and 'tourists as guests'.

Environmentally, slow tourism provides tangible benefits for a more sustainable form of tourism (Klak, 2007). Sharing the ecological goals of ecotourism serves the interests of biodiversity maintenance, endangered species protection and naturalist tourism (Hall, 2006). Further, by advancing local control over the slow tourism product, local populations with a direct stake in the local environment are empowered to limit environmental pressures, particularly since the tourism product and their own livelihoods depend on the local environment (Conway & Timms, 2003). This also spills over to equitable socio-economic development as the local community character is also tied to the slow tourism product.

More concretely, the soft economic growth basis of slow tourism emphasizing quality counters the negative environmental externalities of mass tourism. Capturing a greater share of tourist expenditures from a smaller number of visitors allows local economies to develop while minimizing the resource demands and waste production that a greater number of mass tourists would require. This is particularly important for small island developing states, such as those in the Caribbean and Oceania, but is also applicable to other more marginal locales in mainland areas that are promoted as slow tourism destinations (Timms & Conway, 2011).

In sum, slow tourism's theoretical basis in soft growth promotes a more inclusive concept of sustainability and conviviality and focuses upon countering the loss of local distinctiveness as it relates to leisure, sense of place, hospitality and rest and recuperation (Woehler, 2004). This serves a growing market of travelers that prefer a less alienating tourism experience for both hosts and 'guests' (a preferred identity to that of 'tourist'), and provides a variety of local community tourism leisure-environments that limit, or avoid altogether, negative social, cultural, economic and environmental impacts on local host communities. As such, it focuses on local social, economic and environmental sustainability and is more applicable to developing regions in the Caribbean, Latin America, Oceania, Africa and Asia, while also remaining relevant to destinations in more affluent regions of the world.

Discussion – Similarities and Contradictions

'Slow travel' and 'slow tourism' are both outgrowths of the slow food movement and are underpinned by the same senses of social well-being that are set-up as alternatives to people's alienated existences in contemporary capitalist life-worlds – the time-space pressures they feel at work, home and leisure and the hectic daily grind (Harvey, 1989; Kummer, 2002; Petrini, 2001; Stille, 2001). Like their precursor, 'slow travel' and 'slow tourism' appeal to mature and seasoned visitors' perspectives because they both emphasize 'quality of life' considerations such as leisure and enjoyment of the simple delicacies and profound moments of communal contact.

While they both address the positives accrued to local destinations, slow tourism more explicitly expects there to be sincere societal involvement with the locals they meet and learn from, including the involvement and participation of local stakeholders, who bring their maturity and experience to the provision of quality, culturally-rich services. This consideration lessens undue demands on local host communities and promotes mutual benefits of slow tourism for hosts and guests alike, as the former help determine the pace of life, in accordance with the cultural specificity of local offerings and customs. Hence, slow tourism focuses on reciprocal beneficial returns, be they beneficial experiences for the guests or beneficial financial returns for the hosts and their associated service-providing community stakeholders.

Environmentally, mass tourism's history tells of a now-global industry that has always had a large ecological footprint, making it less than benign in this respect (Hall, 2006). Slow travel's advocacy as a genre of sustainable tourism that lessens this footprint is laudable, therefore. On the other hand, its focus upon local travel itineraries and the preferred use of a wide range of travel modes, such as buses, trains, walking and cycling, brings in the thorny question of how long-distance travel can be attained without resorting to airlines or automobiles? The compromise to such an impasse, after much debate and considerable soul-searching, is for slow travel advocates to allow flying to a distant destination and then undertake, or adopt, slow travel principles and practices on arrival (Dickinson & Lumsdon, 2010). And, a similar debate over high-speed rail use for the long distance part of the journey pertains to slow travel in Europe, and Japan, but not the continental US, where long distance, Amtrack rail journeys are agonizingly slow anyway!

Unfortunately, this compromise of principles would seem to undermine the ecological, sustainability argument that slow travel's footprint is reduced sufficiently. Rather, it stands out as the construct's 'Achilles' heel'. Without unduly stressing the obvious, we can only conclude that destinations for slow travel are mostly limited to wealthier regions of the world due to the emphasis on avoiding long distance air and auto travel. It therefore has limited potential to be offered as an alternative genre for tourism destinations further afield, i.e. the Caribbean, Latin America, Oceania, Asia and Africa.

Conclusions

Proponents of slow travel have claimed that slow travel and slow tourism are one and the same (Dickinson *et al.*, 2011; Lumsdon & McGrath, 2011).

However, we beg to differ. First of all, let us admit that geography definitively differentiates the two genres. Slow travel might very well be an attractive alternative tourism genre undertaken in advanced capitalist countries with nationally integrated and well-developed infrastructures, comprehensive transportation networks and relatively small distances between amenity-rich and environmentally attractive rural and urban destinations – such as those in Europe, New Zealand or Japan. However, the applicability of the slow travel mandate for North America is limited due to a travel infrastructure based on the automobile and airlines.

The last point on North America is also relevant, for somewhat different reasons, to less developed regions of the world – many tropical and sub-tropical tourist venues that are more distant from the global North's heavily populated tourist markets can scarcely meet the slow travel mandate for a low carbon footprint based merely on avoidance of air travel to the destination. Further, once at the destination, other more mundane practical issues intercede, such as safety concerns or even climatic extremes of temperature and humidity in tropical environments that make the slow travel option of cycling or trekking much more unappealing, whereas in the temperate climes of Europe they can be envisaged as a restful, bucolic and enjoyable means of slow travel (Lumsdon, 2000).

Slow travel options, and slow travel mandates, might guide tourist travelers as concerned and ecologically sensitive consumers, but challenges abound in blending the necessity to travel shorter distances, reduce the experiences' carbon footprints, and insist upon a much more explicit focus on the travel experience, as the guiding principals of this alternative tourism. It may not be in the best interest of the providers of tourism experiences in developing regions to adhere to the more restrictive practices that slow travel advocates have formulated in their ideal construct(s). Overlooked, or down-played perhaps, is the essential participatory involvement of local stakeholders – guest-house providers, bed and breakfast owners, naturalist guides, eco-lodge staff and interns, local transport service providers, fishermen, agro-tourism enterprises, and such. The slow tourism experiences they too participate in, and experience, are as much an essential aspect of such a new and emerging genre of sustainable tourism in contemporary times, as are the guests' leisure experiences during their recuperative holidays.

We must provide a concluding caveat, however. Our advocacy of slow tourism as a sustainable development initiative for less developed, or under-developed, marginal locales in far-flung peripheral tourist destinations certainly biases our promotional argument in this research probe since we are less interested in the self-gratifying focus on slow travel in already well-developed venues. We must admit, however, that we have undertaken slow travel leisure trips and indeed have followed its tenets in some of our own recreational pursuits in Britain, Europe, the Caribbean and Latin America (excluding the air travel to reach such destinations of course). On the other hand, pro-poor tourism, ecotourism, nature tourism and heritage tourism could all be undertaken to good effect in developing regions, using slow tourism maxims and following slow tourism best practices (Conway & Timms, 2010).

This entails encouraging new ideas about how to grow locales in more conscious and measured ways so that alternative, more inclusive, community centered, and regional regimes are formed from the existing cultural hearths of local practice and communal/familial knowledge that have always existed in the many overlooked, marginal and out-of-the-way locales (Timms & Conway, 2011). Thus, the unevenness of tourism-driven development that has often favored the most accessible urban and coastal spaces at the expense of rural and interior locales can be countered progressively, and more inclusively, than in times past. In conclusion, then, we see the contradictions outweigh the similarities between slow tourism and slow travel. And while the global greenhouse gas emissions ecological basis of slow travel is indeed creditable, for those who seek ways to develop and promote sustainable tourism we find slow tourism as a less geographically restricted and more comprehensive, alternative path for tourism to follow. Notably, slow tourism is a 'development-from-below' initiative that has ecological sensitivity as well as human, capacity-building potential for peripheral regions of the world, that have been overlooked, or un-developed as local authentic, tourism destinations.

14.2

Advocating and Implementing Slow Tourism

Alison Caffyn

Encouraging visitors to make slower choices when planning and enjoying their holidays has a range of benefits – for the destination's environment and local community and for the visitor themselves. Seeking fewer, more meaningful experiences rather than trying to tick off all the 'must-do' sights in a limited amount of time should mean less traveling and possibly less spending, although hopefully more of the visitor's money may go into local businesses' pockets rather than be lost to the local economy via larger chain businesses. With a slower pace visitors should have time to make connections – with the place, people and local culture they are located in or passing through. Plus, possibly, also make better connections with their companions and themselves at the same time.

Slow visitors will find out more about the natural and built heritage, local cuisine, traditions, and some of the special qualities of their destination, in contrast to those who rush through and move on quickly. They are likely to have a more authentic experience taking time to browse the local market, absorb the atmosphere, people watch, buy something from a craftsperson, chat to some local people, linger over a meal at a typical restaurant, take a guided tour of an archaeological site with a local expert, walk or cycle into the surrounding countryside, watch wildlife or whatever takes their fancy. Slow breaks can also have an element of chilling out, relaxing and 're-creating' in the original sense of the word. Such breaks involve being with people you love, sharing quality time and experiences together and probably seeking some peace and tranquillity during a break from the everyday. Woehler (2004) advocates unshackling tourists from time constraints without denying their wish for self-fulfillment.

Slow tourism and travel should also have fewer environmental impacts – people ideally travel less distance both to the destination and once there use more sustainable methods of travel. By making choices to eat, drink, buy and visit locally fewer resources are used on international products and more money goes into local businesses. The great advantage slow travel and tourism have over green or eco-choices is that slow encompasses a well-being, convivial, even hedonistic, element which should be a much easier sell to a range of markets than trying to sell environmentally sound breaks to what is quite a small, if growing, green tourism market.

More time should be spent working out how to market slow travel and slow tourism and increase the benefits for all, than on arguing over exactly what their definitions are and what the differences between them might be.

Proponents of slow travel place more emphasis on the journey – experiencing a place by traveling by a slow means through a landscape and environment. This suits the types of breaks and holidays where people continue to move around rather than base themselves in one place. With slow tourism more emphasis is usually placed on a single destination. The model might perhaps be of a holiday where visitors venture out from a central base to explore the near locality over several days, perhaps doing a range of activities but probably coming back to the same accommodation. Basing the distinction between slow travel and slow tourism on whether a holiday has one base or a series of bases is not particularly helpful – this is a preference which varies from person to person. Also the feasibility of undertaking a slow journey will be more difficult on some destinations such as peripheral areas, small islands and extreme climates (Conway & Timms, 2010). In fact too much emphasis on the journey and keeping moving could mean a holiday with many short unsatisfying visits to a large number of places – like many cruises.

Similarly, focusing too much attention on the distance traveled to arrive in a holiday destination can divert from the slow holiday itself. Yes, distance and means of travel is an essential issue, however this is down to personal choice – some people will eschew all air travel in their lifetime, others may consider it reasonable to take a long-haul flight every five years, others once a year, others won't see it as an issue at all. As far as slow tourism and travel are concerned the main focus should be to encourage those people who do travel long distances to make it more worthwhile by staying longer. Or to encourage those with limited time to travel shorter distances. There may be some equation that could be worked out which calculates the trade-off between distance traveled and time spent and whether it can be considered a slow trip. It might be considered slower to take a medium-haul flight but stay in one locality for a month than to take a short-haul flight for a weekend city break. Obviously using public transport, slow or non-motorized options (narrow-boats, sailing yachts, canoes, cycling, riding, walking, buses, trains, trams) would be less harmful to the global environment (Dickinson & Lumsdon, 2010) – so the emphasis should be on promoting the choices, encouraging people to stay longer or travel about less when they arrive. People will make their own judgements and have their own time constraints. A general message might be that if you only have a weekend why not spend it not too far from home, but if you are able to take an extended break or unpaid leave from work that's the time to travel to the other side of the world and visit more remote locations. Any more blanket approach to discouraging air travel, for example, would disadvantage more remote destinations and especially island states as the probe points out.

Dickinson *et al.* (2010) point out that many slow travelers focus on the type of holiday they want e.g. wildlife watching, walking, heritage etc., before choosing a destination, rather than the other way round. Destinations and operators may not want to encourage this way of thinking but it could be a valuable tool for travel agents, travel websites or marketing consortia with multiple members.

There are few definitions proposed for slow travel and tourism. Lumsdon and McGrath (2011) suggest slow travel might be best expressed as a group of associated ideas rather than a watertight definition and offer a useful conceptualization. They attempt a definition which places most emphasis on the journey and experience of slow travel itself rather than a destination and experience of place. Matos (2004) identified two essential principles – taking time and attachment to place. Caffyn (2009) only went so far as to apply Honoré's 2004 definition of slow to the phenomenon of tourism which thus 'involves making real and meaningful connections with people, places, culture, food, heritage and the environment'.

The emphasis should essentially be on the word slow – rather than travel or tourism. Arguably one could have a slow day trip or even a slow afternoon out. Slow principles have been applied to many fields through a range of organizations and writers. It varies somewhat according to the topic and the context. So those of us working in the travel and tourism field should not get too hung up on definitions but examine the slow approach to see what it may offer to our own context. Slow could be regarded as an amalgamation, made up of a wide range of elements, perhaps even a continuum – the more elements that are present in any holiday or location the slower that trip or destination will be.

Caffyn (2009) outlined some principles or elements of slow tourism (slightly amended here):

- Minimizing travel distance (at least by car/plane).
- Maximizing the time available for the trip.
- Relaxing, refreshing the mind and body.
- Exploring the local area in depth – seeking out distinctiveness.
- Contact with local people, culture, heritage and community.
- Eating at local restaurants, buying in local markets or direct from producers, trying local drinks, beer, wine.
- Creative and unstructured play for children.
- Learning a new skill or activity – personal development.
- A minimum of mechanization, little technology.
- Limited commercialization, few global brands, local economic multipliers.
- Quality experiences and authenticity.
- Relatively sustainable and a modest carbon footprint.
- Good for the visitor and their companions themselves.

The list is not exhaustive but the more of these elements which are present, the slower a holiday might be judged. They will vary enormously from location to location. Conway and Timms (2010) point out that slowing down the pace of a holiday also boosts inclusiveness in the locale and combats foreign capital leakages, especially important in less developed countries.

What has been underestimated in much of the literature is that while the elements of slow holidays should be an attractive prospect for many potential visitors the word slow itself is still problematic as a marketing proposition. A quick look in any thesaurus reveals many more negative connotations than positive. When Ludlow became the first Cittaslow in the

UK there were debates in the local papers with local people resenting being labeled slow. There are also many people who think slow food is about using a slow cooker for making stews. Yet there does not seem to be any word or phrase which sums up all the wide ranging aspects of slow which could be used in its stead. In much of the literature there is an assumption that people will understand the concept, but in reality those who grasp the breadth of meaning are a small minority.

There is some evidence that this lack of understanding may be changing. Alastair Sawday Guide books (2008–10) have been the first to use Slow in their titles and appear to be selling well as more and more titles have been published. Certain markets in some countries will understand faster than others. Care will need to be taken in communications and marketing messages about slow tourism products.

There are at least two other connotations of slow approaches which can also be problematic – that slow is elitist or exclusive and that it is not compatible with new technologies. Slow food can certainly give an elitist impression – gourmet food for those who know enough to appreciate it, using the rather abstruse language of convivia, gastronomy and artisan food. However at heart it is about food which is good to taste, produced in a natural way that does not harm the environment, animal welfare or human health and that the producers are paid a fair return. There could be a similar danger that slow tourism becomes considered an exclusive niche for the few rather than an approach which could be applied to almost any tourism product or destination.

As regards whether slow is compatible with new technology, it probably depends on what technology and how it is applied. The least slow holiday might involve some type of hi-tech virtual reality theme park. However there is no reason why a more authentic experience of exploring an ancient site could not be promoted or interpreted using the internet or smart phones. Many visitors might feel they will only really slow down when they leave their mobile phone and laptop behind but in reality it may be more pragmatic to focus on how to use such technology to help people to discover more meaningful experiences – e.g. linking them to local guides, signposting the best eating places off the beaten track or using traditional music and art online to interpret a location.

The concept can be applied in different ways in different locations to suit both the locality and context but also to suit the relevant visitor markets. The word slow itself may be useful but could be replaced by synonyms where circumstances or objectives dictate. A pragmatic approach applying the most relevant elements is likely to be most successful in practice – for visitor destinations, tourism businesses and also for visitors themselves.

This commentary concludes with examples of how each of these could introduce a slow approach.

Destinations going slow could:

- Highlight slow ways of arrival, public transport, links to car share websites.
- Encourage longer stays with suggested itineraries, deals and packages.
- Minimize car travel in the destination by keeping visitors engaged within a short distance of base with attractive propositions and activities.
- Provide slow activities such as walking, cycling and riding routes, rowing boats, guided walks, taster sessions of arts and crafts e.g. learning about bush-craft in a woodland.
- Provide opportunities for relaxing and soaking up the environment such as viewpoints, outdoor cafes and seating, picnic sites, car free and tranquil areas.
- Showcase local heritage and culture by highlighting what is distinctive and vernacular, local traditions and festivals.
- Promote and support local food and drink including producers, retail and catering, events, trails, all accompanied by hospitable service.
- Tackle infrastructural issues which make slow choices more difficult (Dickinson *et al.,* 2010) such as encouraging train companies to allow more bicycles on board.

Most tourism businesses should be able to incorporate slow elements or options within their offers:

- Allow visitors to choose their own pace – flexible timings for meals, options to extend stays or take a slower paced tour.
- Promote the use of slower transport options – discounts for those arriving without a car, information on public transport and bicycle hire, good knowledge of slow options and routes.
- Provide slower environments such as tranquil areas, TV-free or mobile phone-free zones, reducing advertising, use of local imagery or literature to strengthen sense of place.
- Source and promote local food and drink, traditional dishes, tastings, fair trade products.
- Develop specific slow products such as relaxation/well-being breaks, learning traditional skills or cuisine, slow tours, chill out holidays, slow food menus.
- Target specific markets to which slow would especially appeal such as people with stressful lives or older people who may want a slower paced holiday (Caffyn, 2007).
- Invest in their staff to encourage not only skilled and hospitable service but good knowledge of the local area and culture and the skills to share that with visitors like local ambassadors.

Finally visitors themselves can be encouraged to slow down in choosing what sort of holiday they want, where to go, by what means of travel and for how long. As Gardner (2009: 11) states 'Slow travel is about making conscious choices'. Visitors could also explore the locale more, plan rest days, include some personal development activities and turn off their mobile phones.

Slow tourism has several key dimensions. It is about:

- Place (locality, distinctiveness, landscape, heritage, environment, produce).
- People (community, culture, local enterprise, cuisine, hospitality, authenticity).
- Time (pace, relaxation, unhurried, more in-depth).
- Travel (distance, speed, mode, low carbon).
- The Personal (well-being, pleasure, recreation, conviviality, learning, meaning, enjoyment, understanding).

What it looks like in reality will vary from place to place. Certain dimensions will have more emphasis in some destinations or with specific products. The more elements which can be incorporated, the more sustainable the tourism will be.

Slow really could have the potential to offer a 'win win win': a more sustainable form of tourism, keeping more of the economic benefits within the local community and destination and delivering a more meaningful and satisfying experience.

Case Study References

Few destinations have risked branding themselves slow. However, the word is being used more frequently in titles of travel books – most notably Alastair Sawday, including a new series in association with Bradt Travel Guides. Another recent example is *A Guide to Slow Travel in the Marches* by Les Lumsdon which promotes breaks in the English-Welsh Marches using eight towns as bases for exploring the surrounding countryside by train, bus, bicycle and walking. Some holiday companies are marketing slow holidays, most notably Inntravel (inntravel.co.uk) which offers self-guided activity holidays walking, cycling and skiing across much of Europe. Slow Life Company (jorgandolif.com), based in Vancouver, calls itself a bicycle inspired lifestyle brand and promotes slow cycling trips, including routes linking slow food outlets and local food producers.

14.3

Questioning Slow as Sustainable

Rachel Dodds

Although the authors of the lead-off article bring up some very valid points surrounding definitions that seem to be so much of the focus on tourism research, this author questions whether we are debating the right issues. First, this author questions what the difference is between sustainable tourism and ecologically sustainable development? This author believed that ecologically sustainable development was one of the pillars of sustainability: the balance of socio-cultural, environmental and economic factors for the long-term well-being of contemporary and future generations? The authors of the lead article asked 'how much should sustainable tourism and sustainable development goals enter into the assessment of the future?' This author believes the fundamental goals of the two are different. The question should not be *how much* but rather *how*? The issues that call for more sustainable forms of tourism and tourism development arose from the same concerns over general sustainable development over 20 years ago (Bramwell & Lane, 1993; Eber, 1992; Hall & Jenkins, 1995). If the tourism industry is going to carry on into the future, then perhaps the question should be how to implement these goals more successfully rather than debating another definition which possibly outlines the same issue that the definition of sustainable tourism did originally? Since sustainable tourism development was first discussed, there has been an agreed-upon confusion about whose needs and what time frames should be considered (e.g. Butler, 1993, 1998; Sharpley, 2003). Today, however, there seems to be no debate that sustainable tourism is needed and the concepts of sustainable tourism are agreed upon:

- Protect and conserve resources.
- Use a multi-stakeholder approach.
- Be environmentally responsible.
- Maintain the well-being and involvement of the local population or host community.
- Have economic benefit.
- Have a long-term view.

- Be equitable.
- Have a triple bottom line approach.
- Government must play a leadership role (e.g. impose a 'greater good' approach).

Since the debate over sustainable tourism began, ecotourism, green tourism, pro-poor tourism, geotourism, slow tourism, community-based tourism, carbon friendly tourism and many others have entered into the debate. All these specific genres identify specific issues and offer specific solutions. All, however, are sub-genres of the larger sustainable tourism concept. Are we not spending too much time defining different elements rather than focusing on the key issue – how do we make our tourism industry more sustainable as a whole?

As Conway and Timms – following the concept's champions, Dickinson and Lumsdon (2010) and Gardner (2009) – have rightly pointed out, slow travel can be understood as the opposite to fast travel and therefore uses slower, more carbon-friendly forms of transportation. Since there is a genuine concern for tourism's contribution to global greenhouse gas emissions, it makes sense to discuss the need to promote other forms of travel which are not dependant on carbon intensive transportation. Is, however, slow travel potentially similar to what is being discussed as 'stay-cations' or carbon friendly travel? Are we coming up with definition upon definition and getting lost in semantics? For example, stay-cations are defined as 'a vacation that is spent at one's home enjoying all that home and one's home environs have to offer' (Staycation, 2008). The term 'stay-cation', not found much in academic literature, is as of late, the focus on many television programs, radio interviews and online posts and videos in both Europe and North America. The British Office of National Statistics and the BBC have suggested that the recession, lower value of the pound and increased promotion by local tourists boards have been a mini-boom for the UK tourist industry (*BBC News*, 2011; Webber *et al.*, 2010). Is a stay-cation different from slow travel? Slow travel, stay-cation and carbon friendly travel all seem to highlight that that there are many opportunities for enjoyable holidays in our own backyard, and this form of tourism can be just as good as traveling thousands of miles. With the increase of travel prices (including the increase of carbon taxes on travel), increased concern for climate change and consumer's ultimately tightening their purse strings because of the recession, tourism marketing bodies are capitalizing on these trends to increase local travel and therefore ultimately increase profitability to their areas. Tourism is a business and therefore profit is a key motivating factor. Is the trend to stay closer to home a push by consumers for a different experience or is it marketing bodies influencing consumer's travel choices by promoting and offering such experiences?

Another question to debate is whether slow tourism is really an alternative to mass travel? If slow food is preferable to fast food because of health aspects then is slow tourism healthier than mass tourism for the same reasons? With almost one billion travelers globally (980 million, according to the UNWTO (2012)), not including domestic travelers, do we want to do away with mass tourism? This author does not disagree that there are many negative impacts of mass tourism and that 'reducing capital leakage, increasing inputs to small, locally-owned tourism facilities and promoting socio-economic benefits to local communities' are positive. The authors, however, propagate that all tourism should focus on this development aspect. The soft economic growth basis of slow tourism emphasizes that 'capturing a greater share of tourist expenditures from a smaller number of visitors allows local economies to develop while minimizing the resource demands and waste production that a greater number of mass-tourists would require'. Is this feasible for all destinations? Should we do away with mass tourist resorts like Bugibba in Malta or Cancun in Mexico and force tourists to explore the surrounding communities? The idea may sound good but in reality it is not always practical. Slow tourism focuses on 'sense of place, hospitality, and rest and recuperation' and is said to serve a growing number of travelers. It is not always beneficial, however, to encourage all mass tourists to leave their enclaves or abandon them altogether in search of *slower* places, as this can lead to overcrowding of pristine locations and cultural clashes with local people. Rather than advocate that slow tourism should be the alternative for mass tourism, perhaps the focus should be that destinations should focus on how to increase local experiences (such as purchasing or trying local food rather than typical Western fare in resorts or ensuring that the souvenirs they buy are made locally – therefore ensuring more money benefits local economies).

There is an assumption in slow tourism that visitors *want* to experience local culture or have a 'sincere societal involvement with the locals they meet' but many tourists may not want such experiences. Can a destination really offer a *sincere* experience when there are so many tourists? A city in Croatia, Dubrovnik, may at one time have offered such an experience but with thousands of tourists descending on the place like flies, 'profound moments of communal contact' are becoming few and far between, (see Black, 1996 for another example of tourists in Mdina, Malta, or Getz, 1994 for residents' attitudes in Scotland). This author questions whether slow tourism can address or solve all the issues that mass tourism seems to spread? The authors claim that 'slow tourism shifts the balance of social power and decision-making to local stakeholders' yet this may be a difficult task if enterprises are foreign-owned. Additionally, the authors claim that 'slow tourism places a greater priority on supply-side issues and concerns of tourism destinations' but can slow tourism really shift supply and demand without critical mass?

Slow tourism or slow travel, as ecotourism or pro-poor tourism, is just a subset of the larger, broader concept of sustainable tourism – not a replacement. The idea is that slow tourism focuses on the benefits to local stakeholders and the community. However, with increasingly large numbers of tourists traveling, there is no guarantee that a once 'slow travel' destination will not become mainstream, thereby losing the opportunity for local economic benefit as larger more commercial or foreign ownership moves in.

The authors also discuss 'enjoying, relaxing and absorbing the experience' which is a key point for slow tourism, however, many tourists have limited vacations and therefore may not be able to take the time to *absorb* a destination. Who can determine what relaxing or absorbing the experience means for each tourist as this may be subjective and who is to say that this experience is more pleasurable than a packed package holiday itinerary?

The fundamental concepts outlining slow tourism or slow travel are valid; however, do ecotourism, heritage tourism, pro-poor tourism and all others need to 'use slow tourism's maxims'? Is slow tourism as a term important or does the industry need to focus on implementing the concepts – the same concepts outlined by sustainable tourism 20 or more years ago – rather than debating new definitions?

The elements of slow tourism are similar to those of sustainable tourism and therefore the task ahead is to ensure that tourism strives to address poverty-alleviation, protect the environment and conserve and celebrate cultural heritage. As there is an increasing search for authenticity among tourists, destinations should celebrate their unique attractions and culture aiming to preserve this *experience.* It is perhaps the tourism industry that needs to slow down and realize that they can be profitable while providing slow or sustainable benefits to all.

Concluding Remarks

Slow tourism is an extension of the 'slow food' philosophy, which was perpetuated by the people of Bra (Italy) in 1989. The 'slow movement' came into existence as a reaction against how fast life was being lived in modern society. The key word here is slow – slow food, slow travel, slow holidays and slow cities. The culture-conscious Bra community rose against the mono-culture of industrialization, which was in existence, in order to conserve their gastronomic excellence which was gradually disappearing. This led to a change in lifestyle – with people encouraged to slow down in everyday life, seek real experiences, be patient and avoid the hustle and bustle, appreciate their quality of life, eat healthily, avoid clock-watching and to enjoy life at a more human pace. The follower of slow tourism makes meaningful connections with people (their local community), places, culture, food, heritage and the environment. They participate in slow activities (gastronomic food and food-safaris, conservation, canoe trips, cycling, spa-breaks and so on). The slow tourist is closer to drifters, backpackers or volunteer tourists. They are likely to be well travelled, in search of a more culturally-rich experience, seeking R & R and such during their lengthier visits as 'guests'.

In this chapter, Dennis Conway and Benjamin Timms explore the difference between slow travel and slow tourism, the former being focused on the journey, the latter on the destination, and they discuss its sustainability aspects. Conway and Timms believe that alternative genres of slow tourism like pro-poor tourism, eco-tourism or heritage tourism can better flourish in 'developing regions, using slow tourism maxims'. Slow tourism is a low-carbon economy with lesser negative impacts. They suggest some useful tips for implementing slow tourism. Rachel Dodds approves slow tourism practices but she questions its sustainability and asks whether eco-tourism or pro-poor tourism need to use 'slow tourism maxims'. Caffyn provides some useful elements for the organization of slow tourism. Because of length and breadth concerns, none of the contributions presented case studies of what might be considered a 'slow town' or 'slow city'.

Discussion Questions

(1) In what ways are slow tourism and slow travel considered to be different genres of alternative tourism?

(2) In what ways are slow tourism and slow travel similar in concept, qualitative experiences and ecological consequences?

(3) Why, how, and in what different ways do geography, societal development and locational attributes influence the comparative potential and possibilities of slow tourism and of slow travel?

(4) Is the adoption of slow tourism as an alternative visiting experience to counter and ameliorate the effects of the frenzied, fast pace of our contemporary world likely to grow in importance, decline in importance, or just become one more variation among a diverse set of alternatives that is mainly attractive to mature travelers, who have more leisure time than others?

(5) What are slow tourism's ecological and economic characteristics in terms of the impacts?

(6) Discuss if slow tourism is a viable approach or old wine in a new bottle.

Note

(1) Long distance journeys in the United States by automobile or mobile home, interstate highway travel cannot be undertaken as 'slow travel' given the lengthy distances involved. Alternatively, long distance journeys using the extremely limited Amtrak railway services offered in the United States might very well feature among the worst examples of slow travel, such is the decrepit state of the nation's railway tracks, which impose agonizingly slow speeds on the system so that journeys are interminably long and by no means an enjoyable respite from the 'fast life'.

References

BBC News (2011) Rise in popularity of the 'Staycation', 25 August. Online at: http://www.bbc.co.uk/news/business-14661566. Accessed 2 September 2011.

Black, A. (1996). Negotiating the tourist gaze: The example of Malta. In J. Boissevain (ed) *Coping with Tourists: European Reactions to Mass Tourism*. USA: Bergham Books.

Bramwell, B. and Lane, B. (1993). Interpretation and sustainable tourism: The potential and the pitfalls. *Journal of Sustainable Tourism* 1(2): 71–80.

Butler, R.W. (1993). Tourism – An evolutionary perspective. In J. Nelson, R. Butler and G. Wall (eds) *Tourism and Sustainable Development: Monitoring, Planning, Managing* (pp. 27–41). Canada: University of Waterloo.

Butler, R.W. (1998). Sustainable tourism – Looking backwards in order to progress? In C.M. Hall and A. Lew (eds) *Sustainable Tourism: A Geographical Perspective* (pp. 25–34). Harlow: Longman.

Caffyn, A. (2007). Slow tourism. *Tourism Society Journal* 133: 12.

Caffyn, A. (2009). The slow route to new markets. *Tourism Insights* (September 2009).

Campbell, S. (1996). Green cities, growing cities, just cities? Urban planning and the contradictions of sustainable development. *Journal of the American Planning Association* 62(3): 296–312.

Conway, D. and Timms, B.F. (2003). Where's the environment in Caribbean development theory and praxis? *Global Development Studies* 3(1&2): 91–130.

Conway, D. and Timms, B.F. (2010). Re-branding alternative tourism in the Caribbean: The case for 'slow tourism'. *Tourism and Hospitality Research* 10(4): 329–344.

Daly, H.E. (1990). Sustainable growth: An impossibility theorem. *Development* 3/5: 45–47.

Daly, H.E. (1996). *Beyond Growth: The Economics of Sustainable Development*. Boston: Beacon Press.

Dickinson, J.E. and Lumsdon, L.M. (2010). *Slow Travel and Tourism*. London: Earthscan.

Dickinson, J.E., Lumsdon, L.M. and Robbins, D. (2011). Slow travel: Issues for tourism and climate change. *Journal of Sustainable Tourism* 19(3): 281–300.

Dickinson, J.E., Robbins, D. and Lumsdon, L. (2010). Holiday travel discourses and climate change. *Journal of Transport Geography* 18(3): 482–489.

Eber, J. (ed) (1992). Beyond the green horizon: A discussion paper on principles for sustainable tourism. Surrey: Tourism Concern/WWF Goldalming.

Gardner, N. (2009). A manifesto for slow travel. *Hidden Europe* 25(March/April): 10–14.

Getz, D. (1994). Residents' attitudes towards tourism: A longitudinal study in Spey Valley, Scotland. *Tourism Management* 15(4): 247–258.

Hall, C.M. (2006). Tourism, biodiversity and global environmental change. In S. Gössling and C.M. Hall (eds) *Tourism and Global Environmental Change: Ecological, Economic, Social and Political Interrelationships* (pp. 211–226). London: Routledge.

Hall, C.M. and Jenkins, J. (1995). *Tourism and Public Policy*. USA: Routledge.

Harvey, D. (1989). *The Condition of Postmodernity*. Oxford, UK: Basil Blackwell.

Honoré, C (2004). *In Praise of Slow*. London: Orion.

Klak, T. (2007). Sustainable ecotourism development in Central America and the Caribbean: Review of debates and conceptual reformulation. *Geography Compass* 1(5): 1037–1057.

Kummer, C. (2002). *The Pleasures of Slow Food*. San Francisco: Chronicle Books.

Lumsdon, L. (2011). *A Guide to Slow Travel in the Marches*. Herefordshire: Logaston Press.

Lumsdon, L.M. (2000). Transport and tourism: Cycle tourism – A model for sustainable development. *Journal of Sustainable Tourism* 8(5): 361–377.

Lumsdon, L.M. and McGrath, P. (2011). Developing a conceptual framework for slow travel: A grounded theory approach. *Journal of Sustainable Tourism* 19(3): 265–279.

Matos, R. (2004). Can slow tourism bring new life to Alpine regions? In K. Weiermair and C. Mathies (eds) *The Tourism and Leisure Industry: Shaping the Future* (pp. 93–103). New York: Haworth.

Momsen, J.H. (1998). Caribbean tourism and agriculture: New linkages in the global era? In T. Klak (ed) *Globalization and Neoliberalism: The Caribbean Context* (pp. 115–134). Lanham, MD: Rowman & Littlefield.

Petrini, C. (2001). *Slow Food: The Case for Taste.* New York: Columbia University Press.

Renard, Y. (2001). *Practical Strategies for Pro-poor Tourism: A Case Study of the St. Lucia Heritage Tourism Programme*. London: Overseas Development Institute.

Sawday, A. (2008). *Go Slow England*. Bristol: Sawdays.

Sawday, A. (2009). *Go Slow Italy*. Bristol: Sawdays.

Sawday, A. (2010). *Go Slow France*. Bristol: Sawdays.

Sharpley, R. (2003). Tourism, modernization and development on the island of Cyprus: Challenges and policy responses. *Journal of Sustainable Tourism* 11(2&3): 246–265.

Slowfood (2012). Definition of slow food. Online at: http://www.slowfood.com. Accessed 9 January 2012

Staycation (2008). Word of the day, cited on May 23, 2008. Online at: http://www.urbandictionary.com. Accessed 21 January, 2012.

Stille, A. (2001). Slow food: An Italian answer to globalization. *The Nation* (21/27): 11–16.

Timms, B.F. (2006). Caribbean agriculture-tourism linkages in a neoliberal world: Problems and prospects for St. Lucia. *International Development Planning Review* 28(1): 35–56.

Timms, B.F. and Conway, D. (2011). Slow tourism at the Caribbean's geographic margins. *Tourism Geographies, iFirst 2011* On-line: 1–23.

Tippett, J. (2008). *Sorrento-Amalfi-Capri: Car Tours and Walks.* London: Sunflower Books.

Torres, R. and Momsen, J.H. (2004). Challenges and potential for linking tourism and agriculture to achieve pro-poor tourism objectives. *Progress in Development Studies* 4: 294–318.

UNWTO (2012). Online at: http://unwto.org/en. Accessed 16 January 2012.

Wainwright, A. (1955–1966). *A Pictorial Guide to the Lakeland Fells, Books 1–7.* London: Frances Lincoln.

Wainwright, A. (1973). *A Coast to Coast Walk.* London: Frances Lincoln.

Weaver, D.B. (2008). *The Encyclopedia of Ecotourism* (2nd edn). Milton, Australia: John Wiley & Sons.

Webber, D., Buccellato, T. and White, S (2010). The global recession and its impact on tourists' spending in the UK. *Economic and Labour Market Review*. 4(8): 65–73.
Woehler, K. (2004). The rediscovery of slowness, or leisure time as one's own and as self-aggrandizement? In K. Weiermair and C. Mathies (eds) *The Tourism and Leisure Industry: Shaping the Future* (pp. 83–92). New York: Haworth.

Further Reading

Caffyn, A. (2007). Slow tourism. *Tourism Society Journal* 133: 12.
Conway, D. and Timms, B.F. (2010). Re-branding alternative tourism in the Caribbean: The case for 'slow tourism'. *Tourism and Hospitality Research* 10(4): 329–344.
Daly, H.E. (1996). *Beyond Growth: The Economics of Sustainable Development*. Boston: Beacon Press.
Dickinson, J.E. and Lumsdon, L.M. (2010). *Slow Travel and Tourism.* London: Earthscan.
Fullagar, S. (2012) *Slow Tourism: Experiences and Mobilities*. Bristol: Channel View Publications.
Gossling, S. Hall, M. Lane, B. and Weaver, D. (2008) The Helsingborg statement on sustainable tourism. *Journal of Sustainable Tourism* 16(1): 122–124.
Heitmann, S. Robinson, P., Povey, G. (2011). Slow food, slow cities and slow tourism. In P. Robinson, S. Heitmann and P.U.C. Dieke (eds) *Research Themes for Tourism* (pp. 114–127). Wallingford: CABI.
Mayer, H. and Knox, P. (2006). Slow cities – Sustainable places in a fast world. *Journal of Urban Affairs* 28: 321–334.
Molz, J. G. (2009) Representing pace in tourism mobilities: staycations, Slow travel and the Amazing Race. *Journal of Tourism and Cultural Change* 7(4): 270–286.
Pink, S. (2008). Sense and sustainability: The case of slow city movement. *Local Environment* 13: 99–106.
Weiermair, K. and Mathies, C. (2004). *The Tourism and Leisure Industry: Shaping the Future.* New York: The Haworth Hospitality Press.

Conclusion

Tej Vir Singh

This book may appear to be unusual in its structure, presentation and style but I believe that sometimes the unconventional and the grotesque can also attract attention! The book was the result of my obsession with wanting to acquire reasonable responses, from various tourism pundits, on a few partly-answered or unanswered questions such as if tourism can eradicate poverty, or if mass tourism and sustainability can go hand-in-hand, and so on. Readers will have noted that almost all the chapters pose a big question and that experts in that area then respond qualitatively, reflecting on their long experience in that particular field of tourism research. Readers will also have noticed that many questions still need to be answered appropriately. This shows that tourism research is still in its infancy and that more effective and grounded research is needed in order to find the required answers. We need good and replicable research models to help us get on the right path. Some best practices are worth making note of and these will help to lead us towards making more appropriate insights – something which will become a possibility when academics and practitioners work more closely together.

We began by looking at tourism and the idea that there is nothing good, bad or ugly in it. Men, fuelled by greed, initially went about it in the wrong way, which gave the industry a bad name. Post-industrial societies have more leisure time, more disposable income and faster transport and these factors, linked with other forces of modernization and the rise of the new middle class, are responsible for 'crowd tourism', more popularly known as mass tourism, which has since resulted in severe adverse consequences in a number of places – most obviously in the developing world. Mass tourism, as Poon (1993) observes, has created havoc and she pleads for a new form of tourism – but it will not be easy to get rid of mass tourism and bring in a more friendly form, which can be sustainable on the triple bottom line. Mass tourism is worth around one billion dollars (according to the UNWTO), is responsible for millions of jobs and a vast amount of infrastructure, without even considering the power and politics that go on behind the scenes. While the experts admitted there was some ambiguity over the concept of sustainability, they nonetheless agreed that mass tourism should become more closely aligned with sustainability. The little that has been achieved in the hospitality sector and elsewhere was classed as a 'paradigm nudge'. There was a consensus on bridging the gap between academics and practitioners. One of the unfortunate facts is that mass tourism is closely associated with rampant consumerism (Chapter 2). Consumerism is a child of global capitalism, and tourism as an economic sector stimulates ever increasing levels of consumption. Mass tourism leads to mass consumption of goods and services. Opinions were expressed about consumerism promoting growth in the economy and, because of this, it is a challenge for us to practice ethical consumption, despite the fact that resources are now more scarce. This is one more reason we should appeal for smaller types of tourism that will be of benefit to the world's developing nations.

A new form of tourism, popularly known as Alternative Tourism (AT), emerged among a variety of benevolent terms (appropriate, responsible, sustainable). As these grew in popularity, more new types came on the scene; the most well-known among them being ecotourism, adventure tourism, nature tourism, wildlife tourism, and so on. These new forms of tourism were considered inherently sustainable as they were small in size, community-based, low-leakage, locally-controlled and had links with other sectors of the local economy.

Tourism was seen at its best in the form of ecotourism, which flourished in a variety of different forms (Chapter 12) including nature-based tourism, green tourism, nature talks and so on. This type of tourism brings humankind closer to nature's mysteries whether they be of the earth, sea or sky – 'often nature in the raw'. Ecotourism, though a Western phenomenon, closely resembles Himalayan pilgrimages – only the genie is missing from the former. Unfortunately, the core aspects of ecotourism are often ignored or disregarded, which threatens the legitimacy of the industry; green-washing environmental opportunism and eco-exploitation are commonplace practices adopted by the providers of services with limited ethical practices. Not all nature-based tourism is ecotourism though. The phenomenon of mass ecotourism is creating problems in many national and international parks. In fact, mass tourism and ecotourism are incompatible. Yosemite and the Grand Canyon National Parks are living examples of the kinds of damage that mass tourism can do to such natural spaces. Scholars like Weaver believe that the impact of mass ecotourism can be reduced by making use of sustainable practices and more effective management techniques.

Some scholars have misgivings about the continued success of a destination that attracts more and more tourists – the likelihood being that a niche attraction could be turned into a mass tourist destination if stakeholders ignore the capacity threshold (see the figure below). There is also a danger that a popular niche destination may be overtaken by a multinational company or corporation. Since some niche tourism centres provide a mosaic of attractions, they may continue to serve as supplements to mass tourism. This marriage of convenience is an ideal condition for the stability and sustainability of some destinations.

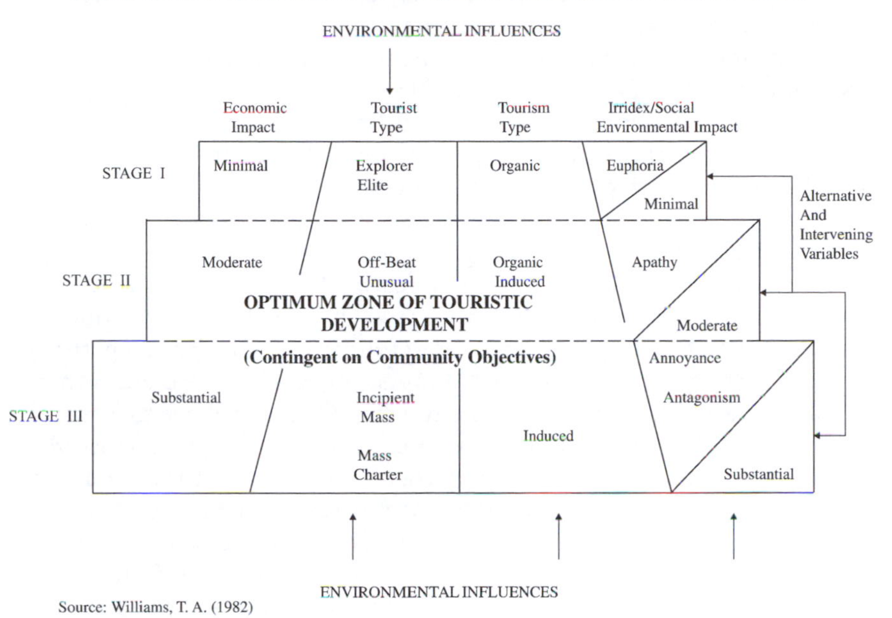

Figure A Three stage model: Impact of tourism development on the environment

The book also reflects the brighter side of the tourism industry in areas such as welfare tourism, tourism and education, tourism and poverty reduction, volunteer tourism, heritage tourism, tourism and community empowerment and tourism and academic reality. There are many more subjects related to 'small' tourism that could not be included in this book for dearth of space and lack of availability of experts in those areas. Since mass tourism was crying out for another probe, being an area that continues to cause grave concern because of its adverse effects, this was discussed at length. Climate change and tourism (Chapter 13) is also a hot topic because of concerns about greenhouse gas (GHG) emissions and the carbon footprints created when travelling, and this was therefore discussed in detail though no workable solution was arrived at. Tourism has been accused of contributing to GHG emissions and was found to be a modest sinner. But humankind has to be careful in the consumption of scarce resources, particularly non-renewable resources, for the health of the planet.

Tourism is a community industry (Chapter 4) and community-based tourism was therefore considered a potential agent for empowering local people through their pro-active participation in tourism projects. It was thought such participation would bring more socio-economic benefits to local communities and while community-based tourism development has yielded promising results, it also has a negative side.

Authenticity (Chapter 10) is a much discussed subject in tourism literature since MacCannell introduced the concept of 'staged authenticity' in 1973. Tourists' insatiable curiosity for authenticity, novelty and serendipity will never die and that is why this persists, even though anthropologists and sociologists, who have largely contributed to this area, say that in this postmodern reality such a desire to seek out 'genuine' experiences is the modern-day equivalent of searching for the holy grail.

The concept of slow tourism, as featured in the final chapter, is a relatively recent arrival and is something which has become quite popular in Europe. In this fast and speedy age, slow tourism sounds appealing, though it is difficult to put into practice. Many scholars argue whether or not it should be classed as part of the tourism sector. Over and above this concept, a new theme has emerged on the scene – the 'staycation'; this is where people love to stay at home and enjoy their leisure time in the surroundings of their own homes. Does this mean that tourism is getting more and more innovative? Or does it mean that people have started to believe that there are no new masterpieces worth seeking out and going travelling for? A few countries, like Japan, have designed prototypes of world-famous attractions, such as the Taj Mahal, so that their people can enjoy the grandeur of such places without the need to visit India at all. The cyber-world also offers more talismanic travel experiences through the creation of virtual reality environments. Such 'surrogates' for travel encourage more staycations, curtail mobility and reduce GHG emission. On the downside, they may also be a threat to tourism as a whole as mobility is the essence of travel and tourism.

References

Poon, A. (1993) *Tourism, Technology and Competitive Strategies*. Walling ford: CABI.

Williams, A.T. (1982). Impacts of domestic tourism on host populations: The evaluation of a model. In T.V. Singh and J. Kaur (eds) *Studies in Tourism Wildlife Parks and Conservation* (pp. 214–223). New Delhi: Metropolitan.

Index